Student Solutions Manual

Elementary Statistics

ELEVENTH EDITION

Robert Johnson
Monroe Community College

Patricia Kuby
Monroe Community College

Prepared by

Robert Johnson
Monroe Community College

Patricia Kuby
Monroe Community College

BROOKS/COLE
CENGAGE Learning

Australia • Brazil • Japan • Korea • Mexico • Singapore • Spain • United Kingdom • United States

© 2011 Brooks/Cole, Cengage Learning

ALL RIGHTS RESERVED. No part of this work covered by the copyright herein may be reproduced, transmitted, stored, or used in any form or by any means graphic, electronic, or mechanical, including but not limited to photocopying, recording, scanning, digitizing, taping, Web distribution, information networks, or information storage and retrieval systems, except as permitted under Section 107 or 108 of the 1976 United States Copyright Act, without the prior written permission of the publisher.

For product information and technology assistance, contact us at
**Cengage Learning Customer & Sales Support,
1-800-354-9706**

For permission to use material from this text or product, submit all requests online at **www.cengage.com/permissions**
Further permissions questions can be emailed to
permissionrequest@cengage.com

ISBN-13: 978-0-840-05388-6
ISBN-10: 0-840-05388-6

Brooks/Cole
20 Channel Center Street
Boston, MA 02210
USA

Cengage Learning is a leading provider of customized learning solutions with office locations around the globe, including Singapore, the United Kingdom, Australia, Mexico, Brazil, and Japan. Locate your local office at: **www.cengage.com/global**

Cengage Learning products are represented in Canada by Nelson Education, Ltd.

To learn more about Brooks/Cole, visit
www.cengage.com/brookscole

Purchase any of our products at your local college store or at our preferred online store
www.cengagebrain.com

Printed in the United States of America
1 2 3 4 5 6 7 15 14 13 12 11

TABLE OF CONTENTS

Chapter 1   Statistics 1

Chapter 2   Descriptive Analysis and Presentation of Single-Variable Data 8

Chapter 3   Descriptive Analysis and Presentation of Bivariate Data 62

Chapter 4   Probability 101

Chapter 5   Probability Distributions (Discrete Variables) 127

Chapter 6   Normal Probability Distributions 150

Chapter 7   Sample Variability 167

Chapter 8   Introduction to Statistical Inferences 182

Chapter 9   Inferences Involving One Population 223

Chapter 10   Inferences Involving Two Populations 273

Chapter 11   Applications of Chi-Square 328

Chapter 12   Analysis of Variance 358

Chapter 13   Linear Correlation and Regression Analysis 380

Chapter 14   Elements of Nonparametric Statistics 413

CHAPTER 1 ∇ STATISTICS

Chapter Preview

The purpose of Chapter 1 is to present:
1. an initial image of statistics that includes both the key role statistics has in the technical aspects of life as well as its everyday applicability,
2. its basic vocabulary and definitions,
3. basic ideas and concerns about the processes used to obtain sample data.

SECTION 1.1 EXERCISES

1.1 a. How often do you eat fruit?
  b. Internet visitors at the Postyour.info website.
  c. 63
  d. 1/63 = 0.01587, 11/63 = 0.1746, 16/63 = 0.25397
  e. No. Only people who visited the site and wanted to answer the question did so.

1.3 a. Americans
  b. length of time before a Wifi user gets antsy and needs to check their messages
  c. 47% of those people surveyed say they get antsy within one hour about checking their email, etc.

1.5 Answers will vary

SECTION 1.2 EXERCISES

> The articles in Applied Example 1.2 gives information about the sample (the number of people surveyed). Be watchful of articles that do not give any of this information. Sometimes not knowing something about the sample or survey size causes a question of credibility.

> <u>Descriptive Statistics</u> - refers to the techniques and methods for organizing and summarizing the information obtained from the sample.
>
> <u>Inferential Statistics</u> - refers to the techniques of interpreting and generalizing about the population based on the information obtained from the sample.

1.7 a. inferential   b. descriptive

1.9 a. married women, ages 25-50, who have 2 or more children
  b. 1,170
  c. how often moms say they have a date night with their spouse
  d. 18% of those surveyed say they have a date night every 4-6 months
  e. (0.18)(1170) = 211

1.11 a. USA teens
  b. 501
  c. what everyday invention they thought would be obsolete in 5 years
  d. 501(0.21) = 105
  e. actual percentage could be 4.3% lower or 4.3% higher than quoted
  f. between 16.7% and 25.3%

> A <u>variable</u> is the characteristic of interest (ex. height), where <u>data</u> is a value for the variable (ex. 5'5"). A variable varies (heights vary), that is, heights can take on different values. Data (singular) such as 5'5" (one person's height) is constant; it does not change in value for a specific subject.
>
> An <u>attribute variable</u> can take on any qualitative or "numerical" qualitative information (ex. kinds of fruit, types of music, religious preference, model year - most answers are in words, although model year would have "numerical" answers such as "2010"). An attribute variable can be <u>nominal</u> (description or name) or <u>ordinal</u> (ordered position or rank; first, second,...).
>
> A <u>numerical variable</u> can take on any quantitative information. This includes any count-type and measurable-type data (ex. number of children in a family, amount of time, age, height, area, volume, miles per gallon). A numerical variable can be <u>discrete</u> or <u>continuous</u>. The domain of a discrete variable has gaps between the possible values; there are numerical values that cannot occur. Theoretically, the domain of a continuous variable has no gaps since all numerical values are possible. Do not be confused by data that has been rounded due to scale being used or for convenience reasons.

1.13 a. 45%   (100% - 55%)   b. The percentages are from different groups.

> <u>Population</u> - the collection of all individuals, objects, or scores whose properties are under consideration.
>
> <u>Parameter</u> - a number calculated from the population of values.
>
> <u>Sample</u> - that part of the population from which the data values or information is obtained.
>
> <u>Statistic</u> - a number calculated from the sample values.
>
> **NOTE:** <u>Parameters</u> are calculated from <u>populations</u>; both begin with *p*.
>   <u>Statistics</u> are calculated from <u>samples</u>; both begin with *s*.

1.15 a. The population is all US adults.
b. A sample is the 1200 randomly selected adults.
c. The variable is "allergy status" for each adult.
d. The statistic is the 33.2% based on the sampled adults.
e. The parameter is the percent of all US adults who have an allergy, in this case, 36%.

1.17 Parameters give the information for the entire population and have one specific value. Statistics come from samples which can vary in size and method of data collection, therefore giving different measurements for each different sample.

1.19 a. Attribute possibilities: marital status, ZIP code, gender, highest level of education
b. Numerical possibilities: annual income, age, distance to store, amount spent

1.21 a. Gender, nominal b. Height, continuous

1.23 a. Severity of side-effects from a particular medicine for a patient
b. attribute(ordinal)

1.25 a. Weight of books and supplies per student
b. numerical (continuous)
c. 10 lbs, 5.67 lbs, 15.2 lbs ...

1.27 a. Average cost of textbooks for the semester per student for all students
b. All students enrolled for this semester
c. The cost of textbooks for the semester for one student
d. The 100 students
e. The average cost of textbooks for the semester per student for the 100 students; add all 100 values and divide the total by 100.

1.29 a. All students currently enrolled at the college
b. finite c. the 10 students selected
d. discrete, continuous (cost rounded to nearest cent), nominal

1.31 a. all 2009 pickup trucks listed on MPGoMatic.com
b. 165, sample = 6 trucks
c. 8 variables
d. Manufacturer, Model, Drive, Transmission
e. all nominal
f. Engine size, Engine Size Displacement, City MPG, Hwy MPG
g. Discrete: Engine Size   Continuous: Engine Size Displacement, City MPG, Hwy MPG

1.33 a. numerical b. attribute c. numerical
d. attribute e. numerical f. numerical

1.35 a. The population contains all objects of interest, while the sample contains only those actually studied.
b. convenience, availability, practicality

SECTION 1.2 EXERCISES

1.37 Group 2, the football players, because their weights cover a wider range of values, probably 175 to 300+, while the cheerleaders probably all weigh between 110 and 150.

1.39 By using a standard weight or measure in conjunction with money, prices between competing product brands can be more easily compared, irrespective of purchase quantity. There is a great deal of variability in container sizes between brands and even within brands of the same product. Problems associated with this variability are simplified by showing the standard unit price in addition to the cash register amount at the point of sale.

1.41 Answers will vary but there is no way to differentiate between the students if everybody attains the same grade. If all students received a 100%, then the test is too easy. If all students received a 0%, then the test was too hard. If the scores are between 40 to 95%, you can distinguish among the students' knowledge about the subject.

SECTION 1.3 EXERCISES

A <u>convenience sample</u> or <u>volunteer sample</u>, as indicated by their very names, can often result in <u>biased samples</u>.

Data collection can be accomplished with <u>experiments</u> (the environment is controlled) or <u>observational studies</u> (environment is not controlled). <u>Surveys</u> fall under observational studies.

Sample designs can be categorized as <u>judgment samples</u> (believed to be typical) or <u>probability samples</u> (certain chance of being selected is given to each data value in the population). The <u>random sample</u> (each data value has the same chance) is the most common probability sample.

Methods (simply defined) to obtain a random sample include:

Single-Stage Methods:
  1. <u>Simple Random Sample</u> using Random Number Table –
      (see Introductory Concepts in Appendix A or the Student Solutions Manual)
  2. <u>Systematic</u> – every kth element is chosen

Multistage Methods:
  1. <u>Stratified</u> – fixed number of elements from each strata (group)
  2. <u>Proportional (Quota)</u> – number of elements from each strata is determined by its size
  3. <u>Cluster</u> – fixed number or all elements from certain strata.

1.43 a. Volunteer
b. Yes, only those that subscribe to USA Today and have strong opinions on the subject will respond.

1.45 Answers will vary but Landers' survey was a volunteer survey, therefore there is a
bias – mostly, only those with strong opinions will respond.

1.47 convenience sampling

1.49 a. (1,1), (1,2), (1,3), (1,4)
(2,1), (2,2), (2,3), (2,4)
(3,1), (3,2), (3,3), (3,4)
(4,1), (4,2), (4,3), (4,4)

b. (1,1,1), (1,1,2), (1,1,3)
    (1,2,1), (1,2,2), (1,2,3)
    (1,3,1), (1,3,2), (1,3,3)

    (2,1,1), (2,1,2), (2,1,3)
    (2,2,1), (2,2,2), (2,2,3)
    (2,3,1), (2,3,2), (2,3,3)

    (3,1,1), (3,1,2), (3,1,3)
    (3,2,1), (3,2,2), (3,2,3)
    (3,3,1), (3,3,2), (3,3,3)

1.51 probability samples

1.53 Statistical methods presume the use of random samples.

1.55 Randomly select an integer between 1 and 25 ($100/x = 100/4 = 25$). Locate the first item by this integer value. Select every 25th data thereafter until the sample is complete.

1.57 A proportional sample would work best since the area is already divided into 35 (different size) listening areas. The size of the listening area determines the size of the subsample. The total for all subsamples would be 2500.

1.59 Only people with telephones and listed phone numbers will be considered, possibly eliminating those with only cell phones.

1.61 a. Fluorescent bulbs use up to 75% less energy than incandescent light bulbs; the average life of compact fluorescent bulbs is up to 10 times as long as incandescent light bulbs.
    b. Yes   c. No
    d. Yes, so one would know which bulb is best   e. part d
    f. collect data on the amount of energy used in each type of bulb for a certain amount of time
    g. collect data on the lifetimes of a sample of each bulb

SECTION 1.4 EXERCISES

1.63 Draw graphs, print charts, calculate statistics

CHAPTER EXERCISES

1.65 Each student's answers will be different. A few possibilities are:
  a. color of hair, major, gender, marital status
  b. number of courses taken, number of credit hours, number of jobs, height, weight, distance from hometown to college, cost of textbooks

1.67 a. T = 3 is a data value - a value from one person.
  b. What is the average number of times per week the people in the sample went shopping?
  c. What is the average number of times per week that people (all people) go shopping?

1.69 a. credit card holders
  b. type/name of credit card, number of months past due on payment, debt amount
  c. name – attribute, number of months, debt amount - numerical

1.71 qualitative, ordinal; responses were descriptive and could be ranked

1.73 a. observational study   b. percent or proportion of sunglass use
  c. proportion of sample that wore sunglasses, 4 out of 10 adults

1.75 Each will have different examples.

1.77 Each will have different examples.

1.79 Answers will vary.

CHAPTER 2 ∇ DESCRIPTIVE ANALYSIS AND PRESENTATION OF SINGLE-VARIABLE DATA

Chapter Preview

Chapter 2 deals with the presentation of data that were obtained through the various sampling techniques discussed in Chapter 1. The four major areas for presentation and summary of the data are:
1. graphical displays,
2. measures of central tendency,
3. measures of variation, and
4. measures of position.

The 2003-2007 American Time Use Survey study by the Bureau of Labor Statistics on how full-time university and college students spend their time on an average weekday is presented in Section 1 of this chapter.

SECTION 2.1 EXERCISES

2.1 Answers will vary

2.3
    a. Answers will vary but both graphs show relative size with respect to the individual answers.
    b. Answers will vary. The circle graph does a better job of representing the relative proportions of the answers to the group as a whole.
    c. The bar graph is more dramatic in representing the relative proportions between the individual answers.

MINITAB - Statistical software
Data is entered by use of a spreadsheet divided into columns and rows. Data for each particular problem is entered into its own column. Each column represents a different set of data. Be sure to name the columns in the space provided above the first row, so that you know where each data set is located. (C1 = Column 1)

EXCEL – Spreadsheet software
Data is entered by use of a spreadsheet divided into columns and rows. Data for each particular problem is entered into its own column. Each column represents a different set of data. If needed, use the first row of a column for a title. (A1 = 1st cell of column A)

TI-83/84 Plus – Graphing calculator
Data is entered into columns called lists. Data for each particular problem is entered into its own list. Each list represents a different set of data. If needed, use the space provided above the first row for a title. Lists are found under STAT > 1:Edit. (L1 = List 1)

Partitioning the circle:

1. Divide all quantities by the total sample size and turn them into percents.

2. 1 circle = 100%
    1/2 circle = 50%
    1/4 circle = 25%
    1/8 circle = 12.5%

3. Adjust other values accordingly.

4. Be sure that percents add up to 100 (or close to 100, depending on rounding).

Computer and calculator commands to construct a Pie Chart can be found in ES11-pp34-35.

The TI-83/84 program 'CIRCLE' and others can be downloaded to your computer through your cengagebrain.com account. The TI-83/84 Plus programs and data files may be in a zipped or compressed format. If so, save the files to your computer and uncompress them using a zip utility. Then download the programs to your calculator using TI-Graph Link Software.
Further details on page 35 in ES11*.
*(ES11 denotes the textbook Elementary Statistics, 11th edition)

2.5 a.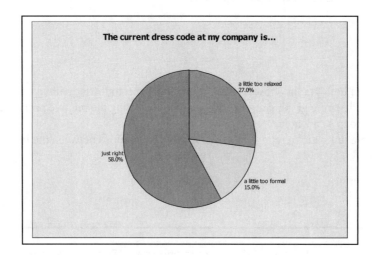

b.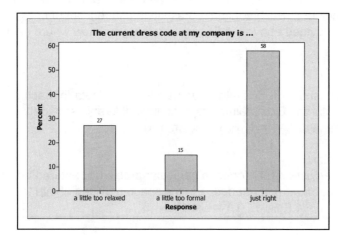

c. Answers will vary.

2.7 a.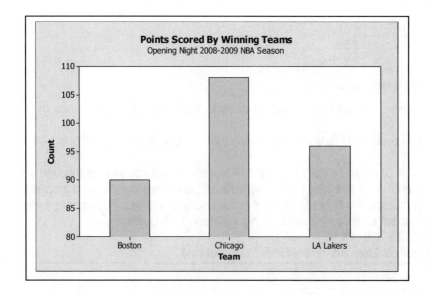

b.

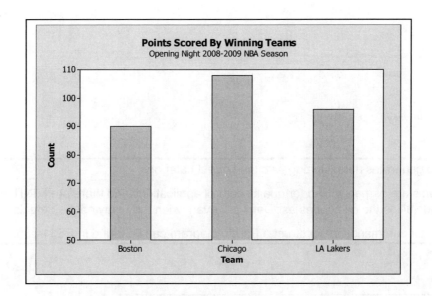

c. The bar graph in 'a' makes the NBA scores appear that they vary more. The Chicago team seems like it outscored the other team by almost 3 times as many points.

d. To make a more accurate representation, begin the vertical scale at zero.

2.9 a.

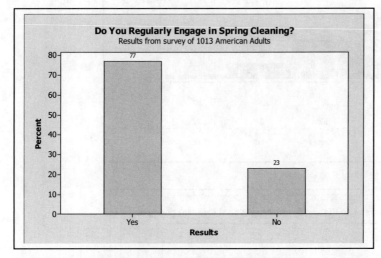

b.

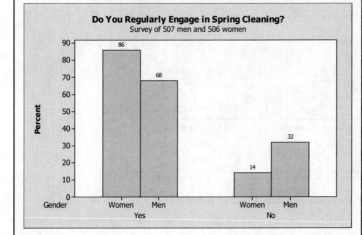

c. Answers will vary.

> The Pareto command generates bars, starting with the largest category.
>
> **NOTE:** Pareto diagrams are primarily used for quality control applications and therefore MINITAB's PARETO command identifies the categories as "Defects", even when they may not be defects.
>
> Computer and calculator commands to construct a Pareto diagram can be found in ES11-p36.

2.11

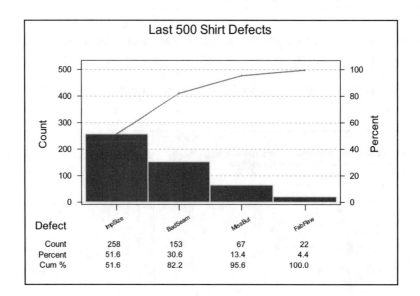

2.13

a.

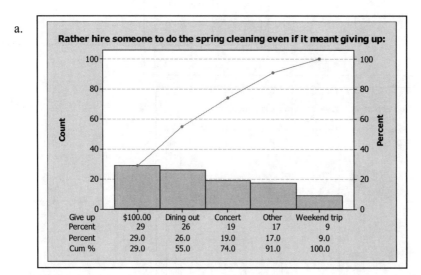

b. It is collection of several answers and needs to be broken down

2.15 a. 150 defects
b. percent of scratch = n(scratch)/150 = 45/150 = 0.30 or 30.0%
c. 90.7% = [37.3 + 30.0 + 15.3 + 8.0]% [round-off error] or (56 + 45 + 23 + 12)/150 = 136/150 = 0.907.   90.7% is the sum of the percentages for all defects that occurred more often than Bend, including Bend.
d. Two defects, Blem and Scratch, total 67.3%. If they can control these two defects, the goal should be within reach.

2.17 a.

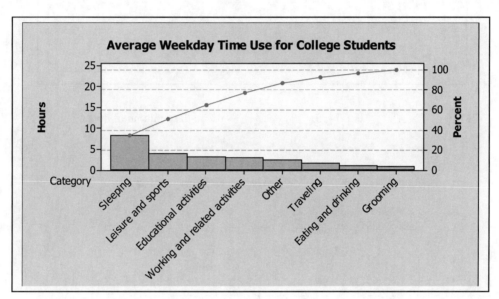

b. sleeping, leisure and sports, educational activities, working and related activities

---

Picking increments (spacing between tick marks) for a dot plot
1. Calculate the spread (highest value minus the lowest value).
2. Divide this value by the number of increments you wish to show (no more than 7 usually).
3. Use this increment size or adjust to the nearest number that is easy to work with (5, 10, etc.).

Computer and calculator commands to construct a dotplot can be found in ES11-pp37&38.
Commands for multiple dotplots are also in ES11-pp41-42.

2.19 Points Scored per Game by Basketball Team

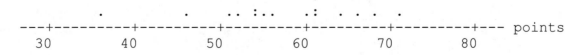

2.21 a.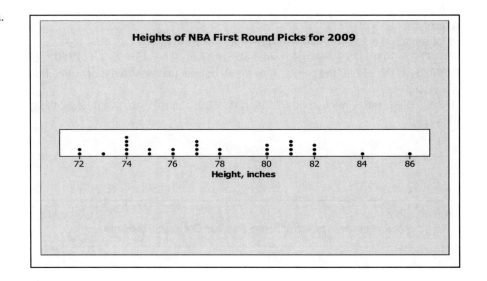

b. 72, 86   c. 74, 5   d. tallest column of dots

2.23   Overall length of commutators

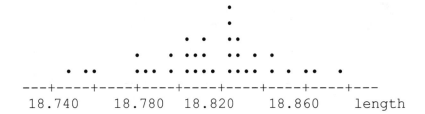

STEM-AND-LEAF DISPLAYS
1. Find the lowest and highest data values.
2. Decide in what "place value" position, the data values will be split.
3. Stem = leading digit(s)
4. Leaf = trailing digit(s) (if necessary, data is first "rounded" to the desired position)
5. Sort stems and list.
6. Split data values accordingly, listing leaves on the appropriate stem.
Computer/calculator commands to construct a stem-and-leaf display can be found in ES11-p39.

2.25 Points scored per game

```
3 | 6
4 | 6
5 | 6 4 5 4 2 1
6 | 1 1 8 0 6 1 4
7 | 1
```

> The column on the left of the stem-and-leaf display is the cumulative count of the data from the top (low-value) down and the bottom (high-value) up until the class containing the median is reached. The number of data values for the median class is in parentheses.

2.27 a. Quik Delivery's delivery charges

```
2. | 0
2. | 9 8 8 9
3. | 1 1
3. | 5 8 8 5 8 6 6 8 7 7 8
4. | 0 3 1 0 0
4. | 5 5 9 6 8 6
5. | 0 4 0 2 4
5. | 6 7
6. | 0 1
6. | 8
7. |
7. | 8
```

  b. The distribution is slightly skewed to the right; the bulk of the distribution is between 2.8 and 5.4, with only one smaller value and 6 larger values that create a longer right-hand tail.

2.29 a. The place value of the leaves is in the hundredths place; i.e., 59|7 is 5.97.
  b. 16
  c. 5.97, 6.01, 6.04, 6.08
  d. Cumulative frequencies starting at the top and the bottom until they reach the class that contains the median. The number in parentheses is the frequency for just the median class.

SECTION 2.2 EXERCISES

> Frequency distributions can be either grouped or ungrouped. Ungrouped frequency distributions have single data values as *x* values. Grouped frequency distributions have intervals of *x* values, therefore, use the class midpoints (class marks) as the *x* values.
>
> Histograms can be used to show either type of distribution graphically. Frequency or relative frequency is on the vertical axis. Be sure the bars touch each other (unlike bar graphs). Increments and widths of bars should all be equal. A title should also be given to the histogram.

Computer or calculator commands to construct a histogram can be found in ES11pp52-54. Note the two methods, depending on the form of your data.

2.31   a.

x	f
0	2
1	5
2	3
3	0
4	2
	12

b. f is frequency, therefore value of 1 occurred 5 times.
c. $2 + 5 + 3 + 0 + 2 = 12$
d. The sum represents the sum of all the frequencies, which is the number of data, or the sample size.

2.33   a. Bar graph, the player's name (number in this case) is qualitative.
b.
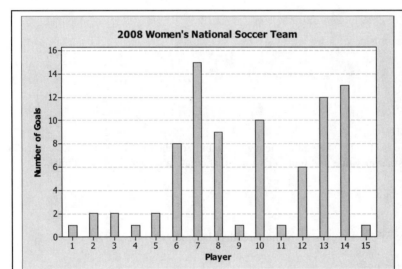
c.
d.
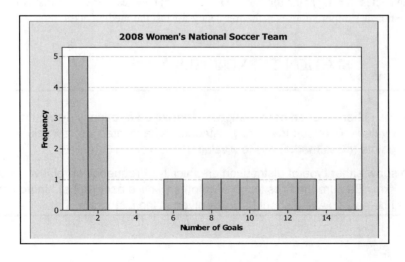

	Relative frequency = frequency / sample size

2.35 a&c.

Height	Count	Rel. Freq.
64	2	0.111
65	4	0.222
66	4	0.222
67	2	0.111
68	4	0.222
69	1	0.056
70	1	0.056

b.

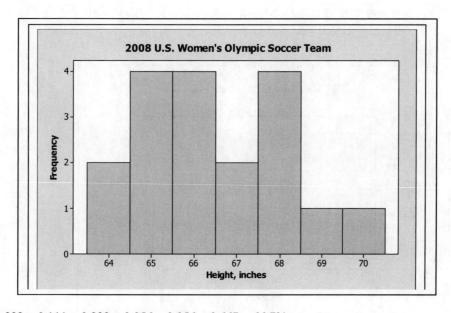

d. $0.222 + 0.111 + 0.222 + 0.056 + 0.056 = 0.667 = 66.7\%$

2.37 a.

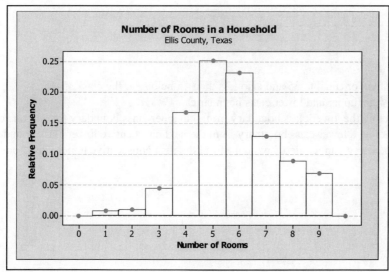

b. mounded, truncated on right due to 9+ class
c. centered on 5; 4 to 7 account for most

An ogive is a line graph of a cumulative frequency or cumulative relative frequency distribution. Start the line at zero for a class below the smallest class. Plot the upper class boundary points from the remaining values of the cumulative (relative) frequency distribution. Connect all of the points with straight line segments. The last point (class) is at the value of one (vertically).

Computer and calculator commands to construct an ogive can be found in ES11p57.

2.39 a.
x	67	68	69	70	71	72	73	74	75	76	77	78	79	80	81	82	83
f	1	8	3	5	10	22	17	28	17	9	9	9	4	1	1	1	1

b.

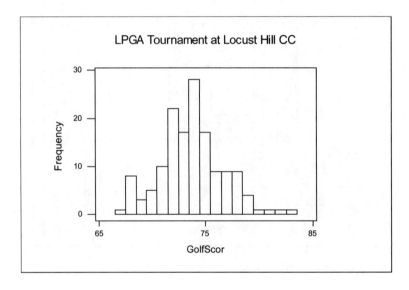

2.41 a. 35-45
b. Values greater than or equal to 35 and also less than 45 belong to the class 35-45.
c. Difference between upper and lower class boundaries.
   i. Subtracting the lower class boundary from the upper class boundary for any one class
   ii. Subtracting a lower class boundary from the next consecutive lower class boundary
   iii. Subtracting an upper class boundary from the next consecutive upper class boundary
d.

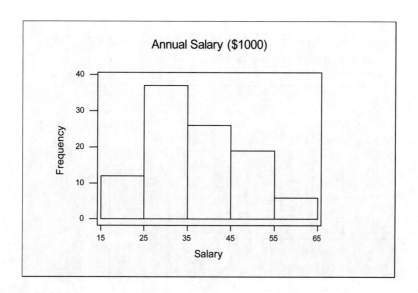

2.43 a. 12 and 16
 b. 2, 6, 10, 14, 18, 22, 26
 c. 4.0
 d. 0.08, 0.16, 0.16, 0.40, 0.12, 0.06, 0.02

 e.
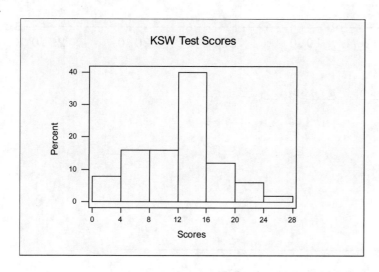

2.45 a.
Class limits	frequency
12 - 18	1
18 - 24	14
24 - 30	22
30 - 36	8
36 - 42	5
42 - 48	3
48 - 54	2

 b. class width = <u>6</u>

 c. class midpoint = (24+30)/2 = <u>27</u>

 lower class boundary = <u>24</u>

 upper class boundary = <u>30</u>

d.

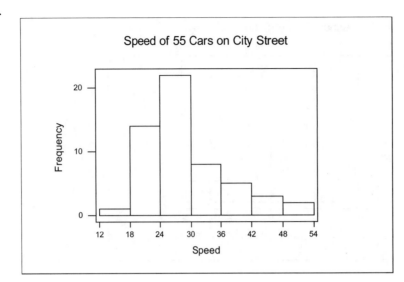

2.47 a.

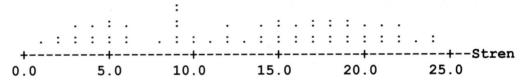

b.
Class boundaries	frequency
1 - 4	6
4 - 7	10
7 - 10	7
10 - 13	6
13 - 16	8
16 - 19	11
19 - 22	10
22 - 25	6
	64

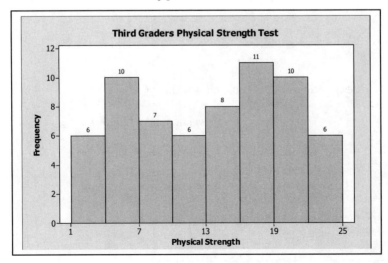

c. | Class boundaries | frequency |
|---|---|
| 0 - 3 | 3 |
| 3 - 6 | 10 |
| 6 - 9 | 4 |
| 9 - 12 | 9 |
| 12 - 15 | 7 |
| 15 - 18 | 11 |
| 18 - 21 | 11 |
| 21 - 24 | 7 |
| 24 - 27 | 2 |
| | 64 |

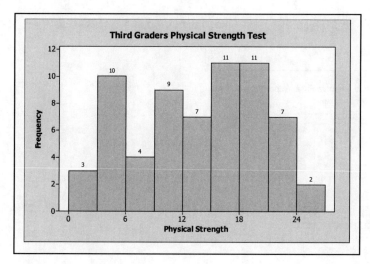

d. | Class boundaries | frequency |
|---|---|
| -2.5 - 2.5 | 3 |
| 2.5 - 7.5 | 13 |
| 7.5 - 12.5 | 13 |
| 12.5 - 17.5 | 15 |
| 17.5 - 22.5 | 17 |
| 22.5 - 27.5 | 3 |
| | 64 |

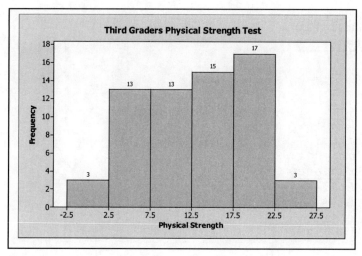

e. Answers will vary.

f. The histograms in parts b and c demonstrate a bimodal distribution whereas the distribution in part d is skewed left. Dotplot shows mode to be 9, which is in the 7-10 class and a cluster centered around 17; while the histogram shows the two modal classes to be 4-7 and 16-22. The mode is not in either modal class.

g. Answers will vary but as the number of classes and the choice of class boundaries change, values will fall into various classes, thereby giving different appearances, all for the same set of data.

2.49 a.
4 – 5	1
5 - 6	9
6 - 7	10
7 – 8	12
8 – 9	4

b. 1

c. 4.5, 5.5, 6.5, 7.5, 8.5,

d.
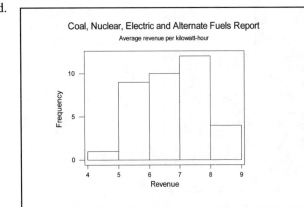

2.51
a. Symmetric: weight of dry cereal per box, breaking strength of certain type of string
b. Uniform: result from rolling a die several hundred times
c. Skewed Right: salaries, high school class sizes
d. Skewed left: hour exam scores
e. Bimodal: heights, weights for groups containing both male and female

2.53 a.
Class Boundaries	Cumulative Frequency
$15 \le x < 25$	12
$25 \le x < 35$	49
$35 \le x < 45$	75
$45 \le x < 55$	94
$55 \le x \le 65$	100

b.
Class Boundaries	Cum. Rel. Frequency
$15 \le x < 25$	0.12
$25 \le x < 35$	0.49
$35 \le x < 45$	0.75

$45 \leq x < 55$ 0.94
$55 \leq x \leq 65$ 1.00

c.

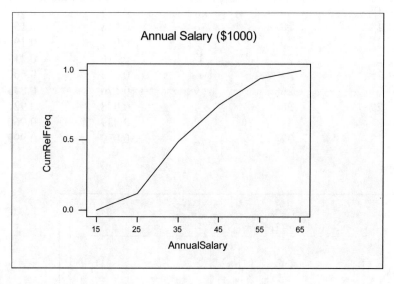

d. $45,000

e. $45,000; They're the same, just asked differently.

2.55   a. Class limits   Cum.Rel.Freq.   (relative frequencies taken from ex. 2.43)

Class limits	Cum.Rel.Freq.
0 - 4	0.08
4 - 8	0.24
8 - 12	0.40
12 - 16	0.80
16 - 20	0.92
20 - 24	0.98
24 - 28	1.00

b.

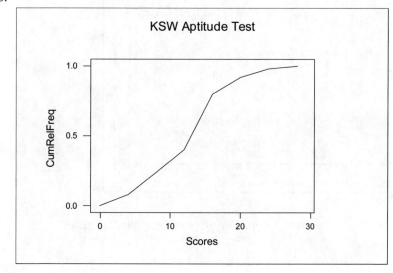

c. $\approx 75\% - 80\%$

2.57 a.

Class Boundar.	Class Midpoints	Freq.	b. Rel. Freq.	d. Cum.Rel Freq
15 – 17	16	2	0.039	0.039
17 – 19	18	6	0.118	0.157
19 – 21	20	2	0.039	0.196
21 – 23	22	11	0.216	0.412
23 – 25	24	13	0.255	0.667
25 – 27	26	9	0.176	0.843
27 – 29	28	4	0.078	0.921
29 – 31	30	2	0.039	0.960
31 – 33	32	2	0.039	0.999

c.

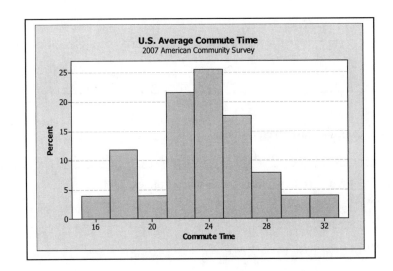

e.

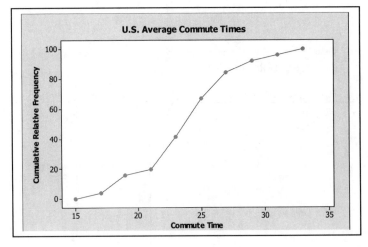

f. ≈ 25 minutes, approx. 70% of the average commute times are less that 25 minutes

SECTION 2.3 EXERCISES

> **NOTE**: A <u>measure of central tendency</u> is a value of the variable. It is that value which locates the "average" value for a set of data. The "average" value may indicate the "middle" or the "center" or the most popular data value.
>
> NOTATION AND FORMULAS FOR MEASURES OF CENTRAL TENDENCY
>
> $\sum x$ = sum of data values
> n = # of data values in the sample
> $\bar{x}$ = sample mean = $\sum x / n$
> $\tilde{x}$ = sample median = middle data value
> $d(\tilde{x})$ = depth or position of median = $(n + 1)/2$
>   mode = the data value that occurs most often
>   midrange = (highest value + lowest value)/2
>
> **NOTE:** REMEMBER TO RANK THE DATA BEFORE FINDING THE MEDIAN.
> $d(\tilde{x})$ only gives the depth or position, not the value of the median. If n is even, $\tilde{x}$ is the
>   average of the two middle values.
>
> See Introductory Concepts (IRM*-Appendix A) for additional information about the Σ (summation) notation.
>
> Computer and calculator commands to find the mean and median can be found in ES11-pp64&66 respectively.
> *(IRM denotes this manual, Instructor's Resource Manual)

2.59 The data resulting from a quantitative variable are numbers with which arithmetic (addition, subtraction, etc.) can be performed. The data resulting from a qualitative variable are 'category' type values such as color. It is not possible to add three colors together, and divide by 3, to obtain a value for the mean color.

2.61 $\bar{x} = \sum x/n = 1429/14 = \underline{\$102.07}$

2.63   a. $\bar{x} = \sum x/n = (16+132+124+191+183+299)/6 = 945/6 = \underline{157.5}$

        Note: For a length of highway with 6 interchanges there are only 5 sections of highway between them.

    b. $\bar{x} = \sum x/n = (16+132+124+191+183+299)/10 = 945/10 = \underline{94.5}$

2.65 $\bar{x} = \sum x/n = [5(425) + 3(750) + 1(1340)]/9 = 5715/9 = \underline{\$635}$

2.67   Ranked data: 70, 72, 73, 74, 76
      $d(\tilde{x}) = (n+1)/2 = (5+1)/2 = 3\text{rd}$; $\tilde{x} = \underline{73}$

2.69   a. $\bar{x} = \sum x/n = 367/10 = \underline{36.7}$
    b.   ranked data: 10 12 20 20 30 35 37 50 53 100
        $d(\tilde{x}) = (n+1)/2 = (10+1)/2 = 5.5\text{th}$; $\tilde{x} = \underline{32.5}$
    c.   the mean is pulled towards the higher values due to the 100

d. $\bar{x} = \Sigma x/n = 267/9 = \underline{29.7}$; $d(\tilde{x}) = (n+1)/2 = (9+1)/2 = $ 5th; $\tilde{x} = \underline{30}$
e. mean

2.71 mode = $\underline{2}$

2.73 a. $\bar{x} = \Sigma x/n = (9+6+7+9+10+8)/6 = 49/6 = 8.166 = \underline{8.2}$
Ranked data: 6, 7, 8, 9, 9, 10
$d(\tilde{x}) = (n+1)/2 = (6+1)/2 = $ 3.5th; $\tilde{x} = \underline{8.5}$
mode = $\underline{9}$
midrange = $(L+H)/2 = (6+10)/2 = 16/2 = \underline{8.0}$
b. Answers will vary. All show centers.

2.75 a. $\bar{x} = \Sigma x/n = (3+5+6+7+7+8)/6 = 36/6 = \underline{6.0}$
b. $d(\tilde{x}) = (n+1)/2 = (6+1)/2 = $ 3.5th; $\tilde{x} = (6+7)/2 = \underline{6.5}$
c. mode = $\underline{7}$
d. midrange = $(H+L)/2 = (8+3)/2 = 11/2 = \underline{5.5}$

2.77 {21, 21, 26, 27, 28, 32, 36, 38, 45, 48}
a. $\bar{x} = \Sigma x/n = 322/10 = \underline{32.2}$ b. $d(\tilde{x}) = (n+1)/2 = (10+1)/2 = $ 5.5th; $\tilde{x} = \underline{30}$
c. midrange = $(H+L)/2 = (21+48)/2 = \underline{34.5}$ d. mode = $\underline{21}$

2.79 a.

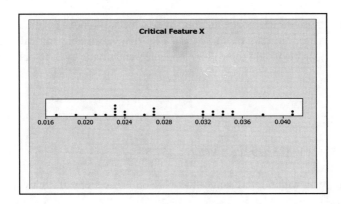

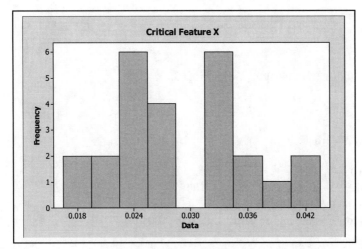

b. $\bar{x} = \sum x/n = 0.714/25 = 0.02856 = \underline{0.0286}$
c. $d(\tilde{x}) = (n+1)/2 = (25+1)/2 = 13\text{th}$; $\tilde{x} = \underline{0.027}$
d. midrange $= (L+H)/2 = (0.017+0.041)/2 = \underline{0.029}$
e. mode $= \underline{0.023}$
f. bimodal distribution; central tendency statistics all fall around the 0.030 center split of the data
g. Answers will vary but problem could be there are two populations.

2.81 a.

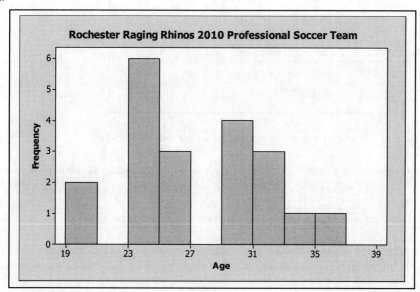

b. skewed right   c. Answers will vary

c. 27.05, 25.5

d. median, mean pulled by few high values

2.83 a.

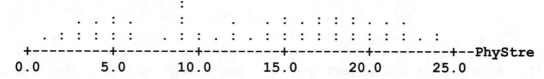

b. mode = $\underline{9}$

c.

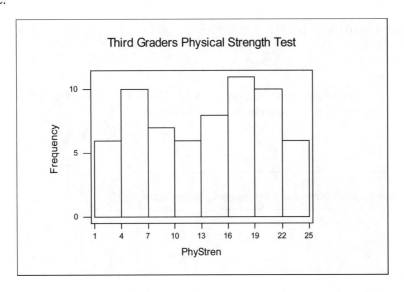

d. The distribution appears to be bimodal. Modal classes are 4-7, and 16-19.

e. Dotplot shows mode to be 9, which is in the 7-10 class; while the histogram shows the two modal classes to be 4-7 and 16-19. The mode is not in either modal class.

f. No. In an ungrouped distribution there is only one numerical value per class.

g. The mode is simply the single data value that occurs most often, while a modal class results from data tending to bunch-up forming a cluster of data values, not necessarily all of one value.

2.85 a. & b.

	Runs at Home	Runs Away
Mean	4.775	4.527
Median	4.735	4.625
Maximum	5.99	5.14
Minimum	3.57	4.00
Midrange	4.78	4.57

c. teams score more runs at home

2.87 a. A quick look at the data listed suggests, Yes. The number of female licensed drivers is larger for most of the states listed.

b. M/F Ratio
0.97970 0.95137 0.97060 1.00864 0.97796 1.08152 0.93919
1.01570 1.02056 0.97678 0.96740 0.92790 1.00073 1.03140
0.98752 0.97085 1.00203 1.01321

c. Near 1.0 means little difference.
Greater than 1.0 means more male drivers.
Less than 1.0 means more female drivers.

d.

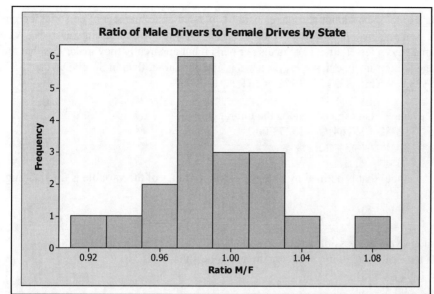

e. The distribution of "Ratio M/F" is mounded and appears to be somewhat normal except for the one value that is somewhat larger than the others, thus making the data skewed to the right.

f. $\overline{x} = \sum x/n = 17.8231/18 = 0.990172 = 0.990$

g. The value to the extreme right means that state has considerably more male drivers than female, approximately 10% more. The value to the extreme left means that state has fewer male drivers than female drivers and since the value is approximately 0.92, there are 8% fewer male drivers.

h. Answers will vary.

i.

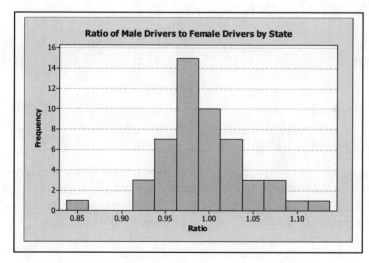

$\overline{x} = \sum x/n = = 50.6136/51 = 0.992423 = 0.992$

j. The results are very similar.

k. Answers will vary.

2.89 a. 'Taxes per capita' means amount of taxes paid per person while 'percent of personal income' is the percent of personal income that is paid in taxes. They are two different variables. A person in Alabama pays a lesser amount of taxes per person but apparently they make less that the percentage of their income that goes for taxes is actually larger than in South Dakota. $2782 is 9.6% of $28,979 while $2842 is 9.1% of $31,231.

b. The only 'average' that can be found is the midrange.
Midrange = ($2782 + $7764)/2 = $5273.00

c. Midrange = (9.1 + 16.6)/2 = 12.85%

d. The only 'average' that is defined by only the extreme values of the variable is the midrange.

2.91 Many different answers are possible.
a. Σx needs to be 500; therefore, need any three numbers that total 330.
<u>100, 100, 130</u> [70, 100]
$\bar{x} = \Sigma x/n = 500/5 = 100$, ck

b. Need two numbers smaller than 70 and one larger.
_, _, 70, _, 100: <u>50, 60, 80</u> [70, 100]
$d(\tilde{x}) = (n+1)/2 = (5+1)/2 = 3rd$; $\tilde{x} = 70$, ck

c. Need multiple 87's. <u>87, 87, 87</u> [70, 100]   mode = 87, ck

d. Need any two numbers that total 140 for the extreme values where one is 100 or larger.
_, _, _, 70, 100
<u>40, 50, 60</u> [70, 100]
midrange = (L+H)/2 = (40+100)/2 = 70, ck

e. Need two numbers smaller than 70 and one larger than 70 so that their total is 330.
_, _, 70, _, 100;
<u>60, 60, 210</u> [70, 100]
$\bar{x} = \Sigma x/n = 500/5 = 100$, ck
$d(\tilde{x}) = (n+1)/2 = (5+1)/2 = 3rd$; $\tilde{x} = 70$, ck

f. Need two numbers of 87 and a third number large enough so that the total of all five is 500.
<u>87, 87, 156</u> [70, 100]
$\bar{x} = \Sigma x/n = 500/5 = 100$, ck; mode = 87, ck

g. Assuming only positive values of x, the mean equal to 100 requires the five data to total 500 and the midrange of 70 requires the total of L and H to be 140; 40, _, 70, _, 100; that is a sum of 210, meaning the other two data must total 290. One of the last two numbers must be larger than 145, which would then become H and change the midrange. Impossible.

h. Assuming only positive values of x, there must be two 87's in order to have a mode, and there can only be two data larger than 70 in order for 70 to be the median. _, 70, 87, 87, 100; Impossible

SECTION 2.4 EXERCISES

NOTE: A <u>measure of dispersion</u> is a value of the variable. It is that value which describes the amount of variation or spread in a data set. A small measure of dispersion indicates data that are closely grouped, whereas, a large value indicates data that are more widely spread.

MEASURES OF DISPERSION - THE SPREAD OF THE DATA

<u>Range</u> = highest value - lowest value

<u>Standard Deviation</u> - s - the average distance a data value is from the mean

$$s = \sqrt{\Sigma(x - \bar{x})^2 / (n - 1)}$$

<u>Variance</u> - s^2 - the square of the standard deviation   (i.e., before taking the square root)

For exercises 2.97-2.101, be sure that the $\Sigma(x - \bar{x}) = 0$.

NOTE: Standard deviation and/or variance cannot be negative. This would indicate an error in sums or calculations.

See Introductory Concepts (IRM-Appendix A) for additional information about Rounding Off.

Computer and calculator commands to find the range and standard deviation can be found in ES11-p78.

If using a non-graphing statistical calculator (one that lets you input the data points) to find the standard deviation of a sample, use the $\sigma(n-1)$ or s_x key. $\sigma(n)$ or σ_x would give the population standard deviation; that is, divide by "n" instead of "n-1".

2.93   a.   range = H - L = $7764 - $2782 = $4982

       b.   range = H - L = 16.6% - 9.1% = 7.5%

2.95   The mean is the 'balance point' or 'center of gravity' to all the data values. Since the weights of the data values on each side of $\bar{x}$ are equal, $\Sigma(x - \bar{x})$ will give a positive amount and an equal negative amount, thereby canceling each other out.

Algebraically: $\Sigma(x - \bar{x}) = \Sigma x - n\bar{x} = \Sigma x - n \cdot (\Sigma x/n) = \Sigma x - \Sigma x = 0$

2.97 a. 1st: find mean, $\bar{x} = \sum x/n = 25/5 = 5$

x	x − $\bar{x}$	(x − $\bar{x}$)²
1	-4	16
3	-2	4
5	0	0
6	1	1
10	5	25
∑ 25	0	46

$s^2 = \sum(x-\bar{x})^2/(n-1)$

$= 46/4 = \underline{11.5}$

b.
x	x²
1	1
3	9
5	25
6	36
10	100
25	171

$SS(x) = \sum x^2 - ((\sum x)^2/n)$

$= 171 - ((25)^2/5)$
$= 171 - 125 = 46$

$s^2 = SS(x)/(n-1) = 46/4 = \underline{11.5}$

c. Both results are the same.

2.99 a. range = H - L = 8 - 3 = <u>5</u>

b. 1st: find mean, $\bar{x} = \sum x/n = 36/6 = 6$

x	x − $\bar{x}$	(x − $\bar{x}$)²
3	-3	9
5	-1	1
6	0	0
7	1	1
7	1	1
8	2	4
∑ 36	0	16

$s^2 = \sum(x-\bar{x})^2/(n-1)$

$= 16/5 = \underline{3.2}$

c. $s = \sqrt{s^2} = \sqrt{3.2} = 1.789 = \underline{1.8}$

2.101 a. 1st: find mean, $\bar{x} = \sum x/n = 104/15 = 6.9$

x	x − $\bar{x}$	(x − $\bar{x}$)²
4	-2.9	8.41
5	-1.9	3.61
5	-1.9	3.61
6	-0.9	0.81
6	-0.9	0.81
6	-0.9	0.81
7	0.1	0.01
7	0.1	0.01
7	0.1	0.01
7	0.1	0.01
8	1.1	1.21
8	1.1	1.21
8	1.1	1.21
9	2.1	4.41
11	4.1	16.81
∑ 104	+0.5*	42.95

$s^2 = \sum(x-\bar{x})^2/(n-1)$

$= 42.95/14$

$= 3.0679$

$= \underline{3.1}$

*The 0.5 is due to the round-off error introduced by using $\overline{x} = 6.9$ instead of 6.933333.

b.
x	x²
4	16
5	25
5	25
6	36
6	36
6	36
7	49
7	49
7	49
7	49
8	64
8	64
8	64
9	81
11	121
Σ 104	764

$SS(x) = \Sigma x^2 - ((\Sigma x)^2/n)$

$= 764 - ((104)^2/15)$

$= 764 - 721.0667 = 42.93333$

$s^2 = SS(x)/(n-1)$

$= 42.9333/14 = 3.0667 = \underline{3.1}$

c. $s = \sqrt{s^2} = \sqrt{3.0667} = 1.751 = \underline{1.8}$

2.103 a. Original data: $n = 6$, $\Sigma x = 37,116$, $\Sigma x^2 = 229,710,344$

$SS(x) = \Sigma x^2 - ((\Sigma x)^2/n) = 229,710,344 - (37,116^2/6) = 110,768.0$

$s^2 = SS(x)/(n-1) = 110,768.0/5 = \underline{22,153.6}$

b. Smaller numbers: $n = 6$, $\Sigma x = 1,116$, $\Sigma x^2 = 318,344$

$SS(x) = \Sigma x^2 - ((\Sigma x)^2/n) = 318,344 - (1,116^2/6) = 110,768.0$

$s^2 = SS(x)/(n-1) = 110,768.0/5 = \underline{22,153.6}$

Both sets of data have the same variance.

2.105 a.

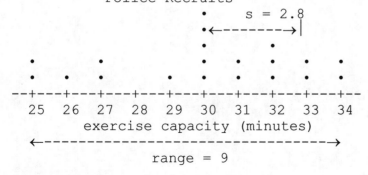

b. $\overline{x} = 601/20 = 30.05$

c. range = H - L = 34 - 25 = $\underline{9}$

d. $n = 20$, $\sum x = 601$, $\sum x^2 = 18{,}209$
   $SS(x) = \sum x^2 - ((\sum x)^2/n) = 18{,}209 - (601^2/20) = 148.95$
   $s^2 = SS(x)/(n-1) = 148.95/19 = 7.83947 = \underline{7.8}$

e. $s = \sqrt{s^2} = \sqrt{7.83947} = 2.7999 = \underline{2.8}$

f. See graph in (a).

g. Except for the value $x = 30$, the distribution looks rectangular. Range is a little more than 3 standard deviations.

2.107 a.

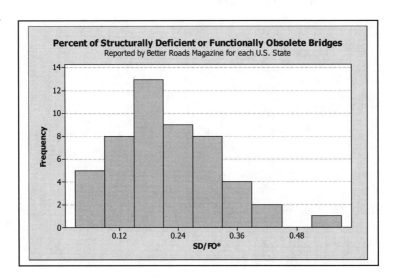

b. The variable "%SD/FO" appears to have a skewed right distribution.

c. $\overline{x} = 0.2176$

d. $\tilde{x} = 0.20$

e. range = H - L = 0.55 – 0.03 = $\underline{0.52}$

f. s = $\underline{0.1038}$

2.109 Set 1: $\bar{x} = 250/5 = 50$ | Set 2: $\bar{x} = 250/5 = 50$

x	x − $\bar{x}$	(x − $\bar{x}$)²		x	x − $\bar{x}$	(x − $\bar{x}$)²
46	−4	16		30	−20	400
55	+5	25		55	+5	25
50	0	0		65	+15	225
47	−3	9		47	−3	9
52	+2	4		53	+3	9
250	0	54		250	0	668

	Σx	Σ(x − $\bar{x}$)	Σ(x − $\bar{x}$)²	Range
Set 1:	250	0	54	9
Set 2:	250	0	668	35

The values of SS(x) [recall SS(x) = Σ(x−$\bar{x}$)²], and range all reflect the fact that there is more variability in the data forming set 2 than in the data of set 1.

2.111 The statement is incorrect. The standard deviation can never be negative. There has to be an error in the calculations or a typographical error in the statement.

2.113 Different answers will result depending on the relative size of the data making up the two sets will determine your answer.

<u>Larger</u>, if the data in one set are larger in value than the data values of the first set; the combined set is more dispersed.
Set I: {4,6,10,14,16} and Set II: {14,16,20,24,26}

Set I: n = 5, Σx = 50, Σx² = 604

SS(x) = Σx² − ((Σx)²/n) = 604 − (50²/5) = 104

s² = SS(x)/(n−1) = 104/4 = 26

s = $\sqrt{s^2}$ = $\sqrt{26}$ = 5.099 = <u>5.1</u>

Set II: n = 5, Σx = 100, Σx² = 2104

SS(x) = Σx² − ((Σx)²/n) = 2104 − (100²/5) = 104

s² = SS(x)/(n−1) = 104/4 = 26

s = $\sqrt{s^2}$ = $\sqrt{26}$ = 5.099 = <u>5.1</u>

Together, Set I and Set II;

n = 10, Σx = 150, Σx² = 2708

SS(x) = Σx² − ((Σx)²/n) = 2708 − (150²/10) = 458

s² = SS(x)/(n−1) = 458/9 = 50.88888

s = $\sqrt{s^2}$ = $\sqrt{50.88888}$ = 7.133644 = <u>7.13</u>

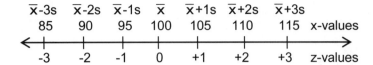

SECTION 2.5 EXERCISES

NOTE: A <u>measure of position</u> is a value of the variable. It is that value which divides the set of data into two groups: those data smaller in value than the measure of position, and those larger in value than the measure of position.

To find any measure of position:

    1. Rank the data - <u>DATA MUST BE RANKED LOW TO HIGH</u>

    2. Determine the depth or position in two separate steps:
        a. Calculate $nk/100$, where n = sample size, k = desired percentile
        b. Determine $d(P_k)$:

            If $nk/100$ = integer $\Rightarrow$ add .5 (value will be halfway between 2 integers)
            If $nk/100$ = decimal $\Rightarrow$ round up to the nearest whole number
    3. Locate the value of P_k

REMEMBER:
$Q_1 = P_{25}$ = 1st quartile - 25% of the data lies below this value

$Q_2 = P_{50} = \tilde{x}$ = 2nd quartile - 50% of the data lies below this value

$Q_3 = P_{75}$ = 3rd quartile - 75% of the data lies below this value

2.115   a.   91 is in the 44th position from the Low value of 39
               91 is in the 7th position from the High value of 98

        b. $nk/100 = (50)(20)/100 = 10.0$; therefore $d(P_{20})$ = 10.5th from L
           $P_{20} = (64+64)/2 = \underline{64}$

          $nk/100 = (50)(35)/100 = 17.5$; therefore $d(P_{35})$ = 18th from L
           $P_{35} = \underline{70}$

        c. $nk/100 = (50)(20)/100 = 10.0$; therefore $d(P_{80})$ = 10.5th from H
           $P_{80} = (88+89)/2 = \underline{88.5}$

          $nk/100 = (50)(5)/100 = 2.5$; therefore $d(P_{95})$ = 3rd from H
           $P_{95} = \underline{95}$

2.117 a.

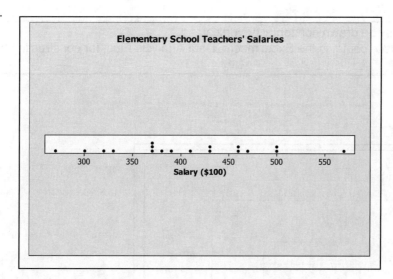

Ranked data:
269 295 317 326 367 367 371 376 391 413 433 434 455 458 471 495 501 574

b. 2nd from the L, 17th from the H

c. nk/100 = (18)(25)/100 = 4.5; therefore d(Q$_1$) = 5th from L
   Q_1 = 367 = $36,700

d. nk/100 = (18)(75)/100 = 13.5; therefore d(Q$_3$) = 14th from L
   Q_3 = 458 = $45,800

2.119 Ranked data:

2.6 2.7 3.4 3.6 3.7   3.9 4.0 4.4 4.8 4.8
4.8 5.0 5.1 5.6 5.6   5.6 5.8 6.8 7.0 7.0

a. nk/100 = (20)(25)/100 = 5.0; therefore d(P$_{25}$) = 5.5th
   Q_1 = P_{25} = (3.7 + 3.9)/2 = 3.8

   nk/100 = (20)(75)/100 = 15.0; therefore d(P$_{75}$) = 15.5th
   Q_3 = P_{75} = (5.6 + 5.6)/2 = 5.6

b. midquartile = (Q$_1$ + Q$_3$)/2 = (3.8 + 5.6)/2 = 4.7

c. nk/100 = (20)(15)/100 = 3.0; therefore d(P$_{15}$) = 3.5th
   P_{15} = (3.4+3.6)/2 = 3.5

   nk/100 = (20)(33)/100 = 6.6; therefore d(P$_{33}$) = 7th
   P_{33} = 4.0

   nk/100 = (20)(90)/100 = 18.0; therefore d(P$_{90}$) = 18.5th
   P_{90} = (6.8+7.0)/2 = 6.9

Box-and-whisker displays may be drawn horizontal or vertical.
The cengagebrain.com website contains the Excel macro, Data Analysis Plus, for constructing box-and-whisker displays.

2.121

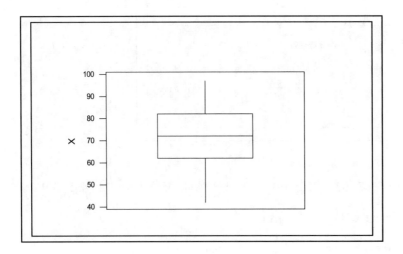

2.123 a.

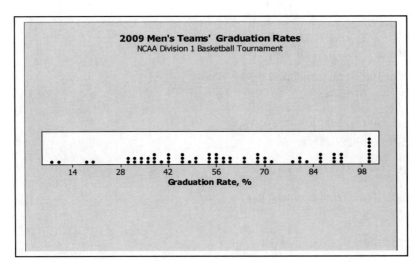

b.
Stem-and-Leaf Display: Graduation Rate, %

```
Stem-and-leaf of Graduation Rate, %
N   = 63
Leaf Unit = 1.0


    1    0   8
    3    1   07
    5    2   09
   15    3   0113466788
   23    4   01225667
  (10)   5   0033355677
   30    6   00347779
   22    7   017
   19    8   002666999
   10    9   122
    7   10   0000000
```

c. 5-number summary: 8, 40, 57, 86, 100

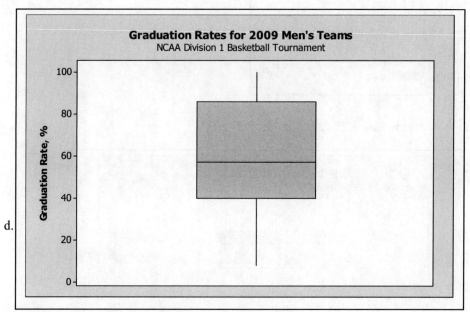

d.

e. Slightly Skewed left, centered somewhat around 60%

f. Answers will vary.

2.125 a.

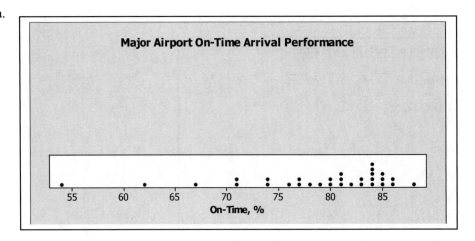

b.
```
Stem-and-leaf of On-Time, %
N   = 31
Leaf Unit = 1.0


  1    5  3
  1    5
  2    6  2
  3    6  6
  7    7  0144
 12    7  67779
(16)   8  0011112333333444
  3    8  668
```

c. 5- number summary: 53.5, 76.2, 81.3, 83.9, 88.2

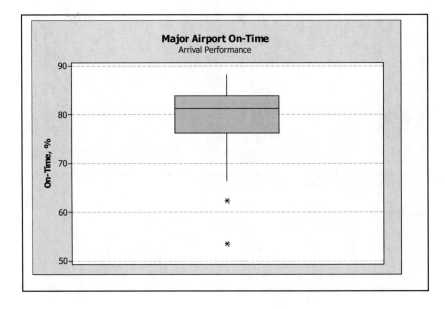

d. $nk/100 = (31)(10)/100 = 3.1$; therefore $d(P_{10}) = 4^{th}$ and $P_{10} = 70.7$

$nk/100 = (31)(20)/100 = 6.2$; therefore $d(P_{20}) = 7^{th}$ and $P_{20} = 74.4$

e. The distribution is skewed to the left with the 2 airports having the lowest rate of on-time performance being quite separated for the others.

f. Travelers more interested in being on time and want the best on-time performance rate. Only the lowest are poor.

g. The airports with the lowest percentages are quite different from the rest of the airports. The two airports are EWR (Newark, NJ) and LGA (LaGuardia, NY).

2.127 The distribution needs to be symmetric

---

z is a measure of position. It gives the number of standard deviations a piece of data is from the mean. It will be <u>positive</u> if x is to the <u>right of the mean</u> (larger than the mean) and <u>negative</u>, if x is to the <u>left of the mean</u> (smaller than the mean). Keep 2 decimal places. (hundredths)

$z = (x - mean)/st. dev.$ $z = (x - \bar{x})/s$

2.129 $z = (x - mean)/st.dev.$

for $x = 92$, $z = (92 - 72)/12 = \underline{1.67}$
for $x = 63$, $z = (63 - 72)/12 = \underline{-0.75}$

2.131 $z = (x - mean)/st.dev.$

a. for $x = 54$, $z = (54 - 74.2)/11.5 = \underline{-1.76}$
b. for $x = 68$, $z = (68 - 74.2)/11.5 = \underline{-0.54}$
c. for $x = 79$, $z = (79 - 74.2)/11.5 = \underline{0.42}$
d. for $x = 93$, $z = (93 - 74.2)/11.5 = \underline{1.63}$

2.133 If $z = (x - mean)/st.dev$; then $x = (z)(st.dev) + mean$

a. for $z = 0.0$, $x = (0.0)(20.0) + 120 = \underline{120}$
b. for $z = 1.2$, $x = (1.2)(20.0) + 120 = \underline{144.0}$
c. for $z = -1.4$, $x = (-1.4)(20.0) + 120 = \underline{92.0}$
d. for $z = 2.05$, $x = (2.05)(20.0) + 120 = \underline{161.0}$

2.135 a. Ranked data:

0.03 0.05 0.05 0.06 0.07 0.10 0.13 0.14 0.14 0.14 0.14 0.14 0.14 0.16 0.16 0.16 0.16
0.17 0.17 0.17 0.19 0.20 0.20 0.20 0.20 0.20 0.21 0.21 0.21 0.22 0.22 0.23 0.23 0.24
0.25 0.29 0.29 0.30 0.30 0.31 0.31 0.32 0.32 0.34 0.35 0.36 0.37 0.39 0.39 0.55

b.

Lowest Value	First Quartile	Median	Third Quartile	Highest Value
0.03	0.14	0.20	0.30	0.55

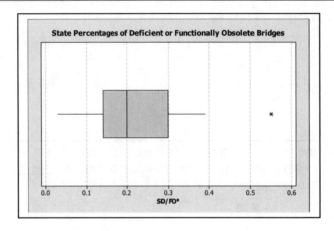

c. Midquartile = (0.14 + 0.30) / 2 = 0.22
Interquartile range = 0.30 − 0.14 = 0.16

d.

	California	Hawaii	Nebraska	Oklahoma	Rhode Is.
z-core	-0.75	1.66	-1.42	0.22	3.20

2.137 for A: z = (85 - 72)/8 = 1.625
for B: z = (93 - 87)/5 = 1.2
Therefore, A has the higher relative position.

SECTION 2.6 EXERCISES

The empirical rule applies to a normal distribution.
    Approximately 68% of the data lies within 1 standard deviation of the mean.
    Approximately 95% of the data lies within 2 standard deviations of the mean.
    Approximately 99.7% of the data lies within 3 standard deviations of the mean.

Chebyshev's theorem applies to any shape distribution.
    At least 75% of the data lies within 2 standard deviations of the mean.
    At least 89% of the data lies within 3 standard deviations of the mean.

2.139 From 175 through 225 words, inclusive.

2.141 Nearly all of the data, 99.7%, lies within 3 standard deviations of the mean.

2.143 a. 97.6 is 2 standard deviations above the mean
    $\{z = (97.6-84.0)/6.8 = 2.0\}$, therefore 2.5% of the time more than 97.6 hours will be required.

    b. 95% of the time the time to complete will fall within 2 standard deviations of the mean, that is $84.0 \pm 2(6.8)$ or from 70.4 to 97.6 hours.

2.145 a. 50%   b. $0.50 - 0.34 = 0.16 =$ 16%
    c. $0.50 + 0.34 = 0.84 =$ 84%   d. $0.34 + 0.475 = 0.815 =$ 81.5%

Chebyshev's theorem
    At least $\left(1 - \dfrac{1}{k^2}\right)\%$ of the data lies within k standard deviations of the mean. (k > 1)

2.147 a. at least 75%   b. at least 89%

2.149 a. at most 11%   b. at most 6.25%

2.151 The interval 5.75 to 132.35 represents the mean plus or minus two standard deviations.

    a. According to Chebyshev's theorem, then we can be sure that at least 75% of the distribution is contained within the interval.

    b. If the distribution is normal, then approximately 95% of the distribution is contained within the interval.

    c. Answers will vary. At three standard deviations both endpoints would be beyond the range of the scale from 0 to 100.

2.153 a.

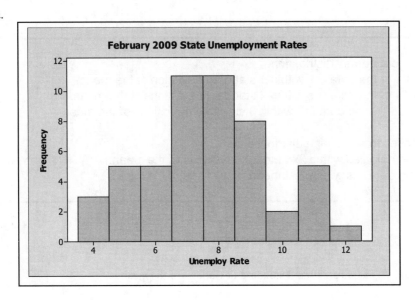

b. Yes, mounded in the middle.

c. $\bar{x} = 7.649$, $s = 1.969$

Ranked data:
3.9 4.2 4.3 4.6 4.9 5.1 5.3 5.4 5.5 5.7 5.9 6.0 6.0 6.5 6.5 6.6 6.6
6.7 6.8 7.0 7.2 7.4 7.4 7.4 7.5 7.7 7.8 7.8 8.0 8.0 8.1 8.2 8.3 8.4
8.4 8.6 9.1 9.1 9.2 9.3 9.4 9.4 9.4 9.9 10.1 10.5 10.5 10.7 10.8 11.0 12.0

d. $\bar{x} \pm s = 7.649 \pm 1.969 = 5.68$ to 9.618
66.7% of the data (34/51) is between 5.68 and 9.618

$\bar{x} \pm 2s = 7.649 \pm 2(1.969) = 9.649 \pm 3.938$ or 3.711 to 11.587
98.0% of the data (50/51) is between 3.711 and 11.587.

$\bar{x} \pm 3s = 7.649 \pm 3(1.969) = 9.649 \pm 5.907$ or 1.742 to 13.556
100% of the data (51/51) is between 1.742 and 13.556.

The empirical rule says approximately 68%, 95%, and 99.7% of the data are within one, two, and three standard deviations, respectively; the 66.7%, 98% and 100% do somewhat agree with the rule; based solely on this information, the distribution can be considered 'approximately' normal and agrees with answer in part b.

e.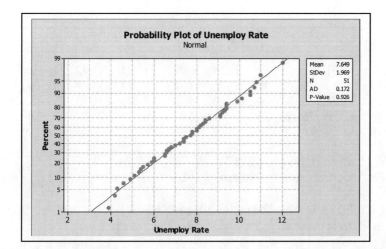

The points closely follow the straight line and the p-value is greater than 0.05, therefore it follows that unemployment rates are approximately normally distributed.

2.155 a.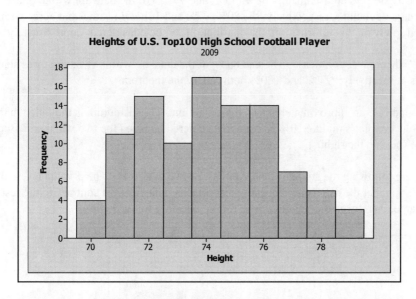

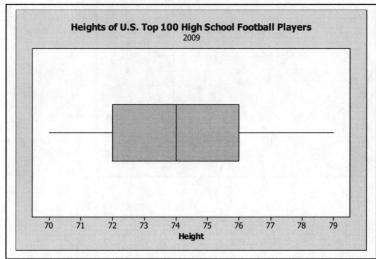

b. $\bar{x} = 74.09$, s = 2.292

c. Ranked data:
    70 70 70 70 71 71 71 71 71 71 71 71 71 71 71 72 72 72 72 72 72 72 72
    72 72 72 72 72 72 72 73 73 73 73 73 73 73 73 73 73 74 74 74 74 74 74
    74 74 74 74 74 74 74 74 74 74 74 75 75 75 75 75 75 75 75 75 75 75 75
    75 75 76 76 76 76 76 76 76 76 76 76 76 76 76 77 77 77 77 77 77 77
    78 78 78 78 78 79 79 79

d. $\bar{x} \pm 1s = 74.09 \pm (2.292)$ or <u>71.798</u> to <u>76.382</u>
    70% of the data (70/100) is between 71.798 and 76.382.

    $\bar{x} \pm 2s = 74.09 \pm 2(2.292) = 74.09 \pm 4.584$ or <u>69.506</u> to <u>78.674</u>
    97% of the data (97/100) is between 69.506 and 78.674.

    $\bar{x} \pm 3s = 74.09 \pm 3(2.292) = 74.09 \pm 6.876$ or <u>67.214 to 80.966</u>
    100% of the data (100/100) is between 67.214 and 80.966.

e. The empirical rule says approximately 68%, 95%, and 99.7% of the data are within one, two, and three standard deviations, respectively; the 70%, 97% and 100% do somewhat agree with the rule; based solely on this information, the distribution can be considered 'approximately' normal.

f. Chebyshev's theorem says at least 75%, and 89%, of the data are within two, and three standard deviations, respectively; 97%, and 100% both satisfy the theorem.

g. The graphs indicate an approximately normal distribution. The histogram is mounded in the center and the boxplot's middle 50% is just about centered between the two whiskers. Both show a slight skewness to the right.

h. The points are following the straight line except at the two extremes. The p-value is less than 0.005 indicating that the data is not normally distributed. This result is contrary to the results found in part e, thereby encouraging various forms of testing for normality.

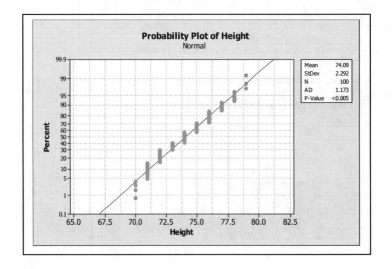

> Helpful hint to use when expecting to count data on histogram:
>
> Minitab: While on the Histogram dialogue box,
> Select: **Labels > Data Labels...**
> Select: **Use y-value labels**
>
> This will direct the computer to print the frequency of each class above its corresponding bar.
>
> Excel: Returns a frequency distribution with the histogram.
> TI-83/84 Plus: Use the TRACE and arrow keys.

2.157 a. Answers will vary.

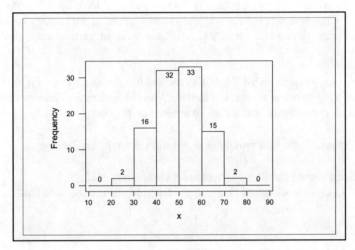

Within one standard deviation, 40 to 60, is 33 + 32 or 65 of the 100 data. 65%

Within two standard deviations, 30 to 70, is 16+32+33+15 or 96 of the 100 data. 96%

Within three standard deviations, 20 to 80, is all 100 of the data, or 100%.

The above results are extremely close to what the empirical rule claims will happen.

b, c, d. Not all sets of 100 data will result in percentages this close. However, expect very similar results to occur most of the time.

SECTION 2.7 EXERCISES

2.159 a. A bar graph; a person's age is used to identify the age group; age is not used as the variable.
        b. Smaller groups, 18 – 19-year olds are different kinds of shoppers than 35 – 39 year olds

2.161 a. Here's a few of the more obvious ones. If you research and find the original information, you'll find more.
1. The ranking information covers an 11-year period while the tuition information covers 35 years. Yet they are shown horizontally as being the same.
2. The units for tuition (share of median income) and ranking (rank number) are totally different, yet the vertical axis treats them as having common units.
3. The ranking graph is placed below the tuition graph creating the impression that cost exceeds quality. Since the vertical scale is meaningless, either line could have been "on top".
4. The sharp "drop" in the ranking graph actually represents an improved ranking. A ranking of "15th best" is not better than a ranking of "6th best", however the vertical scale used makes it look like it is.

        b. The caption under the graph suggests that Cornell's rank has been erratic by varying from 6th to 15th on the national ranking over the 12 years reported. With the hundreds of colleges and universities that exist, to consistently hold a rank like this is quite good.

The "upside-down" scale with the best ranking at the bottom is totally misleading.

2.163 a. The ribs break the circle into equal pie slices. There are 8 slices, each is 12.5%.
        b. Answers will vary. The circle is distorted on the upper left side making the 46% look less than the 24%.

2.165

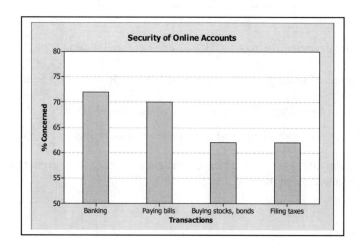

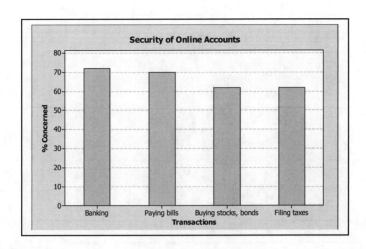

Answers will vary but the 50 to 80 scale gives a misleading conclusion, namely that there is a significant difference among the responses. Starting the percent scale at 0, gives the true perspective of the relationship among the percents.

CHAPTER EXERCISES

2.167 a.

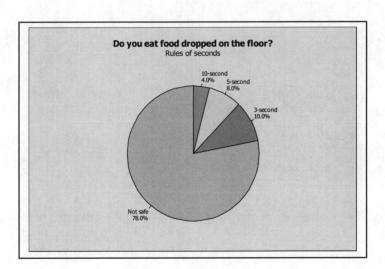

b. (300)(0.10) = 30 respond 'Three-second rule'
(300)(0.08) = 24 respond 'Five-second rule'
(300)(0.04) = 12 respond '10-second rule'
(300)(0.78) = 234 respond 'Not safe'

2.169 a.

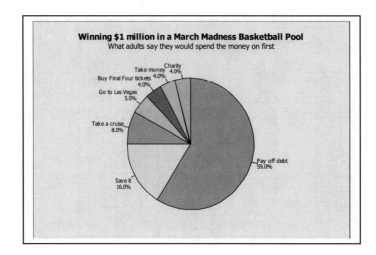

b. Answers may vary. Circle graph demonstrates the smaller percentages differences. It is easier to read.

2.171 a.

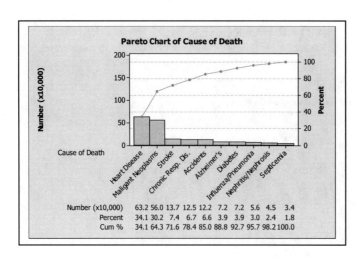

b. Answers will vary. Two leading causes of death are heart disease and malignant neoplasms.

2.173 a. numerical b. attribute c. numerical
d. attribute e. numerical

2.175 a. Mean increased; when one data increases, the sum increases.

b. Median is unchanged; the median is affected only by the middle value(s).

c. Mode is unchanged.

d. Midrange increased; an increase in either extreme value increases the sum H+L.

e. Range increased; difference between high and low values increased.

f. Variance increased; data are now more spread out.

g. Standard deviation increased; data are now more spread out.

2.177 Data summary: n = 8, $\sum x = 36.5$, $\sum x^2 = 179.11$

a. $\bar{x} = \sum x/n = 36.5/8 = 4.5625 = \underline{4.56}$

b. $s = \sqrt{[\sum x^2 - ((\sum x)^2/n)]/(n-1)}$
$= \sqrt{[179.11 - (36.5^2/8)]/7}$
$= \sqrt{1.79696} = 1.3405 = \underline{1.34}$

c. These percentages seem to average very closely to 4%.

2.179 Data summary: n = 118, $\sum x = 2364$

a. $\bar{x} = \sum x/n = 2364/118 = 20.034 = \underline{20.0}$

b. $d(\tilde{x}) = (n+1)/2 = (118+1)/2 = 59.5$th; $\tilde{x} = (17+17)/2 = \underline{17}$

c. mode = $\underline{16}$

d. $nk/100 = (118)(25)/100 = 29.5$; therefore $d(P_{25}) = 30$th
$Q_1 = P_{25} = \underline{15}$

$nk/100 = (118)(75)/100 = 88.5$; therefore $d(P_{75}) = 89$th
$Q_3 = P_{75} = \underline{21}$

e. $nk/100 = (118)(10)/100 = 11.8$; therefore $d(P_{10}) = 12$th
$P_{10} = \underline{14}$

$nk/100 = (118)(95)/100 = 112.1$; therefore $d(P_{95}) = 113$
$P_{95} = \underline{43}$

2.181 Data: 63 67 66 63 69   74 72 70 71 71
72 70 75 85 84   85 85 86 94 91   90 90 95 105 104

Data summary: n = 25, $\sum x = 1{,}997$, $\sum x^2 = 163{,}205$

$\bar{x} = \sum x/n = 1997/25 = 79.88 = \underline{79.9}$

$SS(x) = \sum x^2 - ((\sum x)^2/n) = 163{,}205 - (1{,}997^2/25) = 3684.64$

$s^2 = SS(x)/(n-1) = 3684.64/24 = 153.5267$

$s = \sqrt{s^2} = \sqrt{153.5267} = 12.3906 = \underline{12.4}$

2.183 a. The population is U.S. commercial airline industry; three variables are involved: number of reports, numbers of passengers, number of reports per 1000 airline passengers.
  b. Data; they are values of the variable, number of reports per 1000 airline passengers.
  c. Statistic; it summarizes the data for one month. It is used to estimate the parameter, the value for the whole population.
  d. No. The 20 airline values are a sample of the airline industry, not all are included here.

2.185 Data summary: $n = 100$, $\sum x = 1315$

  a. $\bar{x} = \sum x/n = 1315/100 = \underline{13.15}$

  b. $d(\tilde{x}) = (n+1)/2 = 50.5$th; $\tilde{x} = \underline{13.85}$

  c. mode = $\underline{15.0}$

  d. midrange = $(H+L)/2 = (15.8+10.1)/2 = \underline{12.95}$

  e. range = $H - L = 15.8 - 10.1 = \underline{5.7}$

  f. $nk/100 = (100)(25)/100 = 25$; therefore $d(P_{25}) = 25.5$th
     $Q_1 = P_{25} = \underline{10.95}$

  $nk/100 = (100)(75)/100 = 75$; therefore $d(P_{75}) = 75.5$th
     $Q_3 = P_{75} = \underline{14.9}$

  g. midquartile = $(Q_1 + Q_3)/2 = (10.95+14.9)/2 = \underline{12.925}$

  h. $nk/100 = (100)(35)/100 = 35$; therefore $d(P_{35}) = 35.5$th
     $P_{35} = \underline{12.05}$

  $nk/100 = (100)(64)/100 = 64$; therefore $d(P_{64}) = 64.5$th
     $P_{64} = \underline{14.5}$

  i.
Class limits	freq	rel.fr	k. cum.rel. fr
10.0 - 10.5	15	0.15	0.15
10.5 - 11.0	10	0.10	0.25
11.0 - 11.5	6	0.06	0.31
11.5 - 12.0	3	0.03	0.34
12.0 - 12.5	4	0.04	0.38
12.5 - 13.0	4	0.04	0.42
13.0 - 13.5	2	0.02	0.44
13.5 - 14.0	9	0.09	0.53
14.0 - 14.5	12	0.12	0.65
14.5 - 15.0	11	0.11	0.76
15.0 - 15.5	23	0.23	0.99
15.5 - 16.0	1	0.01	1.00
Σ	100	1.00	

j.

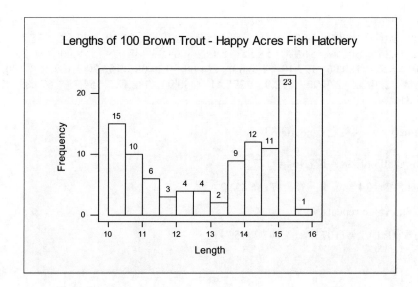

k. Shown in (i) above.

l.

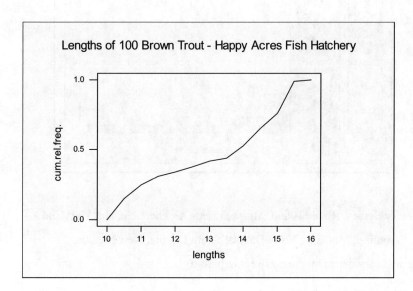

2.187 a. Answers will vary.

b. Answers will vary.

c. Answers will vary.

d. Sum of Area = 3022316

Sum of population = 279583437

Overall density = 279583437/3022316 = 92.506 = 93 people/sq. mile

Densities
5.04   6.13   9.09   9.80   14.95   15.49   18.08   22.16   26.30   32.68   35.28   38.38
41.26   45.04   9.35   50.34   51.99   58.52   59.62   63.35   74.79   78.00   80.29   86.17
86.43   92.10   95.52   100.06   129.19   132.82   134.68   139.05   152.70   167.55   170.72
173.43   213.44   220.20   270.90   272.92   275.41   380.94   382.78   554.79   679.89   768.94
863.52   1073.81

e. Data summary: n = 48, $\sum x$ = 8503.88

   $\bar{x} = \sum x/n = 8503.88/48 = 177.164 = \underline{177.2}$

   $d(\tilde{x}) = (n+1)/2 = 24.5$th; $\tilde{x} = (86.17+86.43)/2 = \underline{86.3}$

   no mode – no value repeats

   midrange = (H+L)/2 = (1073.81+5.04)/2 = $\underline{539.425}$

f.

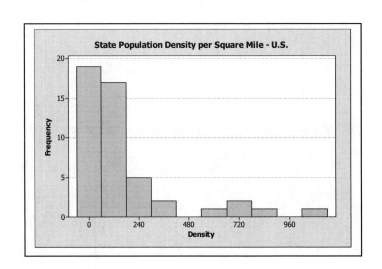

g. Highest: New Jersey, Rhode Island, Massachusetts, Connecticut, and Maryland

   Lowest: Wyoming, Montana, North Dakota, South Dakota, New Mexico

h. Answers will vary depending on answers in parts a – c.

2.189 a. Weight

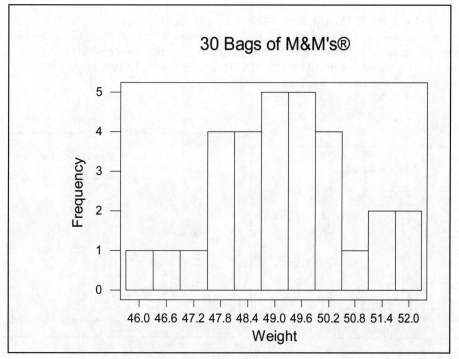

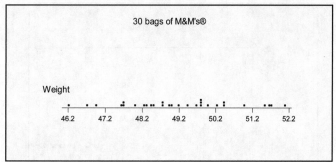

b. $\bar{x} = 49.215$, median = 49.07, s = 1.522, min = 46.22, max = 52.06

c. No, there does not seem to be any inconsistencies in the weight data.

d. Find number per bag: 58, 62, 59, etc.

e.

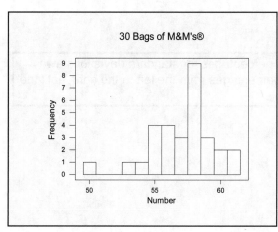
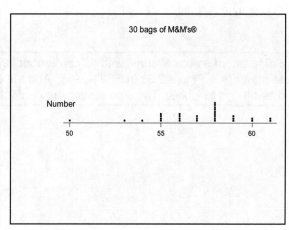

f. $\bar{x} = 57.1$, median = 58, s = 2.383, min = 50, max = 61

g. One bag has "only 50" M&M's in it. This one value appears to be quite different from the rest of the values. Case 14 seems to have a total bag weight that is "typical" (see histogram below). However, its count number is approximately 10% smaller than the "typical" count (see histogram below) – this means that the individual M&M's would have to be 10% larger to make up the weight (see histogram below). Very suspicious!

g. continued

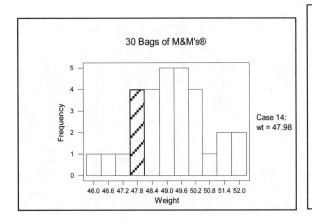
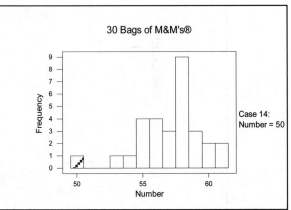
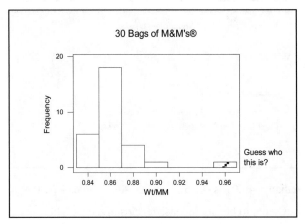

h. After bag 14 was weighed and before the M&M's were counted, someone ate a few of them, approximately 5 of them.

Draw a diagram of a normal curve with its corresponding percentages for standard deviations away from the mean (see Figure 2.32 in ES11-p96). Add the percentages from the left to the right, until the desired z-value is reached. The sum equals the percentile.

2.191 a. 0.50-0.20 = 0.30; therefore, z ≈ -0.8 or -0.9
b. 0.95-0.50 = 0.45; therefore, z ≈ +1.6 or +1.7
c.

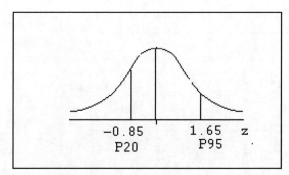

2.193 z-scores must be changed to percentiles in order to make the comparisons. Percentages are obtained from the empirical rule.

z = 2 corresponds to P_{97}

z = 1 corresponds to P_{84}

z = -1 corresponds to P_{16}

z = 0 corresponds to P_{50}

Therefore, Joan has the higher relative score for fitness, agility, and flexibility.
Jean scored highest in posture, while they scored the same in strength.

2.195 Data summary: n = 8, $\sum x$ = 31,825, $\sum x^2$ = 126,894,839

a. $\bar{x} = \sum x/n$ = 31,825/8 = <u>3978.1</u>

b. SS(x) = $\sum x^2 - ((\sum x)^2/n)$ = 126,894,839 - (31,825²/8) = 291,010.88

s^2 = SS(x)/(n-1) = 291,010.88/7 = 41,572.982

s = $\sqrt{s^2}$ = $\sqrt{41572.98}$ = <u>203.9</u>

c. $\bar{x} \pm 2s$ = 3978.1 ± 2(203.9) = 3978.1 ± 407.8 or <u>3570.3</u> to <u>4385.9</u>

2.197 a. Stem-and-Leaf plot:

```
                    50 Service Times
   1 | 8 9
   2 | 1 7 8 7 5 4 9 7 8 3
   3 | 6 5 5 8 5 2 2 8 6 8 5 1 8 3 5 1 8 5 1 2 8 2
   4 | 3 6 5 0 6 8 3 3 3 9 6
   5 | 0 1 2 2 3
```

b.
Summary of data: n = 50, Σx = 1810, Σx² = 69518

18	19	21	23	24	25	27	27	27	28
28	29	31	31	31	32	32	32	32	33
35	35	35	35	35	35	36	36	38	38
38	38	38	38	40	43	43	43	43	45
46	46	46	48	49	50	51	52	52	53

Mean: $\bar{x} = \Sigma x/n = 1810/50 = \underline{36.2}$
Median: $d(\tilde{x}) = (n+1)/2 = (50+1)/2 = 25.5^{th}$
    $\tilde{x} = (35+35)/2 = \underline{35}$
Mode: mode = $\underline{35}$
Range: range = H-L = 53 – 18 = $\underline{35}$
Midrange: midrange = (H+L)/2 = (53+18)/2 = $\underline{35.5}$
Variance:
    $SS(x) = \Sigma x^2 - ((\Sigma x)^2/n) = 69518 - (1810^2/50) = 3996$
    $s^2 = SS(x)/(n-1) = 3996/49 = \underline{81.551}$
Standard deviation:
    $s = \sqrt{s^2} = \sqrt{81.551} = \underline{9.03}$

c.

Lowest Value	First Quartile	Median	Third Quartile	Highest Value
18	31	35	43	53

d. Chebyshev's theorem predicts that no less than 75% of the service times will fall between ±2 standard deviations from the mean. This interval is 36.2 ± 2(9.03) or between 18.14 and 54.26 minutes. In fact, 49 out of 50 heads of hair were cut within this interval, or 98%.

e. Forty minutes should be about right, although this is a judgment call. Less than 30 minutes will make things hectic in the shop and back up customers. Quality would suffer. Over 40 minutes will start to generate excessive slack time between appointments and not keep the barbers busy.

2.199 There are many possible answers for this question; only one of those possibilities is shown.
  a. 70, 77.5, 77.5, 77.5, 85 yields s = 5.30, which is the smallest standard deviation for a sample of 5 data with 70 and 85.
  b. 70, 76, 85, 89, 95 yields s = 10.02.
  c. 70, 85, 90, 99, 110 yields s = 15.02.
  d. In order to increase the standard deviation the data had to become more dispersed.

2.201 a.

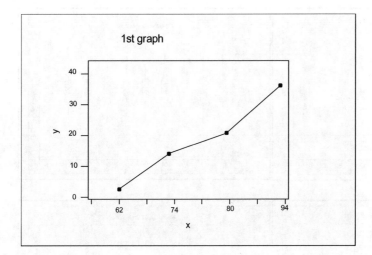

b.

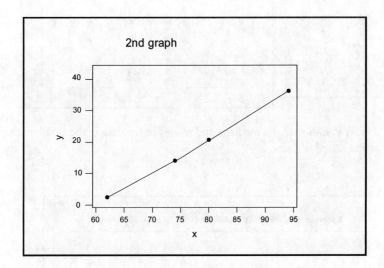

c. Consider the horizontal scale in years. The line graph in (a) suggests an accelerated rate of increase from 1980 to 1995, while the line graph in (b) suggests that the rate of increase has been constant from 1962 to 1995.

2.203 As the sample size increases the distribution of the sample looks more like a normal distribution.

2.205 a.

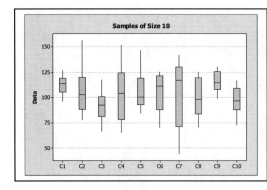
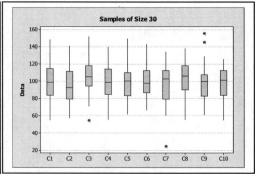
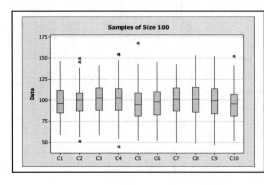
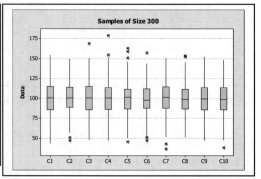

Answers will vary. Greater variation in the medians and ranges of the data in the smaller sample sizes. More outliers in the larger samples sizes.

b.

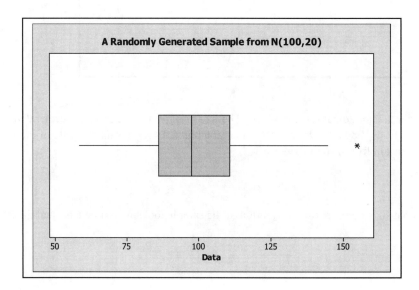

c.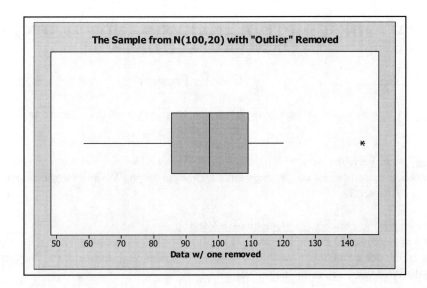

d. The boxplot in part (b) indicated that the largest value was an outlier, then the boxplot in part (c) said the next to the largest data value was now an outlier for the remaining sample. How far could this continue? It doesn't make sense, at this level, to be making decisions about outliers without additional information.

CHAPTER 3 ∇ DESCRIPTIVE ANALYSIS AND PRESENTATION OF BIVARIATE DATA

Chapter Preview

Chapter 3 deals with the presentation and analysis of bivariate (two variables) data. There are three main categories of bivariate data.

1. <u>Two Qualitative Variables</u>

This type of data is best presented in a contingency table and/or bar graph. Variations of the contingency table are given in Section 1 of Chapter 3.

2. <u>One Qualitative Variable and One Quantitative Variable</u>

This type of data can be presented and/or summarized in table form or graphically. More statistical techniques are available because of the one quantitative variable. Dot plots, box plots, and stem-and-leaf diagrams can represent the data for each different value of the qualitative variable.

3. <u>Two Quantitative Variables</u>

Initially, this type of data is best presented in a scatter diagram. If a relationship seems to exist, based on the scatter plot, then linear correlation and regression techniques will be performed.

The relationship between a rainbow trout's length and weight is presented in the chapter opener "Weighing Your Fish with a Ruler."

SECTION 3.1 EXERCISES

3.1 a. Yes   b. Somewhat

Exercises 3.3 - 3.6 present two qualitative variables in the form of contingency tables and bar graphs. A contingency table is made up of rows and columns. Rows are horizontal and columns are vertical. Adding across the rows gives marginal row totals. Adding down the columns gives marginal column totals. The sum of the marginal row totals should be equal to the sum of the marginal column totals, which in turn, should be equal to the sample size.

Computer and/or calculator commands to construct a cross-tabulation table can be found in ES11-p124.

3.3 a.

	On Airplane	Hotel Room	All Other	Marginal total
Business	35.5%	9.5%	5.0%	50%
Leisure	25.0%	16.5%	8.5%	50%
Marginal total	60.5%	26.0%	13.5%	100%

b.

	On Airplane	Hotel Room	All Other	Marginal total
Business	71.0%	19.0%	10.0%	100%
Leisure	50.0%	33.0%	17.0%	100%
Marginal total	60.5%	26.0%	13.5%	100%

The table shows the distribution of ratings for business and leisure separately. For example, 71% of business travelers would like more space on the airplanes while 50% of leisure travelers would like more space on the airplanes.

c.

	On Airplane	Hotel Room	All Other	Marginal total
Business	58.7%	36.5%	37.0%	50%
Leisure	41.3%	63.5%	63.0%	50%
Marginal total	100%	100%	100%	100%

The table shows the distribution of business travelers and leisure travelers for each of the categories. For example, for more space in the hotel room, 36.5% of the responses were from business travelers and 63.5% from leisure travelers.

3.5 a. Population: Adults
Variables: Gender; Age would like to remain rest of life

b.

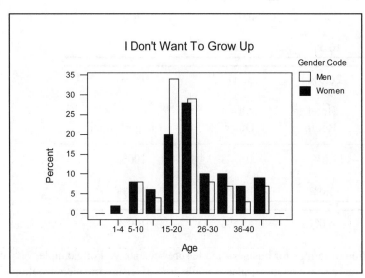

c. There does not seem to be a great difference, however it appears that the women tended to pick ages above 25 more often than did men.

> For percentages based on the marginal total for speed, divide each count by its column total since the columns represent speed.

d. Cross tabulation of vehicle type and maximum interstate highway speed limit (relative frequencies, % of column totals).

	Interstate Highway Speed Limits (mph)					
Vehicle Type	*55*	*60*	*65*	*70*	*75*	*Row Totals*
Cars	0%	25%	47.4%	58.1%	56.5%	50%
Trucks	100%	75%	52.6%	41.9%	43.5%	50%
Column Totals	100%	100%	100%	100%	100%	100%

e. Bar graph of (d):

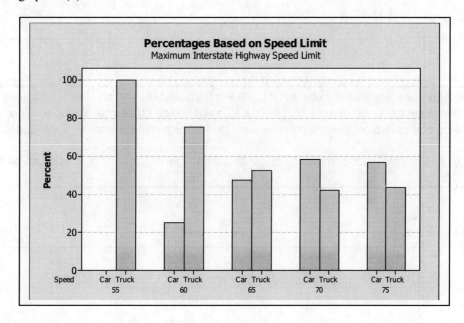

> Before answering any questions concerning data in contingency tables, add all of the rows and columns. Be sure the sum of the row totals = the sum of the column totals = the grand total. Now you are ready to answer all questions easily.

3.7 a. 3350
  b. Two variables, political affiliation and television network preferred, are paired together. Both variables are qualitative.
  c. 880
  d. 46.9% [1570/3350]
  e. 19.2% [203/1060]
  f. 5.9% [197/3350]

3.9 East: $\bar{x} = 9.438, \tilde{x} = 9.65$ West: $\bar{x} = 9.287, \tilde{x} = 9.00$

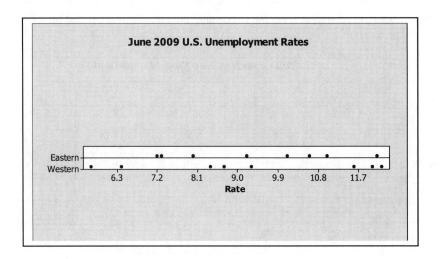

Exercise 3.10 demonstrates the statistical methods that can be used on "one qualitative, one quantitative" type data. Be sure to split the data based on the qualitative variable. The effect is a side-by-side comparison of the quantitative variable for each different value of the qualitative variable.

Computer and/or calculator commands to construct multiple dotplots or boxplots can be found in ES11-p126.

3.11 a.

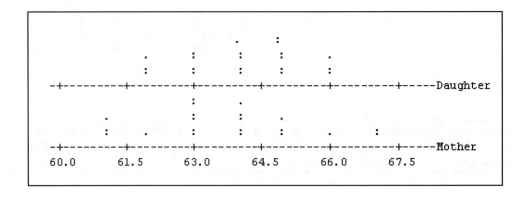

b. The mother heights are more spread out than the daughter heights. No daughters were as short as the shortest mothers and no daughters were as tall as the tallest mothers.

Exercises 3.11c and 3.14-3.24 demonstrate the numerical approach that can be taken now that we have two quantitative variables. A scatter diagram is the first tool we use in determining whether a linear relationship exists between the two variables. Decide which variable is to be predicted. This variable will be the dependent variable.* Let x be equal to the independent variable (input variable) and y be equal to the dependent variable (output variable).

How to construct a scatter diagram:

1. Find the range of the x values and the range of the y values.
2. Based on these, choose your increments for the x-axis (horizontal axis) and then for the y-axis (vertical axis). They will not always be the same.
3. Each point on the scatter diagram is made up of an ordered pair (x,y). (x,y) is plotted at the point that is *x* units on the x-axis <u>and</u> *y* units on the y-axis.
4. Label both axes and give a title to the diagram.

*(ex. the <u>age</u> of a car and the <u>price</u> of a car; price would be the dependent variable if we wish to predict the price of a car based on its age)

Computer and/or calculator commands to construct a scatter diagram can be found in ES11-pp129-130.

c.

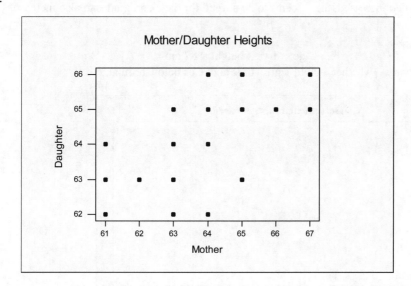

d. As mothers' heights increased, the daughters' heights also tended to increase.

3.13 The input variable most likely would be height. Based on height, weight is often predicted or given in a range of acceptable values depending on the size of a person's frame.

3.15 a.

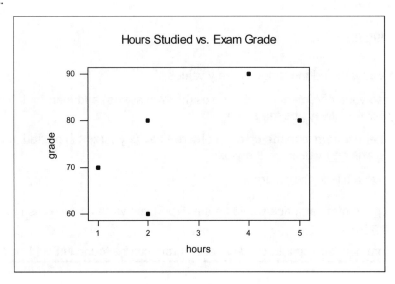

b. As hours studied increased, there seems to be a trend for the exam grades to also increase.

3.17 a. age, height   b. Age = 3 yrs., height = 87 cm.
      c. Answers will vary – whether a child's growth is above or below normal, etc.

3.19

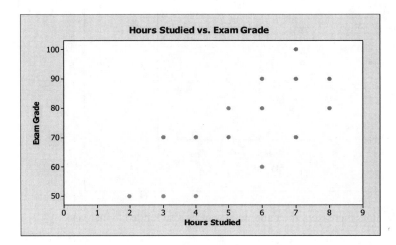

3.21 a. as distance increases, so does commute time

b.
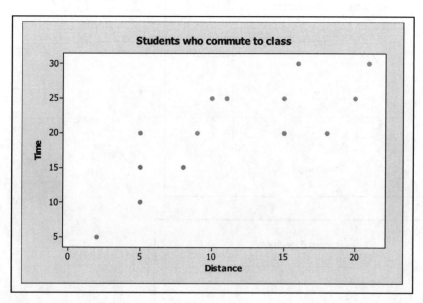

c. yes

3.23 a. Answers will vary, but probably a weak relationship. Or one might expect that a relationship would exist between a stadium's distance from home plate to center field fence and the number of seats; a larger distance would make for a bigger stadium which in turn would mean more seating.

b.
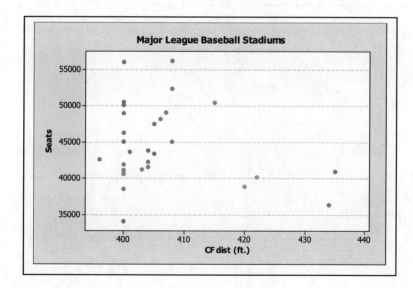

c. Scatter diagram indicates that there is no relationship between the center field distance and the number of seats.

3.25 a.

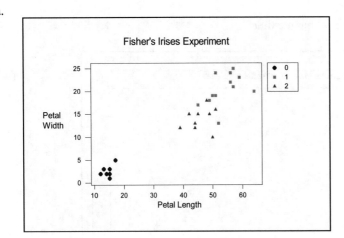

b.

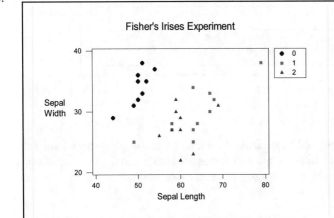

c. Type 0 displays a different pattern than do types 1 and 2, while types 1 and 2 seem to blend together on the scatter diagram.

d.

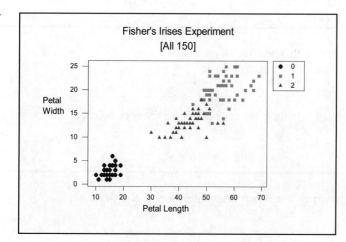

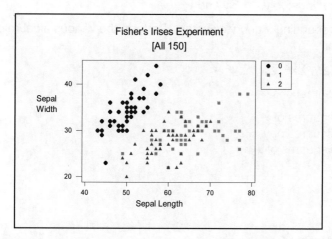

e. Aside from the quantity of points on the scatter diagrams the same patterns are apparent. The random sample of 30 yields the same relationship ideas as the full set of 150. Clearly the 150 are more "impressive" in demonstrating the relationship of the variables as well as the types.

SECTION 3.2 EXERCISES

Based on the scatter diagram, we might suspect whether or not a linear relationship exists. If the y's increase linearly as the x's increase, there exists a relationship called <u>positive</u> <u>correlation</u>. If the y's decrease linearly as the x's increase, there exists a relationship called <u>negative correlation</u>. The measure of the strength of this linear relationship is denoted by *r*, the coefficient of linear correlation.

<u>r = linear correlation coefficient</u>

1. r has a value between -1 and +1, i.e. $-1 \leq r \leq +1$

2. r = -1 specifies perfect negative correlation. All of the data points would fall on a straight line slanted downward.

> 3. r = +1 specifies perfect positive correlation. All of the data points would fall on a straight line slanted upward.
>
> 4. r ≈ 0 indicates little or no consistent linear pattern or a horizontal pattern.

3.27   a. Become closer to a straight line with a positive slope;
           b. Become closer to a straight line with a negative slope

3.29   Coefficient values near zero indicate that there is very little or no linear correlation.

> ### Calculating r - the linear correlation coefficient
>
> Preliminary Calculations:
>
> 1. Set up a table with the column headings: x, y, x^2, xy and y^2.
>
> 2. Insert the bivariate data into corresponding x and y columns. Perform the various algebraic functions to fill in the remaining columns.
>
> 3. Sum all columns, that is, find Σx, Σy, Σx^2, Σxy, Σy^2.
>
> 4. Double check calculations and summations.
>
> 5. Calculate: SS(x) - the sum of squares of x
>   SS(y) - the sum of squares of y
>   SS(xy) - the sum of squares of xy
>   where:
> $SS(x) = \Sigma x^2 - ((\Sigma x)^2/n)$
> $SS(y) = \Sigma y^2 - ((\Sigma y)^2/n)$
> $SS(xy) = \Sigma xy - ((\Sigma x \cdot \Sigma y)/n)$
>
> Final Calculation:
>
> 6. Calculate r:
>
> $$r = \frac{SS(xy)}{\sqrt{SS(x)SS(y)}}$$ (round to the nearest hundredth)
>
> 7. Retain the <u>summations</u> and the <u>sums of squares</u>, as they will be needed for later calculations.
>
> **NOTE:** Remember $SS(x) \neq \Sigma x^2$, $SS(y) \neq \Sigma y^2$ and $SS(xy) \neq \Sigma xy$.
>
> The computer and/or calculator command to calculate the correlation coefficient can be found in ES11-p139.

3.31   a.

x	y	x^2	xy	y^2
2	80	4	160	6400
5	80	25	400	6400
1	70	1	70	4900
4	90	16	360	8100
2	60	4	120	3600
14	380	50	1110	29,400

$SS(x) = \Sigma x^2 - ((\Sigma x)^2/n) = 50 - (14^2/5) = \underline{10.8}$

$$SS(y) = \Sigma y^2 - ((\Sigma y)^2/n) = 29{,}400 - (380^2/5) = \underline{520}$$
$$SS(xy) = \Sigma xy - ((\Sigma x \cdot \Sigma y)/n) = 1110 - (14 \cdot 380/5) = \underline{46}$$

b. $r = SS(xy)/\sqrt{SS(x) \cdot SS(y)} = 46/\sqrt{10.8 \cdot 520} = 0.6138 = \underline{0.61}$

3.33 a.

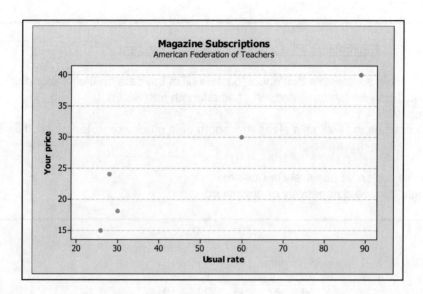

b. $SS(x) = 13877.2 - (232.9^2/5) = 3028.718$
c. $SS(y) = 3616.51 - (126.85^2/5) = 398.326$
d. $SS(xy) = 6951.91 - (232.9 \cdot 126.85/5) = 1043.237$
e. $r = 0.95$

3.35 a. Manatees, powerboats
b. Number of registrations, manatee deaths
c. As one increases, the other does also
d. Answers will vary. Possible: Restrict the number of boat registrations.

Estimating r - the linear correlation coefficient

1. Draw as small a rectangle as possible that encompasses all of the data on the scatter diagram.
    (Diagram should cover a "square window" - same length and width)
2. Measure the width.
3. Let k = the number of times the width fits along the length or in other words: length/width.
4. $r \approx \pm(1 - \frac{1}{k})$
5. Use +, if the rectangle is slanted positively or upward.
   Use -, if the rectangle is slanted negatively or downward.

3.37 a. Estimate r to be near 2/3 or 0.7.

b.

Data	x	y	x^2	xy	y^2
1	2	5	4	10	25
2	3	5	9	15	25
3	3	7	9	21	49
4	4	5	16	20	25
5	4	7	16	28	49
6	5	7	25	35	49
7	5	8	25	40	64
8	6	6	36	36	36
9	6	9	36	54	81
10	6	8	36	48	64
11	7	7	49	49	49
12	7	9	49	63	81
13	7	10	49	70	100
14	8	8	64	64	64
15	8	9	64	72	81
Σ	81	110	487	625	842

$SS(x) = \Sigma x^2 - ((\Sigma x)^2/n) = 487 - (81^2/15) = 49.6$
$SS(y) = \Sigma y^2 - ((\Sigma y)^2/n) = 842 - (110^2/15) = 35.333$
$SS(xy) = \Sigma xy - ((\Sigma x \cdot \Sigma y)/n) = 625 - (81 \cdot 110/15) = 31.0$

$r = SS(xy)/\sqrt{SS(x) \cdot SS(y)} = 31.0/\sqrt{49.6 \cdot 35.333} = \underline{0.741} = \underline{0.74}$

---

* What is the r value of 0.74 in exercise 3.37 telling you? How do your estimated r and calculated r compare? They should be relatively close.   (See the bottom of the next page for answer.)

> ** What is the r value in exercise 3.38 telling you? (See the bottom of the next page for answer.)

3.39 a.

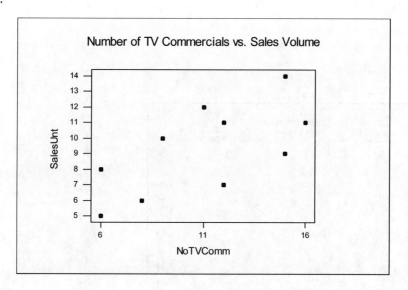

b. From 1/2 to 2/3

c. Summations from extensions table: n = 10, $\Sigma x = 110$, $\Sigma y = 93$, $\Sigma x^2 = 1332$, $\Sigma xy = 1085$, $\Sigma y^2 = 937$

$SS(x) = \Sigma x^2 - ((\Sigma x)^2/n) = 1332 - (110^2/10) = 122.0$
$SS(y) = \Sigma y^2 - ((\Sigma y)^2/n) = 937 - (93^2/10) = 72.10$
$SS(xy) = \Sigma xy - ((\Sigma x \cdot \Sigma y)/n) = 1085 - (110 \cdot 93/10) = 62.00$

$r = SS(xy)/\sqrt{SS(x) \cdot SS(y)} = 62.0/\sqrt{122 \cdot 72.1} = \underline{0.66}$

> *** What is the r value in exercise 3.39 telling you? (See the bottom of this page for answer.)

> *(3.37) There is a positive relationship between the variables, that is, as x increases, y increases. In this case, as the hours studied increased, the grade on the exam increased. What can one deduce from this?
> **(3.38) As age increases, the number of irrelevant answers given during a controlled experiment decreases. Can you make a prediction or generalization based on this information?
> ***(3.39) As the number of television commercials increased, so did the amount of sales. This would be valuable information for the sales department as well as the advertising department.

3.41   Answers will vary: positive vs. negative; nearness to straight line, etc.

3.43   a. 0.95

b. 1.00

c. answers will vary

d. yes, CO_2 will double

3.45   a.

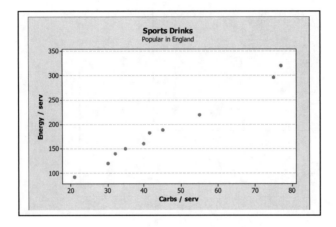

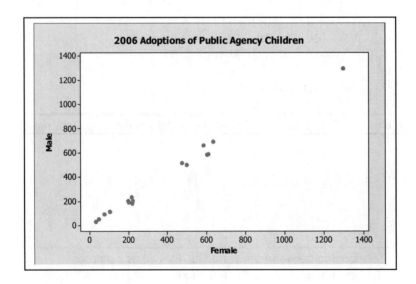

b. Yes, a straight line relationship appears. As carb/serving increase, so does the energy/serving.

c. Summations from extensions tables: n = 10, $\sum x = 451.7$, $\sum y = 1868$, $\sum x^2 = 23528.8$,
$\sum xy = 96642.4$, $\sum y^2 = 397448$

$SS(x) = \sum x^2 - ((\sum x)^2/n) = 23528.8 - (451.7^2/10) = 3125.511$
$SS(y) = \sum y^2 - ((\sum y)^2/n) = 397448 - (1868^2/10) = 48505.6$
$SS(xy) = \sum xy - ((\sum x \cdot \sum y)/n) = 96642.4 - (451.7 \cdot 1868/10) = 12264.84$

$r = SS(xy)/\sqrt{SS(x) \cdot SS(y)} = 12264.84/\sqrt{3125.511 \cdot 48505.6}$
 = <u>0.996</u>

d. There is a strong, almost perfect, positive correlation between the carbs/serving and the energy/serving for sports drinks.

e.

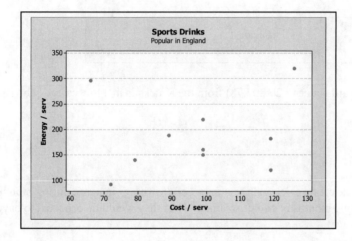

Appears to have no linear relationship between the cost/serving of a sports drink and the energy/serving.

Summations from extensions tables: n = 10, $\sum x = 967$, $\sum y = 1868$, $\sum x^2 = 97303$,
$\sum xy = 182680$, $\sum y^2 = 397448$

$SS(x) = \sum x^2 - ((\sum x)^2/n) = 97303 - (967^2/10) = 3794.1$
$SS(y) = \sum y^2 - ((\sum y)^2/n) = 397448 - (1868^2/10) = 48505.6$
$SS(xy) = \sum xy - ((\sum x \cdot \sum y)/n) = 182680 - (967 \cdot 1868/10) = 2044.4$

$r = SS(xy)/\sqrt{SS(x) \cdot SS(y)} = 2044.4/\sqrt{3794.1 \cdot 48505.6}$
 = 0.1507 = <u>0.15</u>

There is little or no correlation between the cost/serving of a sports drink and the energy/serving.

3.47 a.

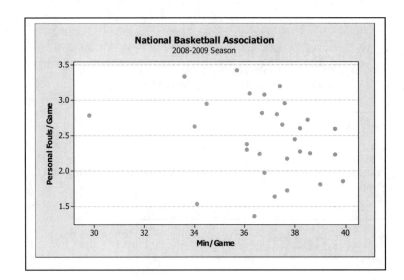

b. no upward or downward pattern; (29.8, 2.78) from the Knicks is in a low category by itself

c. r = -0.261

d. Yes

3.49 No. Both the amount of ice cream sales and the number of drownings increase during months of warmer weather and decrease during months of cooler weather. Thus they would be expected to have a positive correlation.

SECTION 3.3 EXERCISES

If a linear relationship exists between two variables, that is,

1. its scatter diagram suggests a linear relationship
2. its calculated r value is not near zero

the techniques of linear regression will take the study of bivariate data one step further. Linear regression will calculate an equation of a straight line based on the data. This line, also known as the line of best fit, will fit through the data with the smallest possible amount of error between it and the actual data points. The regression line can be used for generalizing and predicting over the sampled range of x.

FORM OF A LINEAR REGRESSION LINE

$$\hat{y} = b_0 + b_1 x$$

where $\hat{y}$ (y hat) = predicted y
b_0 (b sub zero) = y intercept
b_1 (b sub one) = slope of the line
x = independent data value.

3.51

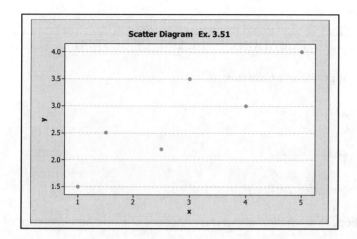

Yes, the scatter diagram shows a linear pattern. The ordered pairs follow a straight line from lower left to upper right.

3.53 a.

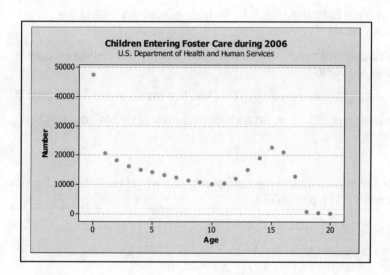

b. The number of children entering foster care increases as they approach the teen year.

c. There appears to be little or no linear correlation.

d. no

e. between the ages of 1 and 10, 11 and 15, 16 and 18

CALCULATING $\hat{y} = b_0 + b_1 x$ - THE EQUATION OF THE LINE OF BEST FIT

1. Retrieve preliminary calculations from previous r calculations.

2. Calculate b_1 where $b_1 = \dfrac{SS(xy)}{SS(x)}$

3. Calculate b_0 where $b_0 = \dfrac{1}{n}(\sum y - b_1 \sum x)$

$\hat{y}$ = predicted value of y (based on the regression line)

NOTE: See Review Lessons for additional information about the concepts of slope and intercept of a straight line.

DRAWING THE LINE OF BEST FIT ON THE SCATTER DIAGRAM

1. Pick two *x*-values that are <u>within</u> the interval of the data *x*-values.
   (one value near either end of the domain)

2. Substitute these values into the calculated $\hat{y} = b_0 + b_1 x$ equation and find the corresponding $\hat{y}$ values.

3. Plot these points on the scatter diagram in such a manner that they are distinguishable from the actual data points.

4. Draw a straight line connecting these two points. This line is a graph of the line of best fit.

5. Plot a third point, the ordered pair $(\bar{x}, \bar{y})$ as an additional check. It should be a point on the line of best fit.

OR:

Computer and/or calculator commands to find the equation of the line of best fit and also draw it on a scatter diagram can be found in ES11-pp152-153.

3.55 a. Summations from extensions table: n = 5, $\sum x = 14$, $\sum y = 380$, $\sum x^2 = 50$, $\sum xy = 1110$, $\sum y^2 = 29{,}400$

$SS(x) = \sum x^2 - ((\sum x)^2/n) = 50 - (14^2/5) = 10.8$
$SS(xy) = \sum xy - ((\sum x \cdot \sum y)/n) = 1110 - (14 \cdot 380/5) = 46$

$b_1 = SS(xy)/SS(x) = 46/10.8 = 4.259$

$b_0 = [\sum y - b_1 \cdot \sum x]/n = [380 - (4.259 \cdot 14)]/5 = 64.0748$

$\underline{\hat{y} = 64.1 + 4.26x}$

b. At x = 1, $\hat{y}$ = 64.1 + 4.26(1) = 68.36; thus (1,68.4)
At x = 3, $\hat{y}$ = 64.1 + 4.26(3) = 76.9; thus (3,76.9)
Points (1,68.4) and (3,76.9) are used to locate the line.

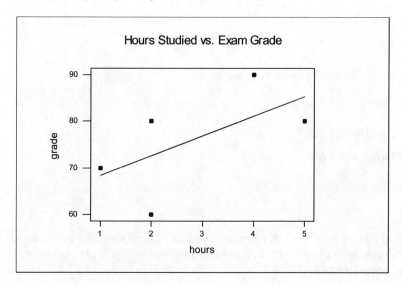

c. Yes, as the hours studied increased, the exam grades appear to increase, also.

3.57 a. $\hat{y}$ = 14.9 + 0.66(20) = <u>28.1</u>

$\hat{y}$ = 14.9 + 0.66(50) = <u>47.9</u>

b. Yes, the line of best fit is made up of all points that satisfy its equation.

3.59 a.

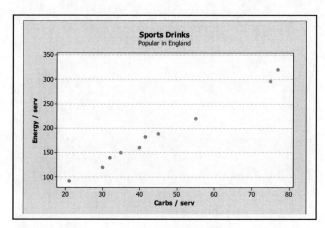

A linear relationship appears between carbs/serving and energy/serving for the sports drinks in the sample. The ordered pairs follow very closely to a straight line.

b. Summations from extensions tables: $n = 10, \Sigma x = 451.7, \Sigma y = 1868, \Sigma x^2 = 23528.8, \Sigma xy = 96642.4$

$SS(x) = \Sigma x^2 - ((\Sigma x)^2/n) = 23528.8 - (451.7^2/10) = 3125.511$
$SS(xy) = \Sigma xy - ((\Sigma x \cdot \Sigma y)/n) = 96642.4 - (451.7 \cdot 1868/10) = 12264.84$
$b_1 = SS(xy)/SS(x) = 12264.84/3125.511 = 3.924$

$b_0 = [\Sigma y - b_1 \cdot \Sigma x]/n = [1868 - (3.924 \cdot 451.7)]/10 = 9.553$

$$\hat{y} = 9.55 + 3.924x$$

c. $\hat{y} = 9.55 + 3.924(40) = 166.51$

d. $\hat{y} = 9.55 + 3.924(65) = 264.61$

3.61 a. The y-intercept of $23.65 is the amount of the total monthly telephone cost when x, the number of long distance calls, is equal to zero. That is, when no long distance calls are made there is still the monthly phone charge of $23.65.

b. The slope of $1.28 is the rate at which the total phone bill will increase for each additional long distance call; it is related to average cost of the long distance calls.

3.63 a. The slope of 4.71 indicates that for each increase in height of one inch, college women's weight increased by 4.71 pounds.

b. The scale for the y-axis starts at y = 95 and the scale for the x-axis starts at x = 60. The y-intercept of –186.5 occurs when x = 0, so the x-axis would have to include x = 0 and the y-axis would have to be extended down.

3.65 $\hat{y} = 7.31 - 0.01x$ when $x = 50$ is
$\hat{y} = 7.31 - 0.01(50) = 6.81$

The predicted value is 6.81(10,000) or $68,100

3.67 a. $\hat{y} = -5359 + 0.9956x$, when x = 500,000
$\hat{y} = -5359 + 0.9956(500,000) = \underline{492,411}$ or $\underline{\$492,411,000}$

b. $\hat{y} = -5359 + 0.9956x$, when x = 1,000,000
$\hat{y} = -5359 + 0.9956(1,000,000) = \underline{990,241}$ or $\underline{\$990,241,000}$

c. $\hat{y} = -5359 + 0.9956x$, when x = 1,500,000
$\hat{y} = -5359 + 0.9956(1,500,000) = \underline{1,488,041}$ or $\underline{\$1,488,041,000}$

3.69 The vertical scale shown on figure 3.27 is located at x = 58 and therefore is not the y-axis; the y = 80 occurs at x = 58.   Remember, the x-axis is the vertical line located at x = 0.

3.71 a. The data would lie on a straight line with slope 1.618

b. The data would be scattered about but generally follow a straight path with slope 1.618

3.73 Answers will vary

3.75 a.

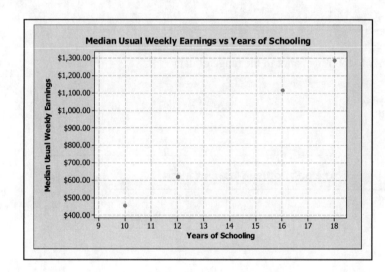

b. Yes, as years of schooling increase, so do median weekly earnings.

c. 0.997   d. yes

e. $\hat{y} = -647.25 + 108.25x$

f. For every additional year of schooling, median weekly earnings increase by $108.25.

g.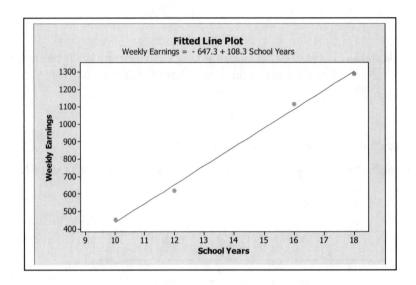

h. -647.25, x = 0 years of schooling is not in the range of data

3.77 a. increasing amounts

b.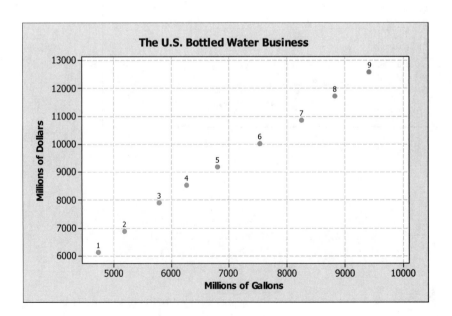

c. yes

d. $\hat{y} = 42.8 + 1.33x$ (Millions of $ = 43 + 1.33 Millions of Gallons)

e. An increase of $1.33 millions of dollars for each additional million gallons of bottled water sold

3.79 a.

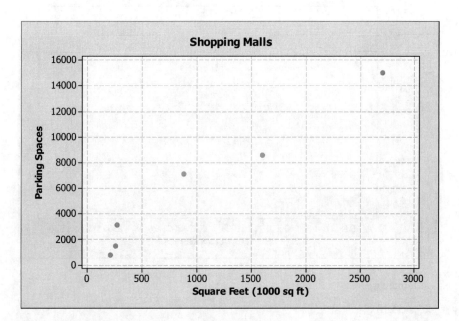

b. Yes, both increase linearly.

c. $\hat{y} = 801 + 5.28x$

d.

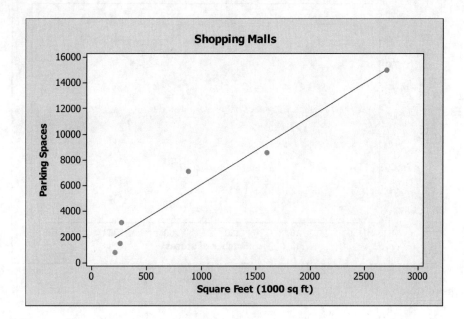

As size of mall increases, so does number of parking spaces.

e. type of store (e.g. Large furniture store – large area but requires less parking)

f.

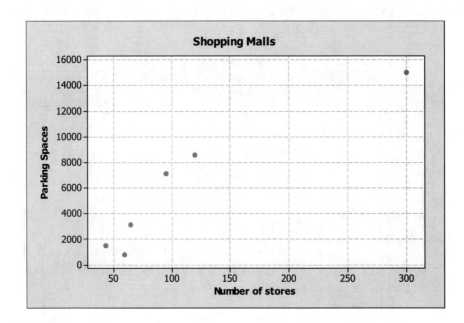

g. Yes, both increase.

h. $\hat{y} = -23 + 53.13x$

i.

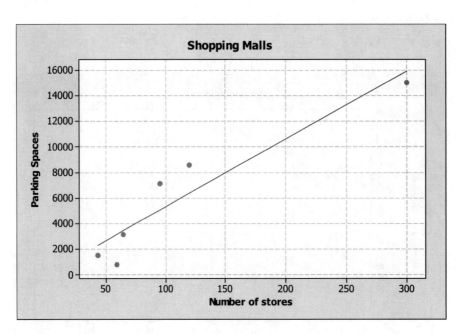

j. type of store

k.

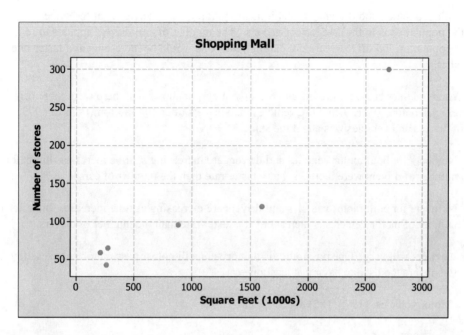

l. Yes, both increase.

m. $\hat{y} = 23.50 + 0.09x$

n.

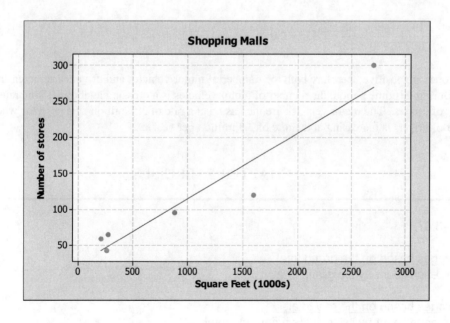

3.81 a. With population and drivers parallel it means that both are increasing at the same rate. Not all of the population is in the category of drivers. The number of non-drivers appears to remain constant. If population and drivers were not parallel, then one would be increasing at a faster rate than the other.

b. When the lines cross for drivers and motor vehicles, it shows that there were more drivers than vehicles before 1971, then more vehicles than drivers after 1973. From 1971 to 1973, the number of drivers and vehicles was about the same.

c. Numbers for both motor vehicles and drivers are increasing as time increases, but before 1973, the number of drivers were increasing at a faster rate than the number of cars.

d. Numbers for both motor vehicles and drivers are increasing as time increases, but after 1973, the number of motor vehicles is increasing at a faster rate than the number of drivers.

e. Answers will vary. One possibility is that drivers will not surpass motor vehicles after 2007. People are too used to having at least one car.

f. Motor vehicles: (1990, 185) & (2002, 227)

$$m = \frac{227-185}{2002-1990} = \frac{42}{12} = 3.5$$

Drivers: (1990, 165) & (2002, 193)

$$m = \frac{193-165}{2002-1990} = \frac{28}{12} = 2.333 = 2.3$$

Both slopes are positive, therefore both the number of motor vehicles and drivers are increasing over time. During this time period, the number of motor vehicles is increasing at about 1.5 the rate of the number of drivers. The motor vehicles are increasing at a rate of 3.5 million cars per year whereas the number of drivers is increasing at the rate of 2.3 million per year.

CHAPTER EXERCISES

Exercises 3.83-3.87

<u>To find percentages based on the grand total</u>
- divide each count by the grand total

<u>To find percentages based on the row totals</u>
- divide each count by its corresponding row total
- each row should add up to 100%

<u>To find percentages based on the column totals</u>
- divide each count by its corresponding column total
- each column should add up to 100%

3.83 a.   Elem.   Jr.H.   Sr.H.   Coll.   Adult   Total

Fear	37	28	25	27	21	138
Do not	63	72	75	73	79	362
Total	100	100	100	100	100	500

b.

	Elem.	Jr.H.	Sr.H.	Coll.	Adult	Total
Fear	7.4%	5.6%	5.0%	5.4%	4.2%	27.6%
Do not	12.6%	14.4%	15.0%	14.6%	15.8%	72.4%
Total	20.0%	20.0%	20.0%	20.0%	20.0%	100.0%

c.

	Elem.	Jr.H.	Sr.H.	Coll.	Adult	Total
Fear	37%	28%	25%	27%	21%	27.6%
Do not	63%	72%	75%	73%	79%	72.4%
Total	100%	100%	100%	100%	100%	100%

d.

	Elem.	Jr.H.	Sr.H.	Coll.	Adult	Total
Fear	26.8%	20.3%	18.1%	19.6%	15.2%	100%
Do not	17.4%	19.9%	20.7%	20.2%	21.8%	100%
Total	20%	20%	20%	20%	20%	100%

e.

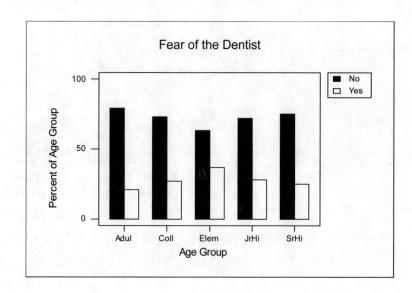

3.85 a.

Year	Lab. Retr.	Golden Retr.	German Sheph.	Beagle	Dach-shunds	Row Totals
2004	146,692	52,550	46,046	44,555	40,770	330,613
2005	137,867	48,509	45,014	42,592	38,566	312,548
Col.Total	248,559	101,059	91,060	87,147	79,336	643,161

b.

Year	Lab. Retr.	Golden Retr.	German Sheph.	Beagle	Dach-shunds	Row Totals
2004	22.81%	8.17%	7.16%	6.92%	6.34%	51.40%
2005	21.44%	7.54%	7.00%	6.62%	6.00%	48.60%
Col Total	44.25%	15.71%	14.16%	13.54%	12.34%	100.00%

c.

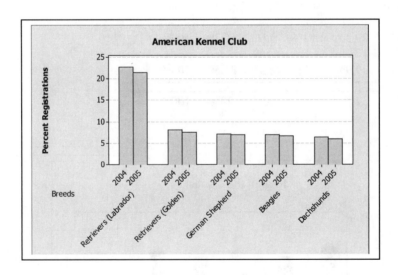

d.

Year	Lab. Retr.	Golden Retr.	German Sheph.	Beagle	Dach-shunds	Row Totals
2004	44.37%	15.89%	13.93%	13.48%	12.33%	100.00%
2005	44.11%	15.52%	14.40%	13.63%	12.34%	100.00%
Col Total	44.24%	15.71%	14.16%	13.55%	12.34%	100.00%

e.

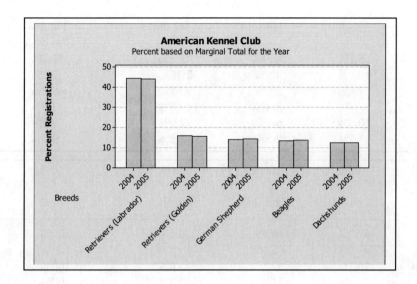

3.87 a.

	1st shift	2nd shift	3rd shift	Total
Sand	87	110	72	269
Shift	16	17	4	37
Drop	12	17	16	45
Corebrk.	18	16	33	67
Broken	17	12	20	49
Other	8	18	22	48
Total	158	190	167	515

b.

	1st shift	2nd shift	3rd shift	Total
Sand	16.9%	21.4%	14.0%	52.2%
Shift	3.1%	3.3%	0.8%	7.2%
Drop	2.3%	3.3%	3.1%	8.7%
Corebrk.	3.5%	3.1%	6.4%	13.0%
Broken	3.3%	2.3%	3.9%	9.5%
Other	1.6%	3.5%	4.3%	9.3%
Total	30.7%	36.9%	32.4%	100%

c.

	1st shift	2nd shift	3rd shift	Total
Sand	55.1%	57.9%	43.1%	52.2%
Shift	10.1%	8.9%	2.4%	7.2%
Drop	7.6%	8.9%	9.6%	8.7%
Corebrk.	11.4%	8.4%	19.8%	13.0%
Broken	10.8%	6.3%	12.0%	9.5%
Other	5.1%	9.5%	13.2%	9.3%
Total	100%	100%	100%	100%

d.

	1st shift	2nd shift	3rd shift	Total
Sand	32.3%	40.9%	26.8%	100%
Shift	43.2%	45.9%	10.8%	100%
Drop	26.7%	37.8%	35.6%	100%
Corebrk.	26.9%	23.9%	49.3%	100%
Broken	34.7%	24.5%	40.8%	100%
Other	16.7%	37.5%	45.8%	100%
Total	30.7%	36.9%	32.4%	100%

e.

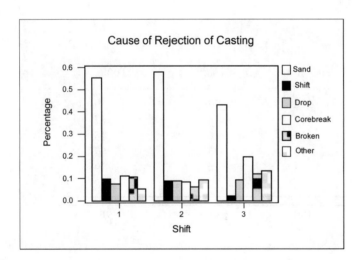

3.89 a. The purpose of a correlation analysis is to determine whether two variables are linearly related or not. The result of correlation analysis is the numerical value of the linear correlation coefficient, r.

b. The purpose of a regression analysis is to determine the equation of the line of best fit that describes the linear relationship of the two variables. The result is the equation.

3.91 a.

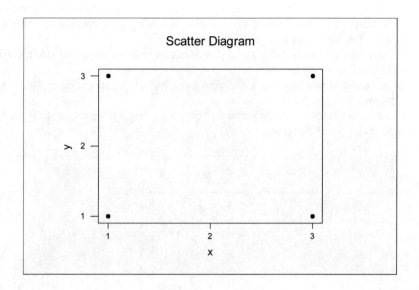

**Looking at the scatter diagram in exercise 3.91a, what would you expect for the correlation coefficient? (See the bottom of this page for answer.)

b. Summations from extensions table: $n = 4$, $\Sigma x = 8$, $\Sigma y = 8$, $\Sigma x^2 = 20$, $\Sigma xy = 16$, $\Sigma y^2 = 20$

$SS(x) = \Sigma x^2 - ((\Sigma x)^2/n) = 20 - (8^2/4) = 4.0$
$SS(y) = \Sigma y^2 - ((\Sigma y)^2/n) = 20 - (8^2/4) = 4.0$
$SS(xy) = \Sigma xy - ((\Sigma x \cdot \Sigma y)/n) = 16 - (8 \cdot 8/4) = 0.0$

$r = SS(xy)/\sqrt{SS(x) \cdot SS(y)} = 0.0/\sqrt{4.0 \cdot 4.0} = \underline{0.00}$

At this point, is there much sense in calculating the line of best fit?
No, the r value shows that no linear relationship exists. Regression analysis would be useless.
NOTE: The line of best fit is a horizontal line (indicating a lack of a linear relationship).

c. $b_1 = SS(xy)/SS(x) = 0.0/4.0 = 0.0$

$b_0 = [\Sigma y - b_1 \cdot \Sigma x]/n = [8 - (0.0 \cdot 8)]/4 = 2.0$

$\hat{y} = \underline{2.0 + 0.0x}$

*(3.90) If all of the points fall exactly on a straight line, perfect positive or negative correlation has occurred. The r value will be either +1 or -1, depending on the upward or downward trend of the data points.
**(3.91) There is no linear trend or pattern to the points in any direction. Therefore, the r value will probably be close to 0.00.

3.93 a. The points will be scatter across both the x and y intervals leaving a "scatter-gun" appearance.
   b. The points will be scatter across both the x and y intervals leaving a pattern that stretches from lower left to upper right and is somewhat elongated.
   c. The points will be scatter across both the x and y intervals leaving a pattern that stretches from lower left to upper right and is quite elongated.
   d. The points will be scatter across both the x and y intervals leaving a pattern that stretches from upper left to lower right and is somewhat elongated.
   e. The points will be scatter across both the x and y intervals leaving a pattern that stretches from upper left to lower right and is quite elongated.

3.95 a.
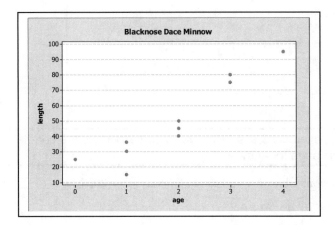

b. Summations from extensions table: $n = 10$, $\Sigma x = 19$, $\Sigma y = 491$, $\Sigma x^2 = 49$, $\Sigma xy = 1196$, $\Sigma y^2 = 30{,}221$

$SS(x) = \Sigma x^2 - ((\Sigma x)^2/n) = 49 - (19^2/10) = 12.9$
$SS(y) = \Sigma y^2 - ((\Sigma y)^2/n) = 30221 - (491^2/10) = 6112.9$
$SS(xy) = \Sigma xy - ((\Sigma x \cdot \Sigma y)/n) = 1196 - (19 \cdot 491/10) = 263.1$

$r = SS(xy)/\sqrt{SS(x) \cdot SS(y)} = 263.1/\sqrt{12.9 \cdot 6112.9} = \underline{0.937}$

c. $b_1 = SS(xy)/SS(x) = 263.1/12.9 = \underline{20.40}$

$b_0 = [\Sigma y - b_1 \cdot \Sigma x]/n = [491 - (20.40 \cdot 10)]/10 = 10.34$

$\underline{\hat{y} = 10.34 + 20.40x}$

d. There is a strong correlation between the age of a blacknose dace minnow and its length. Generally, if the age increases so does the length.

3.97 a.

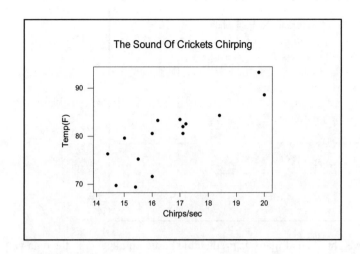

b. There is a strong linearly increasing pattern.

c. Summations from extensions table: n = 15, $\sum x = 249.8$, $\sum y = 1200.60$, $\sum x^2 = 4200.56$, $\sum xy = 20127.5$

(slight variation may occur due to number of decimal places used)

$SS(x) = \sum x^2 - ((\sum x)^2/n) = 4200.56 - (249.8^2/15) = 40.5573$
$SS(xy) = \sum xy - ((\sum x \cdot \sum y)/n) = 20127.5 - (249.8 \cdot 1200.6/15) = 133.508$

$b_1 = SS(xy)/SS(x) = 133.508/40.5573 = 3.2918$

$b_0 = [\sum y - b_1 \cdot \sum x]/n = [1200.6 - 3.2918 \cdot 249.8)]/15 = 25.22$

$\underline{\hat{y} = 25.2 + 3.29x}$

d. x = 14, temperature (F) = 25.2 + 3.29(14) = 71.26 or 71°F
x = 20, temperature (F) = 25.2 + 3.29(20) = 91.0 or 91°F

e. It seems reasonable, temperatures do range from 70 to 90 degrees on summer nights.

f. x = 16, temperature (F) = 25.2 + 3.29(16) = 77.84 or 78°F

3.99 a.

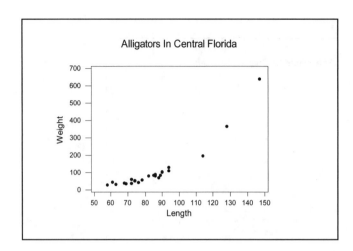

b. Yes, it looks like the weight of an alligator is predictable based on its length – the data very definitely follows a tight pattern along a line. Just not a straight line.

c. No, the pattern is curved. It looks like one side of a parabola.

d. The techniques described in this chapter are for straight line relationships, and this definitely is not a straight line.

e. Summations from extensions table: $n = 25$, $\Sigma x = 2124$, $\Sigma y = 2705$, $\Sigma x^2 = 190282$, $\Sigma xy = 287819$, $\Sigma y^2 = 702127$

$SS(x) = \Sigma x^2 - ((\Sigma x)^2/n) = 190282 - (2124^2/25) = 9826.96$
$SS(y) = \Sigma y^2 - ((\Sigma y)^2/n) = 702127 - (2705^2/25) = 409446$
$SS(xy) = \Sigma xy - ((\Sigma x \cdot \Sigma y)/n) = 287819 - (2124 \cdot 2705/25) = 58002.2$

$r = SS(xy)/\sqrt{SS(x) \cdot SS(y)}$
$= 58002.2/\sqrt{9826.96 \cdot 409446} = \underline{0.914}$

f. The pattern of the data is very elongated and shows an increasing relationship. This overall pattern dominates and the nonlinear nature of the relationship does not effect the calculation of r.

3.101 a. The values for New York City are about 4 times as large as the next smaller piece of data.

b.
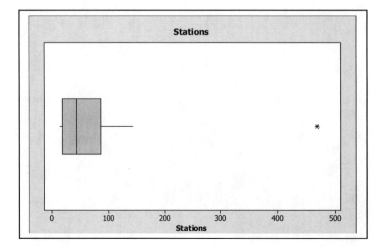

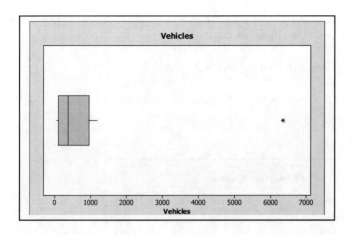

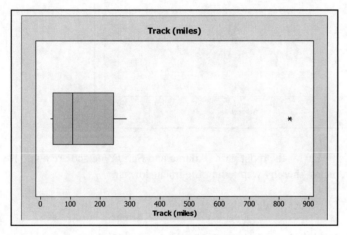

Variable	Mean	Median
Stations	86.8	43.0
Vehicles	961.0	371.0
Track (miles)	196.8	108.0

The boxplots for stations, vehicles and track all denote NYC as an outlier. For all three variables, the mean is at least twice the median, showing that the NYC outlier is pulling the mean in its direction – therefore skewing the distribution to the right.

c.

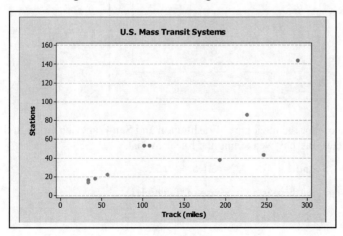

d. The relationship is linear for the most part. Atlanta and San Francisco are low in their number of stations, thereby weakening the linear format.

e. Stations = 4.2 + 0.335 Track (miles) or $\hat{y} = 4.2 + 0.335x$

f. For about every ten miles, there is, on average, about three stations. Or there is a station about every 3 miles.

g.
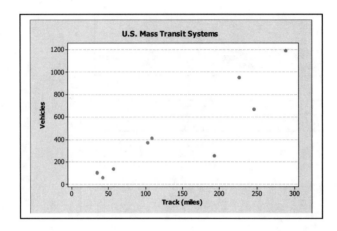

h. The relationship is linear for the most part. Atlanta and San Francisco are again low, this time in their number of vehicles, thereby weakening the linear format.

i. Vehicles = - 59 + 3.63 Track (miles) or $\hat{y} = -59 + 3.63x$

j. For every mile of track, there is on average, about 3 or 4 vehicles.

k.
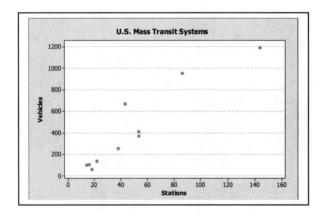

l. The relationship is linear for the most part. Washington and San Francisco are high this time in their number of vehicles, thereby weakening the linear format.

m. Vehicles = - 21.6 + 9.15 Stations   or $\hat{y} = -21.6 + 9.15x$

n. For every additional station, there is an increase of 9 vehicles.

o. Stations = 4.2 + 0.335 Track (miles); x = 50 miles of track;
   = 4.2 + 0.335(50)
   = 20.95 = 21 stations

   Vehicles = - 59 + 3.63 Track (miles); x = 50 miles of track;
   = -59 + 3.63(50)
   = 122.5 = 122 to 123 vehicles

p. At varying miles, the y-intercept will have varying degrees of effect on the final answers.

q. Stations = 4.2 + 0.335 Track (miles); x = 100 miles of track;
   = 4.2 + 0.335(100)
   = 37.7 = 38 stations

   Vehicles = - 59 + 3.63 Track (miles); x = 100 miles of track;
   = -59 + 3.63(100)
   = 304 vehicles

3.103 a.

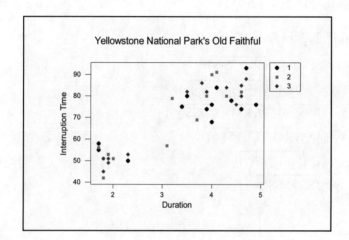

b. There is an overall linear pattern that shows a fairly strong positive relationship, with two separate clusters.

c. Yes, the data from each of the three days is totally inter-dispersed with each day's data showing the same pattern as each other and as the total pattern.

d. Following a 4 minute eruption, the graph shows the length of wait to next eruption to be between 70 and 90 minutes.

e. Summations from extensions table: $n = 39$, $\Sigma x = 130.9$, $\Sigma y = 2734$, $\Sigma x^2 = 487.75$, $\Sigma xy = 9754.7$

$SS(x) = \Sigma x^2 - ((\Sigma x)^2/n) = 487.75 - (130.9^2/39) = 48.3959$
$SS(xy) = \Sigma xy - ((\Sigma x \cdot \Sigma y)/n) = 9754.7 - (130.9 \cdot 2734/39) = 578.274$

$b_1 = SS(xy)/SS(x) = 578.274/48.3959 = 11.94$

$b_0 = [\Sigma y - b_1 \cdot \Sigma x]/n = [2734 - (11.94 \cdot 130.9)]/39 = 30.027$

$\hat{y} = 30.0 + 11.9x$

f. Wait time to next eruption = 30.0 + 11.9(4) = 77.6 or, approximately 78 minutes

g. The distinctive clustering, based on the information given here, does not appear to have any effect on the relationship – the shorter eruptions are followed by the shorter waits and the longer eruptions are followed by the longer waits. There are no short eruptions followed by long waits, etc.

3.105 a. Summations from extensions table: $n = 5$, $\Sigma x = 12$, $\Sigma y = 36$, $\Sigma x^2 = 38$, $\Sigma xy = 101$, $\Sigma y^2 = 286$

$SS(x) = \Sigma x^2 - ((\Sigma x)^2/n) = 38 - (12^2/5) = 9.2$
$SS(y) = \Sigma y^2 - ((\Sigma y)^2/n) = 286 - (36^2/5) = 26.8$
$SS(xy) = \Sigma xy - ((\Sigma x \cdot \Sigma y)/n) = 101 - (12 \cdot 36/5) = 14.6$

$r = SS(xy)/\sqrt{SS(x) \cdot SS(y)} = 14.6/\sqrt{9.2 \cdot 26.8} = \underline{0.9298}$

$b_1 = SS(xy)/SS(x) = 14.6/9.2 = 1.5870$

$b_1 \sqrt{SS(x)/SS(y)} = 1.5870 \cdot \sqrt{9.2/26.8} = \underline{0.9298}$

b. $b_1 \sqrt{SS(x)/SS(y)} = [b_1] \cdot \sqrt{SS(x)/SS(y)}$

$\quad = [SS(xy)/SS(x)] \cdot [\sqrt{SS(x)/SS(y)}]$

$\quad = [SS(xy) \cdot 1/SS(x)] \cdot [\sqrt{SS(x)} \cdot \sqrt{1/SS(y)}]$

$\quad = SS(xy) \cdot [1/SS(x)] \cdot [\sqrt{SS(x)}] \cdot [\sqrt{1/SS(y)}]$

$\quad = SS(xy) \cdot [1/\sqrt{SS(x)}] \cdot [1/\sqrt{SS(y)}]$

$\quad = SS(xy) \cdot [1/\sqrt{SS(x) \cdot SS(y)}]$

$\quad = SS(xy)/\sqrt{SS(x) \cdot SS(y)}$

$\quad = r$

CHAPTER 4 ∇ PROBABILITY

Chapter Preview

Chapter 4 deals with the basic theory and concepts of probability. Probability, in combination with the descriptive techniques in the previous chapters, allows us to proceed into inferential statistics in later chapters.

Some history of M&M's Milk Chocolate Candies is presented in this chapter's Section 4.1, Sweet Statistics.

SECTION 4.1 EXERCISES

4.1 a. Most: yellow, blue and orange ;   Least: brown, red, and green
   b. Not exactly, but similar

4.3
Color	Count	(multiply each Table 4.2 percent by 40)
Brown	5	
Yellow	6	
Red	6	
Blue	9	
Orange	8	
Green	6	
	40	

4.5 P'(5) = 9/40 = <u>0.225</u>

4.7 a. 2/13 = 15.4%   b. 7/13 = 53.8%   c. 6/13 = 46.2%

4.9 a. 11733/78410 = 0.150,   b. (8605+7563)/78410 = 0.206
   c. (13001+12598)/78410 = 0.326   d. (12598+12514+12396)/78410 = 0.478

4.11 a. (2+2+6) = 10% = 0.10   b. 8+14+5 = 27% = 0.27
   c. 9+11+12+17 = 49% = 0.49   d. 0% = 0.00

4.13 a. {0, 1, 2, 3, 4, 5, 6, 7, 8, 9}   b. 0.1   c.   5/10 = 0.5

4.15 M = n(milk), D = n(dark)
   a. All but 42 are milk, the 42 are D >   <u>D = 42</u>
      All but 35 are dark, the 35 are M >   <u>M = 35</u>
   b. 42 + 35 = <u>77 candies</u>
   c. P(M) = 35/77 = <u>5/11</u> or 0.4545 = <u>0.45</u>
   d. P(M or D) = (35 + 42)/77 = 77/77 = <u>1.0, every candy is either milk or dark chocolate</u>
   e. P(M and D) = 0/77 = <u>0.0, no candy is both milk and dark chocolate at the same time</u>

Computer and/or calculator commands to generate random integers and tally the findings can be found in ES11-pp89-91.
Variations in the quantity of random integers and the interval are necessary for exercises 4.17 and 4.18.

4.17 Note: Each will get different results. MINITAB results on one run were:
  a. Relative frequency for: 1 - 0.22, 2 - 0.16, 3 - 0.14, 4 - 0.22, 5 - 0.16, 6 - 0.10

  b. Relative frequency for: H - 0.58, T - 0.42

See the possible outcomes for rolling a pair of dice in ES11-p175. (Also shown in numerical form in ES11-p205)

4.19 P(5) = 4/36; P(6) = 5/36; P(7) = 6/36; P(8) = 5/36;
P(9) = 4/36; P(10) = 3/36; P(11) = 2/36; P(12) = 1/36

4.21 Each student will get different results.
These are the results I obtained: [Note: 12 is an ordered pair (1,2)]

```
12 65 15 32 54   12 52 63 64 62
66 44 42 45 42   35 54 66 54 32
31 12 23 33 26   33 32 23 46 64
63 63 35 54 52   55 56 26 11 44
11 61 46 11 45   55 15 33 43 11
```

  a. P'(white die is odd) = 27/50 = 0.54
  b. P'(sum is 6) = 7/50 = 0.14
  c. P'(both dice show odd number) = 14/50 = 0.28
  d. P'(number on color die is larger) = 16/50 = 0.32

  e. The answers in Exercise 4.20 were determined by using a sample space and are therefore the theoretical probabilities. The answers above were obtained experimentally and are expected to be similar in value, but not exactly the same.

PROPERTIES OF PROBABILITY
1. $0 \leq P(A) \leq 1$ The probability of an event must be a value between 0 and 1, inclusive.
2. $\sum P(A) = 1$ The sum of all the probabilities for each event in the sample space equals 1.

4.23 The three success ratings (highly successful, successful, and not successful) appear to be non intersecting, and their union appears to be the entire sample space. If this is true, none of the three sets of probabilities are appropriate.

Judge A has a total probability of 1.2. The total must be exactly 1.0.

Judge B has a negative probability of -0.1 for one of the events. All probability numbers are between 0.0 and 1.0.

Judge C has a total probability of 0.9. The total must be exactly 1.0.

In exercise 4.25, add the rows and columns first, to find marginal totals.

4.25 a.

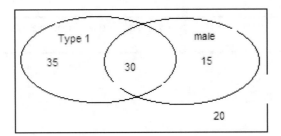

Both representations show that 30 patients are both male and have Type 1 diabetes. There are 65 Type 1 diabetics and 45 males. The 20 female Type 2 diabetics are neither Type 1 nor male.
b. (35+20)/100 = 0.55
c. (15+20)/100 = 0.35

4.27 Let U = used part and D = defective part
Given info: P(U) = 0.60, P(U or D) = 0.61, P(D) = 0.05

P(U and D) = 0.04; 4% are both used and defective

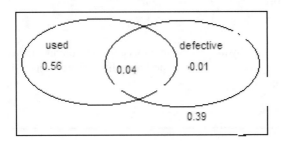

4.29 a. You can expect a 1 to occur approximately 1/6th of the time when you roll a die repeatedly.

b. When one coin is tossed one time, there are two possible outcomes, heads or tails. Each is as likely as the other, therefore each has a probability of occurring of 1/2. 50% of the tosses are expected to be heads, the other 50% tails.

A tree diagram would be helpful to visualize the possible outcomes. Follow the branches for a list of outcomes.
NOTE: See additional information about tree diagrams in Appendix A's Review Lessons, Tree Diagrams.

4.31 a.

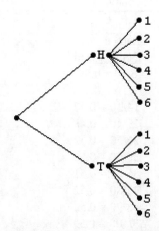

b. {(H,1),(H,2),(H,3),(H,4),(H,5),(H,6),(T,1),(T,2),(T,3),(T,4),(T,5),(T,6)}

NOTE: P' is the notation for an experimental or empirical probability.

4.33 Each student will get different results. These are the results I obtained:

n(heads)/10	P'(head)/set of 10	Cum.P'(head)
6	0.6	6/10 = 0.60
3	0.3	9/20 = 0.45
5	0.5	14/30 = 0.47
5	0.5	19/40 = 0.48
7	0.7	26/50 = 0.52
4	0.4	30/60 = 0.50
6	0.6	36/70 = 0.51
6	0.6	42/80 = 0.52
6	0.6	48/90 = 0.53
5	0.5	53/100 = 0.53
3	0.3	56/110 = 0.51
4	0.4	60/120 = 0.50
7	0.7	67/130 = 0.52
3	0.3	70/140 = 0.50
6	0.6	76/150 = 0.51
3	0.3	79/160 = 0.49
7	0.7	86/170 = 0.51
7	0.7	93/180 = 0.52
4	0.4	97/190 = 0.51
6	0.6	103/200 = 0.52

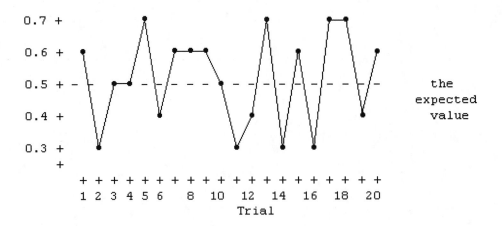

The observed probability varies above and below 0.5, but seems to average approximately 0.5.

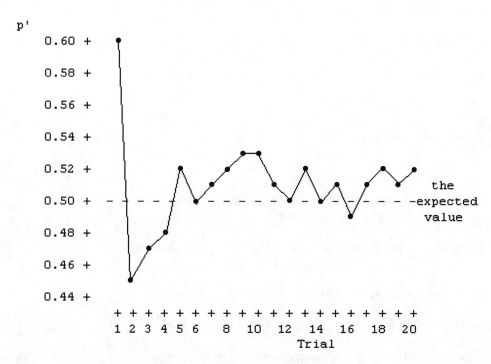

The sum of all the probabilities over a sample space is equal to one. $\Sigma P(x) = 1$
The probability of choosing a red marble plus the probability of choosing a yellow marble plus the probability of choosing a green marble should be equal to the probability of the sample space, which is equal to one.
   P(red) + P(yellow) + P(green) = 1

4.35   P(red) + P(green) + P(blue) + P(yellow) + P(purple) = 1

5 colors all equally likely, therefore

P(red) = P(green) = P(blue) = P(yellow) = P(purple) = 1/5

4.37   P(P) = 4 · P(F)
   P(P) + P(F) = 1
   4·P(F) + P(F) = 1
   5·P(F) = 1
   P(F) = 1/5
   P(P) = 4·P(F) = 4·(1/5) = <u>4/5</u>

4.39   a.   1/7   b.   6:1

4.41 a. P(first place) = 39/195 = 39/(39+156) ⇒ 39:156

b. P(first place) = 39/195 = 0.20

c. P(1st, 2nd, 3rd) = (39+17+28)/195 = 84/(84 + 111) ⇒ 84:111

d. P(1st, 2nd, 3rd) = 84/195 = 0.43

e. placed; twice as likely as 1st place

4.43 a. P(HS as NCAA senior) = 6393/470671 = 6393/(6393+464278) ⇒
6393:464278 = 1:72.6 ≈ 1:73

b. P(NCAA freshman not making pros) = 7619/8219 = 7619/(7619+600) ⇒
7619:600 = 12.7:1 ≈ 38:3

c. P(HS senior being drafted by pros) = 600/134477 = 0.0045

d. P(HS senior playing as NCAA senior) = 6393/134477 = 0.48

4.45 M = n(milk), D = n(dark), S = n(semi-sweet), and W = n(white),

All but 50 are milk	>	50 are not milk	>	D + S + W = 50
all but 50 are dark	>	50 are not dark	>	M + S + W = 50
all but 50 are semi-sweet	>	50 are not semi-sweet	>	M + D + W = 50
all but 60 are white	>	60 are not white	>	M + D + S = 60
				3D + 3M + 3S + 3W = 210

a. D + M + S + W = 70

b. M = 20,   D = 20,   S = 20,   W = 10

c. P(W) = 10/70 = 0.1428 = 0.14

d. P(W or M) = 30/70 = 0.4285 = 0.43

e. P(M and D) = 0/70 = 0.00

f. P(both W) = P(1st is W and 2nd is W) = P(W1) x P(W2) = (10/70) x (9/69) = 0.0186 = 0.019

g. P(one is D and one is S) = P[(1st is D and 2nd is S) or (1st is S and 2nd is D)
= [P(D1) x P(S2)] + [P(S1) x P(D2)]
= (20/70) x (20/69) = (20/70) x (20/69) = 2(0.08281) = 0.1656 = 0.17

h. P(neither are M) = P(1st is not M and 2nd is not M) = (50/70) x (49/69) = 0.507246 = 0.507

4.47 a. probability   b. statistics

4.49 a. statistics   b. probability
c. statistics   d. probability

SECTION 4.2 EXERCISES

4.51 a. (80+55)/300 = 135/300 = 0.45

b. 80/(80+120) = 80/200 = 0.40

c. 55/(55+45) = 55/100 = 0.55

4.53 a. 5146/(5146+9808) = 0.34   b. (1464+2307+156+1723)/14954 = 0.38

c. (1734+7909)/14954 = 0.64   d. (190+227)/14954 = 0.03

e. (190+227+1464)/(190+227+1464+0+6+156) = 0.92

f. (1734+7909)/(2307+958+1734+7909) = 0.75   g. (1734+7909)/9808 = 0.98

4.55 a. 18471/31262 = 0.5908 = 0.59   b. 12791/31262 = 0.4091 = 0.41

c. 10986/31262 = 0.3514 = 0.35   d. 5030/18471 = 0.2723 = 0.27

e. 209/694 = 0.3011 = 0.30   f. 7676/12791 = 0.6001 = 0.60

g. 7676/12791 = 0.6001 = 0.60

h. Same answers, different ways of asking same question.

4.57 a. Some categories would be counted twice.   b. 14,488/139,260 = 0.10

c. 11,139/14,488 = 0.77

d. (6801+949+3954+1258)/139,260 = 0.09

e. 6801/(6801+179+949+3954+1258+5677) = 0.36

SECTION 4.3 EXERCISES

COMPLEMENT - Probability of A complement = $P(\overline{A})$
$P(\overline{A}) = P(\text{not } A) = 1 - P(A)$

4.59 a. 1 – 0.7 = 0.3   b. 1 – 0.78 = 0.22

4.61 1 – 0.63 = 0.37

4.62 P(A or B) = 0.4 + 0.5 - 0.1 = <u>0.8</u>

4.65 P(A or B) = P(A) + P(B) − P(A and B)
0.7 = 0.4 + 0.5 − P(A and B)
0.7 = 0.9 − P(A and B)
−0.2 = −P(A and B)
0.2 = P(A and B)

4.67 Let A = work part-time; $\overline{A}$ = work fulltime;
B = earn more than $20,540
P(A) = 0.37; P($\overline{A}$) = 1 − 0.37 = 0.63; P(B) = 0.50;
P($\overline{A}$ and B) = 0.32
P($\overline{A}$ or B) = 0.63 + 0.50 − 0.32 = 0.81

4.69 Let U = used part and D = defective part
Given info: P(U) = 0.60, P(U or D) = 0.61, P(D) = 0.05

P(U or D) = P(U) + P(D) − P(U and D)
0.61 = 0.60 + 0.05 − P(U and D)

P(U and D) = 0.60 + 0.05 − 0.61

P(U and D) = 0.04; 4% are both used and defective

Same answer as in exercise 4.27, using a Venn diagram.

The <u>probability of event A given event B</u> has occurred is a conditional probability, written as P(A|B).

For any two events, the <u>probability of events A and B</u> occurring simultaneously is equal to:
1. the probability of event A times the probability of event B,
   given event A has already occurred: that is:
$$P(A \text{ and } B) = P(A) \cdot P(B|A)$$
OR
2. the probability of event B times the probability of event A,
   given event B has already occurred: that is:
$$P(A \text{ and } B) = P(B) \cdot P(A|B)$$

4.71 P(A and B) = P(A)· P(B|A) = 0.7 ·0.4 = 0.28

4.73 P(A and B) = P(A)· P(B|A)
0.3 = 0.6 · P(B|A)   (divide both sides by 0.6)
0.5 = P(B|A)

4.75 Given: P(accurate reading) = 0.98; P(clean (not user)) = 0.90
therefore P(user) = 0.10.

P(user and accurate reading (fails test)) = 0.10 · 0.98 = 0.098

4.77 Given: P(extra deductions) = P(E) = 0.10;

P(extra deductions and deny it) = P(E and D) = 0.09

P(D|E) = P(D and E)/P(E)
P(D|E) = 0.09/0.10 = 0.90 or 90%

4.79 a. need help with tree diagram

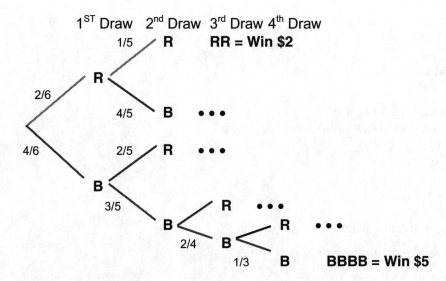

b. 2/5 or 1/5, depending on whether or not the first pick was red or blue

c. $\frac{4}{6} \cdot \frac{3}{5} \cdot \frac{2}{4} \cdot \frac{1}{3} = \frac{1}{15} = 0.067$

d. $P(\$2) = \frac{2}{6} \cdot \frac{1}{5} = 0.067$ Winning $2 and winning $5 have the same probability.

4.81 1st: P(A and B) = P(B) · P(A|B) = 0.4 · 0.2 = 0.08

P(A or B) = P(A) + P(B) - P(A and B)

= 0.30 + 0.40 - 0.08 = 0.62

4.83 1st: P(A or B) = P(A) + P(B) - P(A and B)
0.66 = 0.4 + 0.3 – P(A and B)
0.66 = 0.7 – P(A and B)
-0.04 = -P(A and B)
0.04 = P(A and B)

P(A and B) = P(B and A) = P(B) · P(A|B)
0.04 = 0.3 · P(A|B) (divide both sides by 0.3)
0.133 = P(A|B)

4.85 a. P(B) = 1 – P($\overline{B}$) = 1 – 0.4 = 0.6

b. P(A or B) = P(A) + P(B) - P(A and B)
1.0 = P(A) + 0.6 – 0.3
1.0 = P(A) + 0.3
0.7 = P(A)

c. P(B and A) = P(B) · P(A|B)
0.3 = 0.6 · P(A|B) (divide both sides by 0.6)
0.5 = P(A|B)

4.87 Given: P(A) = 0.5; P(A|B) = 0.25; P(B|A) = 0.2

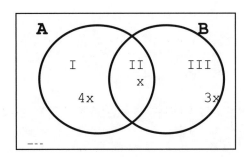

P(B|A) = P(II)/P(A) = x/5x = 0.2 → P(A) = 0.5 = 5x → x = 0.1

a. P(B) = 4x = 0.4

b. P($\overline{A}$) = 1 – P(A) = 1 – 0.5 = 0.5
P(IV) = 1 – P(I + II + III) = 1 – 0.8 = 0.2
P($\overline{B}$|$\overline{A}$) = P(IV)/P($\overline{A}$) = 0.2/0.5 = 0.4

SECTION 4.4 EXERCISES

Mutually Exclusive Events - events that cannot occur at the same time (they have no sample points in common).
Not Mutually Exclusive Events - events that can occur at the same time (they have sample points in common).

4.89 a. Not mutually exclusive. *One head* belongs to both events, therefore the two event intersect.

  b. Not mutually exclusive. All sales that *exceed $1000* also *exceed $100*, therefore $1200 belongs to both events and the two events have an intersection.

  c. Not mutually exclusive. The student selected could be both *male* and *over 21*, therefore the two events have an intersection.

  d. Mutually exclusive. The total cannot be both *less than 7* and *more than 9* at the same time, therefore there is no intersection between these two events.

PROBABILITY - THE ADDITION RULE

P(A or B) = P(A) + P(B) - P(A and B) if A and B are not mutually exclusive

The probability of event A or event B is equal to the probability of event A plus the probability of event B, minus the probability of events A and B occurring at the same time(otherwise that common probability is counted twice).

P(A or B) = P(A) + P(B) if A and B are mutually exclusive

The probability of event A or event B is equal to the probability of event A plus the probability of event B, if A and B have nothing in common (i.e., they cannot occur at the same time).

P(A and B) = 0 if A and B are mutually exclusive

The probability of events A and B occurring at the same time is impossible if A and B are mutually exclusive.

4.91 If two events are mutually exclusive, then there is no intersection. The event, *A and B*, is the intersection. If no intersection, then P(A and B) = 0.0.

4.93 a. P($\overline{A}$) = 1 - 0.3 = <u>0.7</u>

b. P($\overline{B}$) = 1 - 0.4 = <u>0.6</u>

c. P(A or B) = 0.3 + 0.4 = <u>0.7</u>

d. P(A and B) = <u>0.0</u>   (Mutually exclusive events have no intersection.)

4.95 a. Yes, they can not occur at the same time; i.e., a student can not be both male and female.
b. No, they can occur at the same time; i.e., a student can be both male and registered for statistics.
c. No, they can occur at the same time; i.e., a student can be both female and registered for statistics.
d. Yes, the probability of being female at this college plus the probability of being male at this college equals one.
e. No, the two events do not include all of the students.
f. Yes, in both situations there are no common elements shared by the two events.
g. No, two complementary events comprise the sample; two mutually exclusive events do not necessarily make up the whole sample space.

4.97 a. A & C and A & E are mutually exclusive because they cannot occur at the same time.

b. P(A or C) = P(A) + P(C) = 6/36 + 6/36 = <u>12/36</u>
P(A or E) = P(A) + P(E) = 6/36 + 5/36 = <u>11/36</u>
P(C or E) = P(C) + P(E) - P(C and E)
　　　　 = 6/36 + 5/36 - 1/36 = <u>10/36</u>

4.99 a. 'daytime' and 'evening' are mutually exclusive because they cannot occur at the same time.
b. 'preschool' and 'levels' are mutually exclusive because they cannot occur at the same time.
c. 'daytime' and 'preschool' are not mutually exclusive, they can occur at the same time.
d. P(preschool) = (66 + 80)/(145 + 138) = 146/283 = 0.516
e. P(daytime) = 145/283 = 0.5123 = 0.512
f. P(not levels) = 1 – P(levels) = 1 – (69 + 56)/283
　　　　　　　　　 = 1 – 125/283
　　　　　　　　　 = 1 – 0.4417
　　　　　　　　　 = 0.5583 = 0.558
g. P(preschool or evening) = 146/283 + 138/283 – 80/283 = 204/283
　　　　　　　　　　　　　　 = 0.7208 = 0.721
h. P(preschool and daytime) = 66/283 = 0.2332 = 0.233
i. P(daytime|levels) = 69/(69 + 56) = 69/125 = 0.552
j. P(adult & diving|evening) = 2/138 = 0.0144 = 0.014

4.101   (0.62 · 0.58) + (0.43 · 0.42) = 0.3596 + 0.1806 = 0.5402 = 0.54

SECTION 4.5 EXERCISES

Independent Events - when there are independent events, the occurrence of one event **has no effect** on the probability of the other event.

4.103
a. independent
b. not independent
c. independent
d. independent
e. not independent (If you do not own a car, how can your car have a flat tire?)
f. not independent

PROBABILITY - THE MULTIPLICATION RULE

For two independent events:
1. the probability of events A and B occurring simultaneously is equal to the probability of event A times the probability of event B

$$P(A \text{ and } B) = P(A) \cdot P(B)$$

2. the conditional probabilities are equal to the single event probabilities

$$P(A|B) = P(A) \quad \text{AND} \quad P(B|A) = P(B)$$

> **FORMULAS FOR CONDITIONAL PROBABILITIES**
>
> $P(A|B) = \dfrac{P(A \text{ and } B)}{P(B)}$ and $P(B|A) = \dfrac{P(A \text{ and } B)}{P(A)}$
>
> As in Section 4.2, conditionals can also be computed without the formulas above. Suppose $P(A|B)$ is desired. The word *given*,(|), in the conditional tells what the newly reduced sample space is. The number of elements in the reduced sample space, $n(B)$, becomes the denominator in the probability fraction. The numerator is the number of elements in the reduced sample space that satisfy the first event, $n(A \text{ and } B)$. Therefore: $P(A|B) = \dfrac{n(A \text{ and } B)}{n(B)}$

4.105 $P(A \text{ and } B) = P(A) \cdot P(B) = 0.7 \cdot 0.4 = \underline{0.28}$

4.107 $P(A \text{ and } B) = P(A) \cdot P(B)$
$0.3 = 0.6 \cdot P(B)$ (divide both sides by 0.6)
$0.5 = P(B)$

4.109 a. $P(A \text{ and } B) = P(A) \cdot P(B) = 0.3 \cdot 0.4 = \underline{0.12}$

b. $P(B|A) = P(B) = \underline{0.4}$

c. $P(A|B) = P(A) = \underline{0.3}$

4.111 a. $P(A \text{ and } B) = P(B) \cdot P(A|B)$
$0.20 = 0.4 \cdot P(A|B)$; therefore, $P(A|B) = \underline{0.5}$

b. $P(A \text{ and } B) = P(A) \cdot P(B|A)$
$0.20 = 0.3 \cdot P(B|A)$; therefore, $P(B|A) = \underline{0.667}$

c. No, A and B are not independent events.

> Note: A independent of B and A independent of C <u>does not</u> imply that B and C are independent.

4.113 a. P(A) = 12/52 = 3/13 and P(A|B) = 6/26 = 3/13
Therefore, A and B are <u>independent</u> events.

b. P(A) = 12/52 = 3/13 and P(A|C) = 3/13
Therefore, A and C are <u>independent</u> events.

c. P(B) = 26/52 = 1/2 and P(B|C) = 13/13 = 1
Therefore, B and C are <u>dependent</u> events.

4.115 a. P(no life insurance) = 1 − 0.49 = <u>0.51</u>

b. P(18-24 purchase life insurance) = <u>0.15</u>

c. P(no life insurance and 25-34 will purchase life insurance) = (0.51)(0.26) = <u>0.1326</u>

4.117 a. (0.35)(0.35) = 0.1225   b. (0.65)(0.65) = 0.4225

c. (0.35)(0.35)(0.35)(0.35) = 0.0150

4.119 (0.77)(0.77)(0.77) = 0.4565

4.121 Let C = correct decision, I = incorrect decision

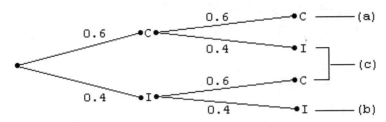

a. P(right decision) = P(C1 and C2) = 0.6·0.6 = <u>0.36</u>

b. P(wrong decision) = P(I1 and I2) = 0.4·0.4 = <u>0.16</u>

c. P(delay) = P[(C1 and I2) or (I1 and C2)] = 0.6 · 0.4 + 0.4 · 0.6 = 0.24 + 0.24 = <u>0.48</u>

4.123 a. P(odd) = <u>3/5</u>

  b. P(neither odd) = (2/5)·(2/5) = 4/25 = <u>0.16</u>

  P(exactly one odd) = [(2/5)·(3/5)] + [(3/5)·(2/5)] = 12/25 = <u>0.48</u>

  P(both odd) = (3/5)·(3/5) = 9/25 = <u>0.36</u>

4.125 Need P(priv | part time), P(publ grad | part time), P(attend private inst.) and P(attend public inst.)

SECTION 4.6 EXERCISES

4.127 a. Two events are mutually exclusive if they cannot occur at the same time or they have no elements in common.

b. Two events are independent if the occurrence of one has no effect on the probability of the other.

c. Mutually exclusive has to do with whether or not the events share common elements; while independence has to do with the effect one event has on the other event's probability.

4.129 a. P(G|H) = P(G and H)/P(H) = 0.1/0.4 = <u>0.25</u>

b. P(H|G) = P(G and H)/P(G) = 0.1/0.5 = <u>0.2</u>

c. P($\overline{H}$) = 1 - P(H) = 1 - 0.4 = <u>0.6</u>

d. P(G or H) = P(G) + P(H) - P(G and H) = 0.5 + 0.4 - 0.1 = <u>0.8</u>

e. P(G or $\overline{H}$) = P(G) + P($\overline{H}$) - P(G and $\overline{H}$) = 0.5 + 0.6 - 0.4 = <u>0.7</u>

f. No. G and H have an intersection, P(G and H) = 0.1; therefore, they are not mutually exclusive.

g. No. P(G|H) does not equal P(G)

4.131 a. P(M and N) = <u>0.0</u> (they are mutually exclusive)

b. P(M or N) = P(M) + P(N) = 0.3 + 0.4 = <u>0.7</u>

c. P(M or $\overline{N}$) = P($\overline{N}$) = 1 - P(N) = 1 - 0.4 = <u>0.6</u>
(M is a subset of $\overline{N}$ since M and N are mutually exclusive.)

d. P(M|N) = <u>0.0</u> (they are mutually exclusive)

e. P(M|$\overline{N}$) = P(M and $\overline{N}$)/P($\overline{N}$) = 0.3/0.6 = <u>0.5</u>

f. No. Mutually exclusive events are disjoint, therefore they must be dependent.

4.133 a. P(satisfied|unskilled) = (150+100)/(250+150) = 0.625

b. P(satisfied|skilled female) = 25/100 = 0.25

c. Compare P(satisfied|skilled female) to P(satisfied|unskilled female)
P(satisfied|skilled female) = 25/100 = 0.25
P(satisfied|unskilled female) = 100/150 = 0.667
Since these two probabilities are not equal, therefore the events are not independent.

CHAPTER EXERCISES

4.135 a. 857/2092 = 0.41   b. 4/2092 = 0.002

c. (9 + 33)/2092 = 0.02   d. without replacement

4.137 P[(med or sh) and (mod or sev)] = (90 + 121 + 35 + 54)/1000 = 0.300

4.139 a. S = {GGG, GGR, GRG, GRR, RGG, RGR, RRG, RRR}

b. P(exactly one R) = 3/8

c. P(at least one R) = 7/8

4.141 P(boy) is approximately equal to 7/8

4.143 a. 133/332 = 0.4006 = 0.40

b. 164/332 = 0.4939 = 0.49

c. 21/332 = 0.0632 = 0.06

d. (164+199-90)/332 = 273/322 = 0.8222 = 0.82

e. 38/95 = 0.40

f. 90/199 = 0.4522 = 0.45

4.145 a. Answers will vary, may include Brazil, Spain, India, etc.

b. The condition is, 'based on countries included.'

c. Total number = 49.8+27.4+42.3+3.8+22.1+13.8+9.8+132.4 = 301.4
132.4/301.4 = 0.4393 = 0.44 = 44%

d. 132.4/301.4 = 0.4393 = 0.44

e. Both are asking the same question but require the answer in a different format, percentage versus decimal form.

4.147 If event A is a subset of event B, then P(A and B) = P(A)
P(A or B) = P(A) + P(B) - P(A and B)
= P(A) + P(B) - P(A) = P(B)

| Note the wording: pink seedless denotes pink <u>and</u> seedless. Use formulas accordingly. |

4.149 a. (seedless) = (10+20)/100 = <u>0.30</u>

  b. (white) = (20+40)/100 = <u>0.60</u>
  c. (pink and seedless) = 10/100 = <u>0.10</u>

  d. (pink or seedless) = (10+20+30)/100 = <u>0.60</u>

  e. (pink|seedless) = 0.10/0.30 = <u>0.333</u>

  f. (seedless|pink) = 0.10/0.40 = <u>0.25</u>

| Rearrange probability formulas in order needed. Remember P(A and B) = P(A) · P(B|A) is the same as P(B and A) = P(B)· P(A|B) since A and B is the same as B and A. |

4.151 a. P(A and B) = P(B)·P(A|B) = (0.36)·(0.88) = <u>0.3168</u>

  b. P(B|A) = P(A and B)/P(A) = 0.3168/0.68 = <u>0.4659</u>

  c. <u>No.</u> P(A) does not equal P(A|B)

  d. <u>No.</u> P(A and B) does not equal 0.0

  e. It would mean that the two events "candidate wants job" and "RJB wants candidate" could not both happen.

4.153 a. P(both damage free) = (10/15)·(9/14) = <u>0.429</u>

  b. P(exactly one) = (10/15)·(5/14) + (5/15)·(10/14) = <u>0.476</u>

  c. P(at least one) = 0.429 + 0.476 = <u>0.905</u>

4.155 P(satisfactory) = P(all good) = p^6

  a. P(satisfactory|p=0.9) = 0.9^6 = <u>0.531</u>

  b. P(satisfactory|p=0.8) = 0.8^6 = <u>0.262</u>

  c. P(satisfactory|p=0.6) = 0.6^6 = <u>0.047</u>

4.157 a. 156006721/307212123 = 0.5078 = 0.508

b. 61944831/307212123 = 0.2016 = 0.202

c. 102665043/307212123 = 0.3341 = 0.334

d. (156006721+16901232)/ 307212123 = 172907953/307212123 = 0.5628 = 0.563

e. 30305704/156006721= 0.1942 = 0.194

f. 102665043/(102665043+103129321) = 102665043/205794364 = 0.4988 = 0.499

g. P(M) is not = P(M|F) since P(M and F) = 0
The events of Male and female are mutually exclusive and complements.

4.159 a. <u>False.</u> If mutually exclusive, P(R or S) is found by adding 0.2 and 0.5.

b. <u>True.</u> 0.2 + 0.5 - (0.2·0.5) = 0.6

c. <u>False.</u> If mutually exclusive, P(R and S) must be equal to zero; there is no intersection.

d. <u>False.</u> 0.2 + 0.5 = 0.7, not 0.6

4.161

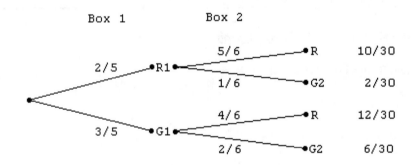

Let G2 = green ball is selected from Box 2 and R1 = red ball is selected from box 1.

P(G2) = P[(R1 and G2) or (G1 and G2)]
= P(R1)·P(G2|R1) + P(G1)·P(G2|G1)
= (2/5)·(1/6) + (3/5)·(2/6) = 2/30 + 6/30 = <u>8/30</u>

4.163 a.

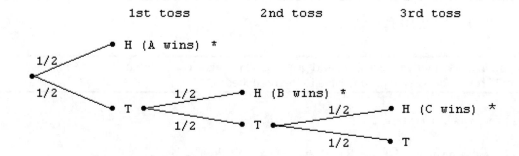

P(A wins on 1st turn) = 1/2

P(B wins on 1st turn) = P(A does not)·P(B wins) = (1/2)·(1/2) = 1/4

P(C wins on 1st turn) = P(A does not)·P(B does not)·P(C wins) = (1/2)·(1/2)·(1/2) = 1/8

b.

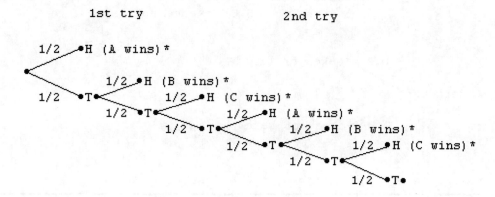

P(A wins on 2nd turn)
 = P(A not on 1st)·P(B not)·P(C not)·P(A wins on 2nd)
 = (1/2)·(1/2)·(1/2)·(1/2) = 1/16

P(A wins on 1st try or 2nd try) = 1/2 + 1/16 = 9/16

P(B wins on 1st try or 2nd try) = 1/4 + 1/32 = 9/32

P(C wins on 1st try or 2nd try) = 1/8 + 1/64 = 9/64

> Exercise 4.164 involves a series of steps. This is a clue to use a tree diagram. Assign probabilities to the branches.

> Exercise 4.165 involves many possibilities and given conditionals. These are clues that a tree diagram should be used. Assign probabilities to the branches.

4.165 See tree diagram on next page.

$$P(\text{wakes on time}) = P(A \text{ or } B \text{ or } C) = P(A) + P(B) + P(C)$$
$$= (0.7)(0.8)(0.9) + (0.7)(0.2)(0.2) + (0.3)(0.2)$$
$$= 0.504 + 0.028 + 0.060 = \underline{0.592}$$

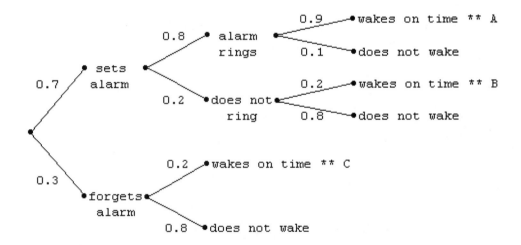

> Notice the different steps required to answer exercise 4.166. First there is a possible locked door, then a possible key, and then a selection of 2 keys. For this type of *steps* problem, a tree diagram is very useful.

4.167 P(2-wk and S) = 3/13 => 12 weeks of the 52 are 2-wk sculptor's, and that is 6 S showings – leaving 14 1-wk S showings – using 26 of the weeks.
Therefore the 22 painters must have 18 1-wk and 4 2-wk showings.

    a. (18x1 + 4x2)/52 = 26/52 = 0.50   b. (14x1 + 6x2)/52 = 26/52 = 0.50

    c. (18+14)/52 = 32/52 = 0.615   d. (4x2 + 6x2)/52 = 20/52 = 0.385

4.169 Let AW = Team A wins game, BW = Team B wins game

a. P(AW) = 0.60; P(A wins a one game series) = <u>0.60</u>

b. P(A wins best of 3 game series) =
= P[(AW1,AW2) or (AW1,BW2,AW3) or (BW1,AW2,AW3)]
= (0.6)(0.6) + (0.6)(0.4)(0.6) + (0.4)(0.6)(0.6)
= <u>0.648</u>

c. P(A wins best of 7 game series)
= P(A wins in 4 games) + P(A wins in 5 games) +
P(A wins in 6 games) + P(A wins in 7 games)
= $1 \cdot (0.6)^4 + 4 \cdot (0.6)^4 (0.4)^1 + 10 \cdot (0.6)^4 (0.4)^2 + 20 \cdot (0.6)^4 (0.4)^3$
= 0.1296 + 0.20736 + 0.20736 + 0.16589 = <u>0.710</u>

d. (a) 0.70   (b) 0.784   (c) 0.874

e. (a) 0.90   (b) 0.972   (c) 0.997

f. The larger the number of games in the series, the greater the chance that the "best" team will win. The greater the difference between the two teams individual chances, the more likely the "best" team wins.

4.171

a. The sample space for tossing 3 coins with no conditions is:
{HHH, HHT, HTH, THH, HTT, TTH, THT, TTT}

The condition, at least one is a head, removes TTT from consideration and leaves the sample space being:
{HHH, HHT, HTH, THH, HTT, TTH, THT}

P(3 heads|at least one of the coins shows a head) = 1/7

b. P(3 heads|at least one of the coins shows a head)
= P(3H and at least one H)/P(at least one H)
= P(3H and at least one H)/(1 - P(0H))
= (1/8)/(1 - 1/8) = (1/8)/(7/8) = 1/7

PROBABILITY

BASIC PROPERTIES: $\quad 0 \leq$ each probability $\leq 1 \quad$ (1)

$$\sum_{\text{over } S} P(A) = 1 \quad (2)$$

Finding probabilities	From an equally likely sample space	By formula, given certain probabilities
$P(A)$, any event A	$P(A) = \dfrac{n(A)}{n(S)} \quad (3)$	-does not apply-
$P(\overline{A})$, complementary event	$P(\overline{A}) = \dfrac{n(\overline{A})}{n(S)} \quad (4)$	$P(\overline{A}) = 1.0 - P(A) \quad (11)$
Any 2 events, no special conditions or relations known: $P(A\|B)$, conditional event $P(A \text{ or } B)$, union of 2 events $P(A \text{ and } B)$, intersection of 2 events	$P(A\|B) = \dfrac{n(A \text{ and } B)}{n(B)} \quad (5)$ $P(A \text{ or } B) = \dfrac{n(A \text{ or } B)}{n(S)} \quad (6)$ $P(A \text{ and } B) = \dfrac{n(A \text{ and } B)}{n(S)} \quad (7)$	$P(A\|B) = \dfrac{P(A \text{ and } B)}{P(B)} \quad (12)$ $P(A \text{ or } B) = P(A) + P(B) - P(A \text{ and } B) \quad (13)$ $P(A \text{ and } B) = P(A) \cdot P(B\|A) \quad (14)$
2 events, known to be mutually exclusive $P(A \text{ or } B)$ $P(A \text{ and } B)$ $P(A\|B)$	$P(A \text{ or } B) = \dfrac{n(A) + n(B)}{n(S)} \quad (8)$	$P(A \text{ or } B) = P(A) + P(B) \quad (15)$ $P(A \text{ and } B) = 0 \quad (16)$ $P(A\|B) = 0 \quad (17)$
2 events, known to be independent $P(A \text{ and } B)$ $P(A\|B)$	$P(A \text{ and } B) = \dfrac{n(A \text{ and } B)}{n(S)} \quad (9)$ $P(A\|B) = \dfrac{n(A \text{ and } B)}{n(B)} = \dfrac{n(A)}{n(S)} \quad (10)$	$P(A \text{ and } B) = P(A) \cdot P(B) \quad (18)$ $P(A\|B) = P(A) \quad (19)$

Resulting Properties:
(20)   If $P(A) + P(B) = P(A \text{ or } B)$; then A and B are mutually exclusive.
(21)   If $P(A) \cdot P(B) = P(A \text{ and } B)$; then A and B are independent.
(22)   If $P(A|B) = P(A)$, then A and B are independent.
(23)   If $P(A \text{ and } B) = 0$, then A and B are mutually exclusive.
(24)   If $P(A \text{ and } B) \neq 0$, then A and B are not mutually exclusive.

<u>The Relationship between Independence and Mutually Exclusive</u>
(25)   If events are independent, then they are NOT mutually exclusive.
(26)   If events are mutually exclusive, then they are NOT independent.

CHAPTER 5 ∇ PROBABILITY DISTRIBUTIONS (DISCRETE VARIABLES)

Chapter Preview

Chapter 5 combines the "ideas" of a frequency distribution from Chapter 2 with probability from Chapter 4. This combination results in a discrete probability distribution. The main elements of this type distribution will be covered in this chapter. The elements include:
1. discrete random variables
2. discrete probability distributions
3. the mean and standard deviation of a discrete probability distribution
4. binomial probability distribution
5. the mean and standard deviation of a binomial distribution.

SECTION 5.1 EXERCISES

5.1 a. 22%   b. 1 vehicle
c. Number of vehicles per household
d. Yes, the events (one, two, three, …eight) are non-overlapping.

Random Variable - a numerical quantity whose value depends on the conditions and probabilities associated with an experiment.

Discrete Random Variable - a numerical quantity taking on or having a finite or countably infinite number of values.

Use x to denote a discrete random variable.(x is often a count of something; ex. the number of home runs in a baseball game)

Continuous Random Variable – a numerical quantity taking on or having an infinite number of values. (often a measurement of something; ex. a person's height)

5.3 One of the random variables is the number of siblings that a classmate has. The possible values for the random variable are x = 0, 1, 2, 3, ..., n.
The other random variable is the length of the last conversation a classmate had with their mother. The random variable will be a numerical value between 0 and 60 minutes for most classmates.

5.5 a. The variable "number of siblings" is discrete because it is a count.
   The variable "length of conversation" is continuous because time is a measurable value.
b. The variable "number of dinner guests" is discrete because it is a count of people.
c. The variable "number of miles" is continuous because distance is a measurable value.

5.7 a. The random variable is the *number of new jobs at a given company*.
b. The random variable is discrete because it represents a count.

5.9 The random variable is the *distance from center to arrow*.
x = 0 to n, where n = radius of the target, measured in inches, including all possible fractions. The variable is continuous.

5.11 a. The random variable is the *average amount of time spent on various activities*.
b. The random variable is continuous because time is a measurable value

SECTION 5.2 EXERCISES

> Probability distributions look very much like frequency distributions. The probability P(x) takes the place of frequency. The frequency *f* column contains integers (counts), whereas the probability *P(x)* column contains fractions or decimals between 0 and 1. The probability P(x) relates to relative frequency, the frequency divided by the size of the data set.
>
> The two main properties of a probability distribution are:
> 1. $0 \le$ each P(x) ≤ 1, each probability is a number between 0 and 1 inclusive.
> 2. Σ P(x) = 1, the sum of all the probabilities should be equal to 1.
>
> Remember to always: 1. make sure each entry in the P(x) column is between 0 and 1, and
>
>   2. sum your P(x) column and check that it is equal to 1.
>
> Both properties <u>must</u> exist.

5.13

x	0	1
P(x)	1/2	1/2

> **Function Notation**
>
>   P(x) $\Rightarrow$ an equation with x as its variable, which assigns probabilities to the corresponding or given values.
>
>   P(0) $\Rightarrow$ replace x on the right side of the equation with 0 and evaluate.
>
>   P(3) $\Rightarrow$ replace x on the right side of the equation with 3 and evaluate.
>
> ex.: $P(x) = \dfrac{x+1}{26}$ $P(0) = \dfrac{0+1}{26} = \dfrac{1}{26}$ $P(3) = \dfrac{3+1}{26} = \dfrac{4}{26}$
>
> **NOTE:** Only evaluate P(x) for the x values in its domain, otherwise P(x) = 0. The domain of a variable is the specified set of replacements (x-values).

5.15 a. The values of x in a probability distribution form a set of mutually exclusive events because they can never overlap. Each possible outcome is assigned a unique numerical value.

  b. All possible outcomes (values of the random variable) are accounted for.

> *What shape distribution does the histogram in exercise 5.16 depict? (See the bottom of the next page for answer.)

5.17 a.

x	P(x)
1	0.12
2	0.18
3	0.28
4	0.42
Σ	1.00

P(x) is a probability function:

1. Each P(x) is a value between 0 and 1.

2. The sum of the P(x)'s is 1.

b.

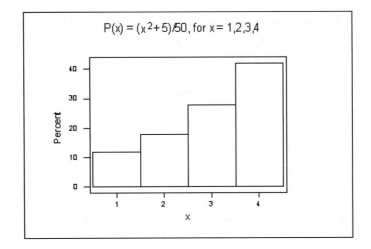

$P(x) = (x^2+5)/50$, for $x = 1,2,3,4$

** What shape distribution does the histogram in exercise 5.17b depict? (See the bottom of this page for answer.)

*(5.16) A uniform, rectangular, or constant distribution.
**(5.17b) A J-shaped distribution or skewed left distribution.

5.19

x	0	1	2	3
P(x)	0.20	0.30	0.40	0.10

Notice that each P(x) is a value between 0.0 and 1.0, and the sum of all P(x) values is exactly 1.0.

> The mean and standard deviation of a probability distribution are μ and σ respectively. They are parameters since we are using theoretical probabilities.
>
> $$\mu = \Sigma[xP(x)] \qquad \sigma^2 = \Sigma[x^2P(x)] - \{\Sigma[xP(x)]\}^2 \Rightarrow \qquad \sigma = \sqrt{\sigma^2}$$
> $$\text{OR}$$
> $$\sigma^2 = \Sigma[x^2P(x)] - \mu^2 \qquad \Rightarrow \qquad \sigma = \sqrt{\sigma^2} \text{ (easier formula)}$$

5.21 $\sigma^2 = \Sigma[(x - \mu)^2 \cdot P(x)]$
  $= \Sigma[(x^2 - 2x\mu + \mu^2) \cdot P(x)]$
  $= \Sigma[x^2 \cdot P(x) - 2x\mu \cdot P(x) + \mu^2 \cdot P(x)]$
  $= \Sigma[x^2 \cdot P(x)] - 2\mu \cdot \Sigma[x \cdot P(x)] + \mu^2 \cdot [\Sigma P(x)]$
  $= \Sigma[x^2 \cdot P(x)] - 2\mu \cdot [\mu] + \mu^2 \cdot [1]$
  $= \Sigma[x^2 \cdot P(x)] - 2\mu^2 + \mu^2$
  $= \Sigma[x^2 \cdot P(x)] - \mu^2 \text{ or } \Sigma[x^2 \cdot P(x)] - \{\Sigma[x \cdot P(x)]\}^2$

5.23 The sum of the number values, once each. Nothing of any meaning.

5.25

x	P(x)	xP(x)	x^2P(x)
0	0.2	0.0	0.0
1	0.2	0.2	0.2
2	0.2	0.4	0.8
3	0.2	0.6	1.8
4	0.2	0.8	3.2
Σ	1.0 ck	2.0	6.0

$\mu = \Sigma[xP(x)] = 2.0$

$\sigma^2 = \Sigma[x^2P(x)] - \{\Sigma[xP(x)]\}^2 = 6.0 - \{2.0\}^2 = 2.0$

$\sigma = \sqrt{\sigma^2} = \sqrt{2.0} = 1.4$

5.27

x	P(x)	xP(x)	x²P(x)
1	0.209	0.209	0.209
2	0.213	0.426	0.852
3	0.241	0.723	2.169
4	0.194	0.776	3.104
5	0.143	0.715	3.575
Σ	1.0 ck	2.849	9.909

$\mu = \Sigma[xP(x)] = 2.849$

$\sigma^2 = \Sigma[x^2P(x)] - \{\Sigma[xP(x)]\}^2 = 9.909 - \{2.849\}^2 = 1.7922$

$\sigma = \sqrt{\sigma^2} = \sqrt{1.7922} = 1.3387 = 1.34$

5.29 a. Yes, variable "number of dogs" is discrete.

b.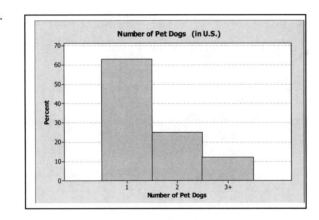

c.

x	P(x)	xP(x)	x²P(x)
1	0.63	0.63	0.63
2	0.25	0.50	1.00
3	0.12	0.36	1.08
Σ	1.0 ck	1.49	2.71

$\mu = \Sigma[xP(x)] = 1.49$

$\sigma^2 = \Sigma[x^2P(x)] - \{\Sigma[xP(x)]\}^2 = 2.71 - \{1.49\}^2 = 0.4899$

$\sigma = \sqrt{\sigma^2} = \sqrt{0.4899} = 0.6999 = 0.70$

d. Average number of dogs per household is 1.49.

e. Mean (average number) is actually higher and standard deviation larger.

5.31 a.

x	P(x)	xP(x)	x²P(x)
1	0.6	0.6	0.6
2	0.1	0.2	0.4
3	0.1	0.3	0.9
4	0.1	0.4	1.6
5	0.1	0.5	2.5
Σ	1.0 ck	2.0	6.0

$\mu = \Sigma[xP(x)] = 2.0$

$\sigma^2 = \Sigma[x^2P(x)] - \{\Sigma[xP(x)]\}^2 = 6.0 - \{2.0\}^2 = 2.0$

$\sigma = \sqrt{\sigma^2} = \sqrt{2.0} = 1.4$

b. $\mu - 2\sigma = 2.0 - 2(1.4) = -0.8$

$\mu + 2\sigma = 2.0 + 2(1.4) = 4.8$

The interval from -0.8 to 4.8 encompasses the numbers 1, 2, 3 and 4.

c. The total probability associated with these values of x is 0.9.

* How does this value of 0.9 in exercise 5.31 compare with Chebyshev's theorem? (See the bottom of the next page for answer.)

5.33 a.

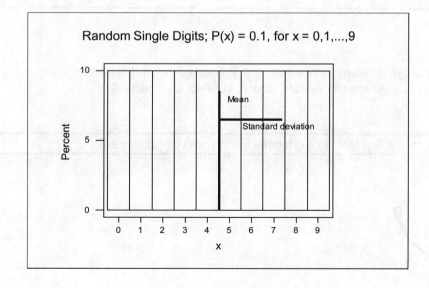

> * What shape distribution does the histogram in exercise 5.33a depict? Is the mean where you would expect it? (See the bottom of the next page for the answer.)

b.

x	P(x)	xP(x)	x²P(x)
0	0.1	0.0	0.0
1	0.1	0.1	0.1
2	0.1	0.2	0.4
3	0.1	0.3	0.9
4	0.1	0.4	1.6
5	0.1	0.5	2.5
6	0.1	0.6	3.6
7	0.1	0.7	4.9
8	0.1	0.8	6.4
9	0.1	0.9	8.1
Σ	1.0 ck	4.5	28.5

> *(5.31) Chebyshev's theorem states that for any shape distribution, at least 75% of the data is within 2 standard deviations of the mean. 90% is well over the minimum limit of 75%.

$$\mu = \Sigma[xP(x)] = 4.5$$

$$\sigma^2 = \Sigma[x^2P(x)] - \{\Sigma[xP(x)]\}^2 = 28.5 - \{4.5\}^2 = 8.25$$

$$\sigma = \sqrt{\sigma^2} = \sqrt{8.25} = 2.87$$

c. See graph in part (a).

d. $\mu \pm 2\sigma = 4.5 \pm 2(2.87) = 4.5 \pm 5.74$ or -1.24 to 10.24
   The interval from -1.24 to 10.24 contains all the x-values of this probability distribution; 100%

> ** How does the 100% from exercise 5.33d compare to Chebyshev's Theorem? (See bottom of this page for answer.)

5.35 The percentages sum to exactly 1.00 indicating all options are included, however the variable is not a random variable, it is an attribute variable, random variables are numerical.

> Note: The higher the probability, the more often the number will be generated.

*(5.33a) A uniform distribution. The mean is exactly in the center since all outcomes are equally likely.

**(5.33d) Chebyshev's theorem states that for any shape distribution, at least 75% of the data is within 2 standard deviations of the mean. 100% is well over the minimum limit of 75%.

For more information on the computer commands to generate discrete data according to a probability distribution, see ES11-p235.

5.37   a.   Everyone's generated values will be different. Listed below are the results obtained from one sample.

Value	Count
1	32
2	36
3	20
4	12

b. & c.

Sample obtained		Given distribution	
x	rel.freq.	x	rel.freq.
1	0.32	1	0.40
2	0.36	2	0.30
3	0.20	3	0.20
4	0.12	4	0.10
ALL	1.00	ALL	1.00

d.

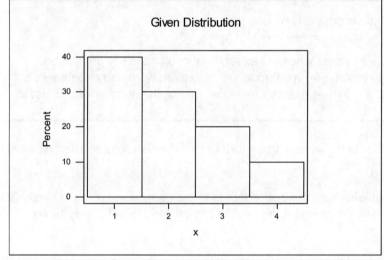

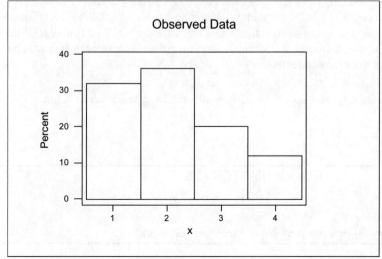

e. Results will vary, but expect: 3 or 4 is almost always the least frequent number, 1 or 2 is almost always the most frequent number, the histograms seem to vary but yet almost always look somewhat like the histogram of the given distribution.

f. Results will vary, but expect: 1 is almost always the most frequent number, 4 is almost always the least frequent number, the histograms vary some but almost always look similar to the given distribution.

SECTION 5.3 EXERCISES

<u>BINOMIAL EXPERIMENTS</u> must have:

1. n independent repeated trials
   a) n - the number of times the trial is repeated
   b) independent - the probabilities of the outcomes remain the same throughout the entire experiment

2. two possible outcomes for each trial
   a) success - the outcome or group of outcomes that is the focus of the experiment
   b) failure - the outcome or group of outcomes not included in success

3. p = probability of success on any one trial
   q = probability of failure on any one trial (q = 1 - p)

4. x = number of successes when the experiment of all n trials is completed. x can range in value from 0 through n. However, when the experiment is completed, x will have exactly one value, that is, the number of successes that occurred.

5.39 a. Each question is in itself a separate trial with its own outcome having no effect on the outcomes of the other questions.

b. There are four different ways that one correct and three wrong answers can be obtained in four questions, each with the same probability. The sum of the 4 probabilities is the same value as 4 times one of them.

c. The 1/3 is the probability of success for each question, i.e., the probability of choosing the right answer from the 3 choices. The 4 is the number of independent trials, i.e., the number of questions.
The expected average would be the sample size times the probability of success; if the probability of guessing one answer correctly is 1/3, then it seems reasonable that on the average one should be able to guess 1/3 of all questions correctly.

5.41 The number of defective items should be fairly small and therefore easier to count.

<u>FACTORIALS</u>

$n! = n(n-1)(n-2)\ldots 1$ ex.: $3! = 3 \cdot 2 \cdot 1 = 6$

$0! = 1$ (defined this way so that the algebra of factorials will work)

$$\binom{n}{x} = \binom{n(\text{trials})}{n(\text{successes}) \quad n(\text{failures})} = \frac{n!}{x!(n-x)!}$$

ex.: $\binom{8}{3} = \binom{8\text{trials}}{3\text{successes} \quad 5\text{failures}} = \frac{8!}{3!5!} = \frac{8 \cdot 7 \cdot 6 \cdot 5 \cdot 4 \cdot 3 \cdot 2 \cdot 1}{3 \cdot 2 \cdot 1 \cdot 5 \cdot 4 \cdot 3 \cdot 2 \cdot 1}$ or

$$= \frac{8 \cdot 7 \cdot 6 \cdot 5!}{3 \cdot 2 \cdot 1 \cdot 5!} = 56$$

EXPONENTS

$b^n = b \cdot b \cdot b \cdots$ (n times)   ex.: $.2^3 = (0.2)(0.2)(0.2) = 0.008$

NOTE: See additional information about factorial notation in Review Lessons – Appendix A cengagebrain.com.

5.43   a. $4! = 4 \cdot 3 \cdot 2 \cdot 1 = 24$

b. $7! = 7 \cdot 6 \cdot 5 \cdot 4 \cdot 3 \cdot 2 \cdot 1 = 5{,}040$

c. $0! = 1$ (by definition)

d. $\dfrac{6!}{2!} = \dfrac{6 \cdot 5 \cdot 4 \cdot 3 \cdot 2 \cdot 1}{2 \cdot 1} = 6 \cdot 5 \cdot 4 \cdot 3 = 360$

e. $\dfrac{5!}{3! \cdot 2!} = \dfrac{5 \cdot 4 \cdot 3 \cdot 2 \cdot 1}{3 \cdot 2 \cdot 1 \cdot 2 \cdot 1} = 10$

f. $\dfrac{6 \cdot 5 \cdot 4 \cdot 3 \cdot 2 \cdot 1}{4 \cdot 3 \cdot 2 \cdot 1 \cdot 2 \cdot 1} = 15$

g. $(0.3)^4 = (0.3)(0.3)(0.3)(0.3) = 0.0081$

h. $\dfrac{7 \cdot 6 \cdot 5 \cdot 4 \cdot 3 \cdot 2 \cdot 1}{3 \cdot 2 \cdot 1 \cdot 4 \cdot 3 \cdot 2 \cdot 1} = 35$

i. $\dfrac{5!}{2! \cdot 3!} = \dfrac{5 \cdot 4 \cdot 3 \cdot 2 \cdot 1}{2 \cdot 1 \cdot 3 \cdot 2 \cdot 1} = 10$

j. $\dfrac{3!}{0! \cdot 3!} = \dfrac{3 \cdot 2 \cdot 1}{1 \cdot 3 \cdot 2 \cdot 1} = 1$

> $\binom{4}{1}(0.2)^1(0.8)^3 = \binom{4}{1} \cdot (0.2)^1 \cdot (0.8)^3$ The use of the multiplication dot is optional. They are sometimes used to emphasize that each of the three parts to a binomial must be evaluated separately first, then multiplication can take place.

    k. $4 \cdot (0.2)(0.8)(0.8)(0.8) = 0.4096$

    l. $1 \cdot 1 \cdot (0.7)^5 = 0.16807$

5.45   Binomial properties:

n = 100 trials (shirts),
two outcomes (first quality or irregular),
p = P(irregular),
x = n(irregular); any integer value from 0 to 100.

5.47 a. x is not a binomial random variable because the trials are not independent. The probability of success (get an ace) changes from trial to trial. On the first trial it is 4/52. The probability of an ace on the second trial depends on the outcome of the first trial; it is 4/51 if an ace is not selected, and it is 3/51 if an ace was selected. The probability of an ace on any given trial continues to change when the experiment is completed without replacement.

    b. x is a binomial random variable because the trials are independent. n = 4, the number of independent trials; two outcomes, success = ace and failure = not ace; p = P(ace) = 4/52 and q = P(not ace) = 48/52; x = n(aces drawn in 4 trials) and could be any number 0, 1, 2, 3 or 4. Further, the probability of success (get an ace) remains 4/52 for each trial throughout the experiment, as long as the card drawn on each trial is replaced before the next trial occurs.

5.49 a.

Trial 1	Trial 2	Trial 3	b.	c.
S	S	S	p^3	x = 3
S	S	F	p^2q	x = 2
S	F	S	p^2q	x = 2
S	F	F	pq^2	x = 1
F	S	S	p^2q	x = 2
F	S	F	pq^2	x = 1
F	F	S	pq^2	x = 1
F	F	F	q^3	x = 0

    d. $P(x) = \binom{3}{x} p^x q^{3-x}$, for x = 0, 1, 2, 3

BINOMIAL PROBABILITY FUNCTION

$$P(x) = \binom{n}{x} p^x q^{n-x} \text{ for } x = 0,1,2,...n$$

where: $P(x)$ = probability of x successes;   n = the number of independent trials

$\binom{n}{x} = \dfrac{n!}{x!(n-x)!}$ = binomial coefficient = the number of combinations of successes and failures that result in exactly x successes in n trials.

p^x = probability of x successes, that is, p·p·p···, x times. Remember x is the number of successes, therefore every time a success occurs, the probability p is multiplied in.

q^{n-x} = probability of (n-x) failures, that is, q·q·q···, (n-x) times. This is the probability for "all of the rest of the trials."

Check: The sum of the exponents should equal n.

5.51 $P(x) = \binom{3}{x}(0.5)^x(0.5)^{3-x}$

$P(0) = \binom{3}{0}(0.5)^0(0.5)^3 = 1(1)(0.125) = 0.125$

$P(2) = \binom{3}{2}(0.5)^2(0.5)^1 = 3(0.25)(0.5) = 0.375$

$P(3) = \binom{3}{3}(0.5)^3(0.5)^0 = 1(0.125)(1) = 0.125$

5.53   a. 0.3585   b. 0.0159   c. 0.9245

Exercise 5.55, parts a and c show more detailed solutions. Review factorials in Appendix A of this manual or at cengagebrain.com, if necessary.

5.55   a. $\binom{4}{1}(0.3)^1(0.7)^3 = \dfrac{4!}{1!3!}(0.3)^1(0.7)^3 = 4(0.3)(0.343) = 0.4116$

b. $\binom{3}{2}(0.8)^2(0.2)^1 = 0.384$

c. $\binom{2}{0}(1/4)^0(3/4)^2 = \dfrac{2!}{0!2!}(1/4)^0(3/4)^2 = 1(1)(9/16) = 0.5625$

d. $\binom{5}{2}(1/3)^2(2/3)^3 = 0.329218$

e. $\binom{4}{2}(0.5)^2(0.5)^2 = 0.375$

f. $\binom{3}{3}(1/6)^3(5/6)^0 = 0.0046296$

Use Table 2, the Binomial Probability Table, to find the needed probabilities. Locate n and p, then the particular x or x's. If more than one x is needed, add the probabilities.

5.57 By inspecting the function we see the binomial properties:
1. n = 5,
2. p = 1/2 and q = 1/2 (p + q = 1),
3. The two exponents x and 5-x add up to n = 5, and
4. x can take on any integer value from zero to n = 5; therefore it is binomial.

By inspecting the probability distribution:

x	T(x)
0	1/32
1	5/32
2	10/32
3	10/32
4	5/32
5	1/32
Σ	32/32 = 1.0

It is a probability distribution.

1. Each T(x) is between 0 and 1.

2. $\Sigma T(x) = 1.0$

$$T(x) = \binom{5}{x}(1/2)^x(1/2)^{5-x} \text{ for } x = 0, 1, \ldots, 5$$

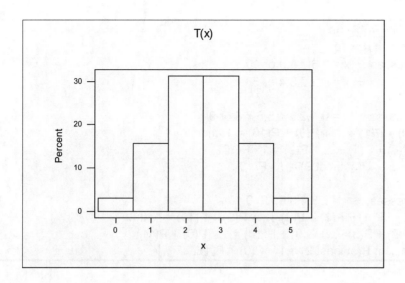

Notation for a Binomial Probability Distribution
$P(x \mid B(n,p))$ where x = whatever values (# of successes) are needed for the problem (can be more than one, just list possibilities) B = binomial distribution n = number of trials p = probability of success
Use Table 2, (Appendix B, ES11-pp713-715), the Binomial Probabilities table, to find needed probabilities.

5.59 0.143

5.61 a. 0.088 b. 0.039 c. 0.00000154

5.63 P(shut down) = $P(x \geq 2)$, where x represents the number defective in the sample of n = 10.

$P(x \geq 2) = 1.0 - [P(x = 0) + P(x = 1)]$

$$P(x = 0) = \binom{10}{0}(0.005)^0(0.995)^{10} = 0.9511$$

$$P(x = 1) = \binom{10}{1}(0.005)^1(0.995)^9 = 0.0478$$

$P(x \geq 2) = 1.0 - [0.9511 + 0.0478] = 0.0011$

5.65 $P(x = 4 \mid B(n = 5, p = P(surv) = 0.8)) = 0.410$

> Terminology in probability problems
> For a binomial problem with **n = 10**:
>   at least 5 successes $\Rightarrow$ x = 5,6,7,8,9 or 10
>   at most 5 successes $\Rightarrow$ x = 0,1,2,3,4 or 5
>
>   at most 9 successes $\Rightarrow$ x = 0,1,2,3,4,5,6,7,8 or 9;
>   Since **P(0) + P(1) + ... + P(9)** + P(10) = 1, then
>   P(0) + P(1) + ... + P(9) = 1 - P(10).
>   Therefore use: P(at most 9) = 1 - P(10).
>
>   at least 2 successes $\Rightarrow$ x = 2,3,4,5,6,7,8,9,10
>   Since P(0) + P(1) + **P(2) + P(3) +...+ P(9) + P(10)** = 1, then
>   P(2) + P(3) +...+ P(9) + P(10) = 1 - [P(0) + P(1)].
>   Therefore use: P(at least 2) = 1 - [P(0) + P(1)].

5.67 a. $P(x = 5 | B(n = 5, p = 0.90)) = 0.590$

b. $P(x = 4, 5 | B(n = 5, p = 0.90)) = 0.328 + 0.590 = 0.918$

5.69 a. $P(x = 0 | B(n = 10, p = 0.40)) = 0.006$

b. $P(x = 3 | B(n = 10, p = 0.40)) = 0.215$

c. $P(x \geq 4 | B(n = 10, p = 0.40)) = 1 - [0.006 + 0.040 + 0.121 + 0.215]$
$= 1 - 0.382 = 0.618$

d. $P(x \leq 2 | B(n = 10, p = 0.40)) = 0.006 + 0.040 + 0.121 = 0.167$

5.71 $P(x = 1, 2, 3, 4, 5, 6 | B(n = 6, p = 0.5)) = 1 - P(x = 0)$
$$= 1 - \binom{6}{0}(0.5)^0(0.5)^6 = 0.984$$

> For a <u>binomial distribution</u> the <u>mean</u> can be calculated using **µ = np** and
> the <u>standard deviation</u> by using **σ = $\sqrt{npq}$**.
>
> Remember for <u>any type</u> discrete probability distribution:
> $\mu = \Sigma[xP(x)]$ and $\sigma = \sqrt{\sigma^2}$ where $\sigma^2 = \Sigma[x^2 P(x)] - [\Sigma[xP(x)]]^2$
>
> Both formulas work for the binomial, however np and $\sqrt{npq}$ are quicker and less prone to computational error.

5.73 $\mu = np = 30 \cdot 0.6 = 18$
$\sigma = \sqrt{npq} = \sqrt{30 \cdot 0.6 \cdot 0.4} = \sqrt{7.2} = 2.68 = 2.7$

5.75 a. Use extension table in exercise 5.74

$\mu = \Sigma[xP(x)] = 0.549 = 0.55$

$\sigma^2 = \Sigma[x^2P(x)] - \{\Sigma[xP(x)]\}^2 = 0.819 - \{0.549\}^2 = 0.5176$

$\sigma = \sqrt{\sigma^2} = \sqrt{0.5176} = 0.7194 = 0.72$

b. The mean and standard deviation of the probability distribution round to exactly the mean and standard deviation of the given binomial distribution.

5.77 a. $\mu = np = 50 \cdot (0.5) = 25.0$

$\sigma = \sqrt{npq} = \sqrt{50 \cdot (0.5) \cdot (0.5)} = 3.5355 = 3.5$

b. $\mu = np = 40 \cdot (0.11) = 4.4$

$\sigma = \sqrt{npq} = \sqrt{40 \cdot 0.11 \cdot 0.89} = 1.9789 = 1.98$

c. $\mu = np = 400 \cdot 0.06 = 24.0$

$\sigma = \sqrt{npq} = \sqrt{400 \cdot 0.06 \cdot 0.94} = 4.7497 = 4.7$

d. $\mu = np = 50 \cdot 0.88 = 44.0$

$\sigma = \sqrt{npq} = \sqrt{50 \cdot 0.88 \cdot 0.12} = 2.298 = 2.3$

5.79 a. x might be approximated by using a binomial random variable. There are only two possible outcomes for each trial of the experiment, there are n = 12 repeated trials. But are the trials independent? [This might depend on whether you are traveling alone or with someone.] Probably not, but the binomial distribution might be a reasonable estimator. x is the count of people and can take on values from zero to 12.

b. $P(x = 4,5 | B(n = 12, p = 0.41)) = [(0.2054 + 0.2284)] = 0.4338$

c. $\mu = np = 12 \cdot 0.41 = 4.92$
$\sigma = \sqrt{npq} = \sqrt{12 \cdot 0.41 \cdot 0.59} = 1.70376 = 1.7$

d.

5.81 a. P(x =

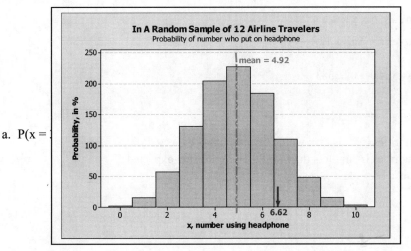

5.83

x	P(x)	x	P(x)	x	P(x)
1*	0.0000	8	0.1009	15	0.0351
2	0.0003	9	0.1328	16	0.0177
3	0.0015	10	0.1502	17	0.0079
4	0.0056	11	0.1471	18	0.0031
5	0.0157	12	0.1254	19	0.0010
6	0.0353	13	0.0935	20	0.0003
7	0.0652	14	0.0611	21*	0.0001

* any other probabilities are each less than 0.00005

5.85 $P(x \geq 20 | B(n = 25, p = 0.90)) = 0.9666$

Computer and/or calculator commands found in ES11-pp251-252 will save time in answering Exercise 5.87

5.87 a. $P(x \leq 5 | B(n = 20, p = 0.48)) = 0.03132$

b. $P(x \geq 3 | B(n = 20, p = 0.48)) = 1.0000 - 0.00038 = 0.99962$

5.89 $\mu = np = 200$ and $\sigma = \sqrt{npq} = 10$, therefore:
$npq = 100$
$200q = 100$
$q = 100/200 = 0.5$
$p = 1 - q = 0.5$
$n = 200/0.5 = 400$

5.91 Given that n = 15 and p = 0.4

$\mu = np = 15(0.4) = 6$

$\sigma = \sqrt{npq} = \sqrt{15 \cdot 0.4 \cdot 0.6} = 1.897 = 1.9$

$\mu + 2\sigma = 6.0 + 2(1.9) = 6.0 + 3.8 = 9.8$
greater than 9.8 gives x = 10,11,12,13,14,15

$P(x \geq 10 | B(n = 15, p = 0.4)) = 0.03383$

5.93 a. Percentage of minorities is "less than would be reasonably expected"
b. Percentage of minorities is "not less than would be reasonably expected";
c. 0.96, "not less than";
d. 0.035, "less than"

5.95 a.

x	P(x)
0	0.886385
1	0.107441
2	0.005969
3	0.000201
4*	0.000005

*the rest are each less than 0.0000005

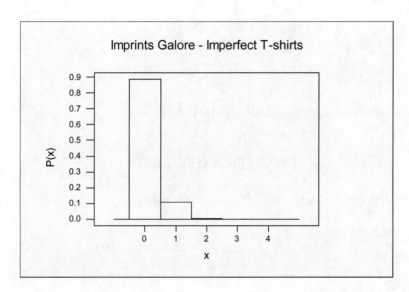

b. $P(x = 0 | B(n = 12, p = 0.01)) = 0.886385$

c.
x	cumP(x)
0	0.88638
1	0.99383
2	0.99979
3*	1.00000

$P(x = 0,1 | B(n = 12, p = 0.01)) = 0.99383$

*the rest are each 1.00000

d. $\mu = np = 12 \cdot (0.01) = 0.12$

$\sigma = \sqrt{npq} = \sqrt{12 \cdot (0.01) \cdot (0.99)} = 0.344674 = 0.345$

e. $\mu - \sigma = 0.12 - 0.345 = -0.225$ and
$\mu + \sigma = 0.12 + 0.345 = 0.465$ includes only x = 0, therefore
the proportion of the distribution is 0.88638

f. $\mu - 2\sigma = 0.12 - 2(0.345) = -0.57$ and
$\mu + 2\sigma = 0.12 + 2(0.345) = 0.81$ includes only x = 0,
therefore the proportion of the distribution is 0.88638

g. The two percentages do not agree with the empirical rule; the shape is skewed right, not a normal shape. Part (f) does agree with Chebyshev's proportion of at least 0.75. Chebyshev's works for any shape distribution.

h.

C4	Count	Simulation	Expected
0	177	177/200 = 0.885	0.886385
1	21	21/200 = 0.105	0.107441
2	2	2/200 = 0.01	0.005969
3	0	0/200 = 0.00	0.000201
4	0	0/200 = 0.00	0.000005
N =	200		

	Simulation	Expected
Mean	0.125	0.12
St. Dev.	0.36059	0.345

i. All simulations were very close to the expected probabilities.

CHAPTER EXERCISES

5.97 1. Each probability, P(x), is a value between zero and one inclusive.

2. The sum of all the P(x) is exactly one.

5.99 a.

x	f(x)	
0	(3/4)/[(0!)(3!)] = 0.125	f(x) is a probability
1	(3/4)/[(1!)(2!)] = 0.375	function since:
2	(3/4)/[(2!)(1!)] = 0.375	
3	(3/4)/[(3!)(0!)] = 0.125	i) $0 \leq$ each $f(x) \leq 1$
Σ	1.000	ii) $\Sigma f(x) = 1.0$

b.

x	f(x)	
9	0.25	f(x) is a probability
10	0.25	function since:
11	0.25	
12	0.25	i) $0 \leq$ each $f(x) \leq 1$
Σ	1.00	ii) $\Sigma f(x) = 1.0$

c.

x	f(x)	
1	1.00	f(x) is NOT a probability
2	0.50	function since:
3	0.00	
4	-0.50	f(x=4) is not between 0 and 1
Σ	1.00	

d.

x	f(x)	
0	1/25	f(x) is NOT a probability
1	3/25	function since:
2	7/25	
3	13/25	$\Sigma f(x) = 24/25$, not 1.0
Σ	24/25	

5.101   a. P(exactly 14 arrive) = P(x = 14) = 0.1

b. P(at least 12 arrive) = P(x = 12, 13 or 14)
$$= P(12) + P(13) + P(14)$$
$$= 0.2 + 0.1 + 0.1 = 0.4$$

c. P(at most 11 arrive) = P(x = 10 or 11)
$$= P(10) + P(11)$$
$$= 0.4 + 0.2 = 0.6$$

5.103

x	P(x)	xP(x)	x^2P(x)
1	0.10	0.10	0.10
2	0.15	0.30	0.60
3	0.25	0.75	2.25
4	0.35	1.40	5.60
5	0.15	0.75	3.75
Σ	1.0	3.30	12.30

a. $\mu = \Sigma[xP(x)] = 3.30 = 3.3$

b. $\sigma^2 = \Sigma[x^2P(x)] - \{\Sigma[xP(x)]\}^2 = 12.30 - \{3.30\}^2 = 1.41$

$\sigma = \sqrt{\sigma^2} = \sqrt{1.41} = 1.1874 = 1.187$

5.105   No, the variable is attribute, not numerical.

5.107   a. P(x = 0, 1, 2|B(n = 10, p = 0.10)) = P(0) + P(1) + P(2)
$$= 0.349 + 0.387 + 0.194 = 0.930$$

b. P(x = 2, 3, 4, ... , 10|B(n = 10, p = 0.10))

$$= 1 - P(x = 0, 1|B(n = 10, p = 0.10))$$

$$= 1.000 - [0.349 + 0.387] = 0.264$$

5.109   P(x = 5|B(n = 10, p = 0.70)) = 0.103

5.111   a. P(x > 4|B(n = 15, p = 0.70)) = 1 − P(x ≤ 4)
$$= 1 - (0+ + 0.001) = 1 - 0.001 = 0.999$$

b. P(x = 10|B(n = 15, p = 0.7)) = 0.206

c. P(x < 10|B(n = 15, p = 0.70)) = (0+ + 0.001 + 0.003 +
0.012 + 0.035 + 0.081 + 0.147) = 0.279

5.113 If x is the random variable n(defective), then *success* is *defective*. On the first selection, P(defective) = 3/10.
However, the P(defective) changes for the next selection: it is either 3/9 or 2/9 depending on whether or not the first selection resulted in a defective or not. Since the probability of defective changes, the trials are not independent. Thus, the experiment is not binomial.

5.115 a. P(accepted) = P(x = 0, 1|B(n = 10,p = 0.05))
= P(0) + P(1) = 0.599 + 0.315 = 0.914

b. P(not accepted) = P(x = 2, 3, ... ,10|B(n = 10,p = 0.20))
= 1 - P(x = 0, 1|B(n = 10,p = 0.20))
= 1 - [0.107 + 0.268] = 0.625

b. Even though the P(defective) changes from trial to trial, if the population is very large, the probabilities are very similar. For example, suppose the population has 10 thousand items and 50 are defective. P(defective) on the first trial is 50/10,000 = 0.0050; if after 10 trials 45 defectives have been selected, p(defective) will be 45/9990 = 0.0045.

5.117 P(bridge in poor or fair condition) = 4706/13268 = 0.355

a. P(x = 0|B(n = 5, p = 0.355)) = 0.1116

b. P(x = 1 or 2|B(n = 5, p = 0.355)) = 0.3072 + 0.3382 = 0.6454 = 0.645

c P(x = 5|B(n = 5, p = 0.355)) = 0.0056 = 0.006

5.119 $\mu = np = (20)(0.4) = 8.0$ and $\sigma^2 = npq = (20)(0.4)(0.6) = 4.8$
$\sigma^2 = \sum x^2 P(x) - \mu^2$
$4.8 = \sum x^2 P(x) - 8^2$ or $\sum x^2 P(x) = 68.8$

5.121 a. Binomial with n = number of seeds per row,
p = P(germination)
b. Number of seeds planted per row.
c.
1. The total number of germinating seeds was 50.
2. The fewest possible total number of seeds planted per row would be 4, since 1 row had 4 seeds geminate and all rows had the same number of seeds planted.

3. There are a limited number of possible values for the germination rate:
a) If 4 planted in each of 50 rows means 200 seeds planted, therefore, 50/200 = 0.25
b) If 5 planted in each of 50 rows means 250 seeds planted, therefore, 50/250 = 0.20
c) If 6 planted in each of 50 rows means 300 seeds planted, therefore, 50/300 = 0.1667

x	B(4, 0.25)	Expected	Actual
0	0.316406	16	17
1	0.421875	21	20
2	0.210938	11	10
3	0.046875	2	2
4	0.003906	0	1

The binomial distribution B(4, 0.25) is a fairly good fit, therefore if 4 seeds were planted per row the germination rate would be 0.25 and that would be the highest possible rate.

5.123 a.

x	P(x)
0	0.107
1	0.268
2	0.302
3	0.201
4	0.088
5	0.026
6	0.006
7	0.001
8	0+
9	0+
10	0+
Σ	0.999 (due to P(x) being near 0 for x = 8, 9, 10 but being treated as exactly 0)

b. P(x=7) = 0.001

c. P(x=6,7,8,9,10) = 0.007

d. No. The probability of his accomplishing such an event is extremely small. So small that the truth of his statement must be doubted.

CHAPTER 6 ∇ NORMAL PROBABILITY DISTRIBUTIONS

Chapter Preview

Chapter 6 continues the presentation of probability distributions started in Chapter 5. In this chapter, the random variable is a continuous random variable (versus a discrete random variable in Chapter 5); therefore, the probability distribution is a continuous probability distribution. There are many types of continuous distributions, but this chapter will limit itself to the most common, namely, the normal distribution. The main elements of a normal probability distribution to be covered are:
1. how probabilities are found
2. how they are represented
3. how they are used.

Intelligence and aptitude test score measuring are presented in this chapter's opening section, Intelligence Scores.

SECTION 6.1 EXERCISES

6.1 a. It's a quotient defined by [100 x (Mental Age/Chronological Age)]

b. I.Q.: 100, 16; SAT: 500, 100; Standard score: 0, 1

c. $z = (I.Q. - 100)/16$; $z = (SAT - 500)/100$

d. 2, 132, 700

e. The percentages are the same (other than for round-off)

6.3 a. Proportion b. Percentage c. Probability

SECTION 6.2 EXERCISES

Continuous Random Variable - a numerical quantity that can take on values on a certain interval

STANDARD NORMAL DISTRIBUTION

- bell shaped, symmetric curve
- $\mu = 0$, $\sigma = 1$
- distribution for the standard normal score z

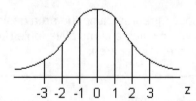

$$z = 0 \Rightarrow \mu$$

$z = 1 \Rightarrow \mu + 1\sigma$ $z = -1 \Rightarrow \mu - 1\sigma$
$z = 2 \Rightarrow \mu + 2\sigma$ $z = -2 \Rightarrow \mu - 2\sigma$
$z = 3 \Rightarrow \mu + 3\sigma$ $z = -3 \Rightarrow \mu - 3\sigma$

- area under the curve = 1

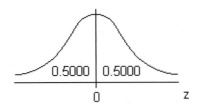

- also symmetric, therefore

The Empirical Rule and Table 3

≈ 68% of the data lies within 1 standard deviation of the mean.
- Note z = 1, gives 0.8413 and z = -1 gives 0.1587. Finding the area from z = -1 to z = 1 would be 0.8413 − 0.1587 = 0.6826 ≈ 68%.

≈ 95% of the data lies within 2 standard deviations of the mean.
- Note z = 2, gives 0.9773 and z = -2 gives 0.0228. Finding the area from z = -2 to z = 2 would be 0.9773 − 0.0228 = 0.9545 ≈ 95%.

≈ 99.7% of the data lies within 3 standard deviations of the mean.
- Note z = 3, gives 0.9987 and z = -3 gives 0.0014. Finding the area from z = -3 to z = 3 would be 0.9987 − 0.0014 = 0.9973 ≈ 99.7%.

6.5 a. A bell-shaped distribution with a mean of 0 and a standard deviation of 1.

  b. The variable is z; the standard score and this distribution is the *standard* or reference used to determine the probabilities for all other normal distributions.

Draw a picture of a normal distribution and shade in the section representing the area desired. Remember right tail areas require subtraction from 1.000 when using a cumulative table like Table 3. Area between two z-values requires the larger area minus the smaller area.

6.7 a. 0.0968   b. 0.0052   c. 0.0007   d. 0.2611

6.9 a. 0.9821   b. 0.8849   c. 0.9994   d. 0.7612

6.11 a. 1 − 0.3192 = 0.6808   b. 1 − 0.1563 = 0.8437   c. 1 − 0.0004 = 0.9996

6.13 a. 1 − 0.9993 = 0.0007   b. 1 − 0.9671 = 0.0329   c. 1 − 0.7734 = 0.2266

6.15 0.5000 − 0.0823 = 0.4177

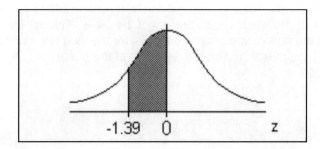

6.17 0.8907 − 0.0336 = 0.8571

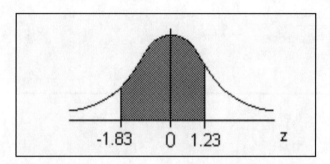

6.19
a. 0.9394 − 0.5000 = 0.4394
b. 1 − 0.9394 = 0.0606
c. 0.9394
d. 0.9394 − 0.0606 = 0.8788

6.21
a. 1.0000 − 0.5000 = 0.5000
b. 1 − 0.8531 = 0.1469
c. 1 − 0.0107 = 0.9893
d. 0.9452
e. 0.0548

6.23
a. 0.9906 − 0.5000 = 0.4906
b. 0.9904 − 0.0179 = 0.9725
c. 1 − 0.5517 = 0.4483
d. 0.9306

6.25
a. 0.7704 − 0.5000 = 0.2704
b. 0.9738 − 0.1210 = 0.8528
c. 1 − 0.8944 = 0.1056
d. 0.9599

6.27 0.9878 − 0.7734 = 0.2144

6.29
a. 0.4129 − 0.1151 = 0.2978
b. 0.0618 − 0.0401 = 0.0217
c. 0.9951 − 0.9032 = 0.0919
d. 0.9998 − 0.6368 = 0.3630

Look at the *inside* of Table 3 (Appendix B, ES11-pp716-717), and get as close as possible to the probability desired. Locate the position (row and column) on the outside of the table. This will be the corresponding z value. Remember that the table is a cumulative table – for the following normal curve diagrams, you will have to add 0.5000 to probabilities on the right side of the mean or subtract left non-tail probabilities from 0.5000.

NOTE: NORMAL CURVE   TABLE 3
- inside ⇒ probabilities
- outside ⇒ z-values

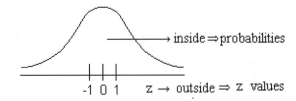

6.31   a. 1.14   b. 0.47   c. 1.66

       d. 0.86   e. 1.74   f. 2.23

6.33   a. 1.65   b. 1.96   c. 2.33

6.35   1.28, 1.65, 2.33

6.37   -0.84

6.39   a. 0.84   b. 1.04   c. -0.67 and +0.67

6.41   a. 0.84   b. -1.15 and +1.15

SECTION 6.3 EXERCISES

x is still used to denote a continuous random variable, but *x* is now referred to on an interval, not a single value. (ex.: a < x < b) As before, any letter may be used; *x* is just the most common.

*** If you use the cumulative probability function on a computer or calculator, or use an applet, to find the answers to questions involving probability distributions, you will obtain very accurate answers. The answers below are found using Table 3. Remember, the entries in Table 3 have been rounded off (z in hundredths, probabilities in ten-thousandths), and therefore the results listed here may contain a round-off error that does not occur when using a calculator, computer or applet.***

6.43 a. 0.7606 using applet   or   P(5 < x < 9) = P[(5 - 7)/1.7 < z < (9 - 7)/1.7]
= P[-1.18 < z < 1.18] = 0.8810 - 0.1190 = 0.7620

b. 0.0372 using applet   or   P[2 < x < 4] = P[(2 – 7)/1.7 < z < (4 - 7)/1.7]
= P[-2.94 < z < -1.76] = 0.0392 – 0.0016 = 0.0376

c. 0.2689 using applet   or   P[8 < x < 11] = P[(8 - 7)/1.7 < z < (11 - 7)/1.7]
= P[0.59 < z < 2.35] = 0.9906 + 0.7224 = 0.2682

6.45 $z = (x - \mu)/\sigma = (58 – 43)/5.2 = 2.88$

Applications of the Normal Distribution

1. Draw a sketch of the desired area, noting given μ and σ.

2. Write the desired probability question in terms of the given variable - usually x (ex.: P(x > 10)).

3. Transform the given variable x into a z value using $z = (x - \mu)/\sigma$.

4. Rewrite the probability question using z (ex.: P(z > *), where * = calculated z value).

5. Use Table 3 (Appendix B, ES11-pp716-717) and find the probability.

Remember: Probabilities (areas) given in Table 3 are cumulative, starting on the far left tail of the curve having a probability of approximately 0, continuing over to the far right tail having a probability of approximately 1.

For an interval probability, subtract the smaller probability from the larger probability.
If the interval is the right tail, subtract the probability given in the table from 1.0000.

Reminder: When calculating z → use 2 decimal places.
When calculating areas or probabilities → use 4 decimal places.

6.47 Use formula $z = (x - \mu)/\sigma$:

a. P[x > 60] = P[z > (60 - 60)/10]
= P[z > 0.00] = 1.0000 – 0.5000 = 0.5000

b. P[60 < x < 72] = P[(60 - 60)/10 < z < (72 - 60)/10]
= P[0.00 < z < 1.20] = 0.8849 – 0.5000 = 0.3849

c. P[57 < x < 83] = P[(57 - 60)/10 < z < (83 - 60)/10]
= P[-0.30 < z < 2.30] = 0.9893 - 0.3821 = 0.6072

d. P[65 < x < 82] = P[(65 - 60)/10 < z < (82 - 60)/10]
= P[0.50 < z < 2.20] = 0.9861 - 0.6915 = 0.2946

e. P[38 < x < 78] = P[(38 - 60)/10 < z < (78 - 60)/10]
= P[-2.20 < z < 1.80] = 0.9641 - 0.0139 = 0.9502

f. P[x < 38] = P[z < (38 - 60)/10]
= P[z < -2.20] = 0.0139

6.49 a. $P(100 < x < 120) = P[(100 - 100)/16 < z < (120 - 100)/16]$
$= P[0.00 < z < 1.25] = 0.8944 - 0.5000 = \underline{0.3944}$

b. $P(x > 80) = P[z > (80 - 100)/16]$
$= P[z > -1.25] = 1.0000 - 0.1057 = \underline{0.8943}$

6.51 Use formula $z = (x - \mu)/\sigma$:

a. $P[7200 < x < 10800] = P[(7200 - 9000)/1800 < z < (10800 - 9000)/1800]$
$= P[-1.00 < z < 1.00] = 0.8413 - 0.1587 = \underline{0.6826} = \underline{68.26\%}$

b. $P[5400 < x < 12600] = P[(5400 - 9000)/1800 < z < (12600 - 9000)/1800]$
$= P[-2.00 < z < 2.00] = 0.9773 - 0.0228 = \underline{0.9545} = \underline{95.45\%}$

c. $P[3600 < x < 14400] = P[(3600 - 9000)/1800 < z < (14400 - 9000)/1800]$
$= P[-3.00 < z < 3.00] = 0.9987 - 0.0014 = \underline{0.9973} = \underline{99.73\%}$

d. $0.6826 \approx 68\%$; $0.9545 \approx 95\%$; $0.9973 \approx 99.7\%$

6.53 a. $P(17 < x < 22) = P[(17 - 47.5)/16.6 < z < (22 - 47.5)/16.6]$
$= P[-1.84 < z < -1.54] = 0.0618 - 0.0329 = \underline{0.0289} = \underline{2.9\%}$

b. $P(x < 25) = P[z < (25 - 47.5)/16.6]$
$= P[z < -1.36] = \underline{0.0869} = \underline{8.7\%}$

c. $P(x > 21) = P[z > (21 - 47.5)/16.6]$
$= P[z > -1.60] = 1.0000 - 0.0548 = \underline{0.9452} = \underline{94.5\%}$

d. $P(48 < x < 68) = P[(48 - 47.5)/16.6 < z < (68 - 47.5)/16.6]$
$= P[0.03 < z < 1.23] = 0.8907 - 0.5120 = \underline{0.3787} = \underline{37.9\%}$

e. $P(x > 75) = P[z > (75 - 47.5)/16.6]$
$= P[z > 1.66] = 1.0000 - 0.9515 = \underline{0.0485} = \underline{4.9\%}$

6.55 a. $P(x > 8) = P[z > (8 - 6.1)/1.4]$
$= P[z > 1.36] = 1.0000 - 0.9131 = \underline{0.0869} = \underline{8.7\%}$

b. $P(x < 4) = P[z < (4 - 6.1)/1.4]$
$= P[z < -1.50] = \underline{0.0668} = \underline{6.7\%}$

6.57 a. $P(x < 20) = P[z < (20 - 21.4)/6.0]$
$= P[z < -0.23] = \underline{0.4090}$

b. $P(18 < x < 24) = P[(18 - 21.4)/6.0 < z < (24 - 21.4)/6.0]$
$= P[-0.57 < z < 0.43] = 0.6664 - 0.2843 = \underline{0.3821}$

c. $P(x > 30) = P[z > (30 - 21.4)/6.0]$
$= P[z > 1.43] = 1.0000 - 0.9236 = \underline{0.0764}$

75th percentile = P_{75} ⇒ 75% of the data is less than this value, therefore shade in the left side of the normal curve (all 0.5000) plus part of the right side (0.2500 of it). Locate the probability 0.7500 in Table 3 and find the corresponding z value. Substitute known values into the z formula and solve for the unknown.

    d. P_{75} corresponds to z = 0.67
        Use formula z = (x - μ)/σ: 0.67 = (x – 21.4)/6.0
                                   (0.67)(6.0) = x – 21.4
                                        x = 21.4 + 4.02 = <u>25.42</u>

6.59 a. P[x > 32.02] = P[z > (32.02 - 32.00)/$\sqrt{0.003}$]
                           = P[z > 0.37]
                           = 1.0000 - 0.6443 = <u>0.3557</u>

       b. 100(0.3557) = 35.57 ≈ <u>36 bottles</u>

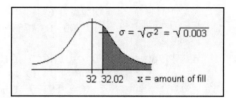

Draw a picture first, filling in the given and calculated probabilities. Based on the probability, determine the z value. Use z = (x - μ)/σ to find x.

6.61 a. Top 8% means 0.9200 lies to the left; corresponding to z = 1.41
        Use formula z = (x - μ)/σ: 1.41 = (x – 72)/12.5
                                     (1.41)(12.5) = x – 72
                                     x = 72 + 17.625 = 89.625 = <u>89.6</u>

       b. Top 28% means 0.7200 lies to the left; corresponds to z = 0.58
        Use formula z = (x - μ)/σ: 0.58 = (x – 72)/12.5
                                   (0.58)(12.5) = x – 72
                                   x = 72 + 7.25 = 79.25 = <u>79.2</u>

       c. Bottom 12% fail means 0.1200 lies to the left, corresponding to z = -1.175
        Use formula z = (x - μ)/σ: -1.175 = (x – 72)/12.5
                                 (-1.175)(12.5) = x – 72
                                 x = 72 – 14.69 = 57.31 = <u>57.3</u>

6.63 Smallest 3% means 0.0300 lies to the left;
       corresponding to z = -1.88
       Use formula z = (x - μ)/σ:
                 -1.88 = (15 – μ)/2.8
           (-1.88)(2.8) = 15 - μ
                     μ = 15 + 5.26 = <u>20.26</u>

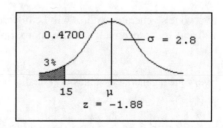

6.65 Shortest 10% means 0.1000 lies to the left;
corresponding to z = -1.28
Use formula $z = (x - \mu)/\sigma$:
$$-1.28 = (6 - \mu)/1.3$$
$$(-1.28)(1.3) = 6 - \mu$$
$$\mu = 6 + 1.664 = \underline{7.664}$$

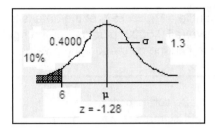

6.67 a. From (300 – 150) to (300 + 100) or <u>from 150N to 400N</u>

b. P[150 < x < 400] = P[(150 – 310)/36 < z < (400 – 310)/36]
= P[-4.44 < z < 2.50]
= 0.9938 - 0.00002 = 0.99378 = <u>99.4%</u>

c. P[x > 250] = P[z > (250 - 310)/36]
= P[z > -1.67]
= 1.0000 - 0.0475 = 0.9525 = <u>95.3%</u>

d. P[260 < x < 360] = P[(260 – 310)/36 < z < (360 – 310)/36]
= P[-1.39 < z < 1.39]
= 0.9177 - 0.0823 = <u>0.8354</u> = <u>83.5%</u>

6.69 a. Everyone's results will be different. Computer and calculator commands can be found in ES11-pp. 283-284.

b. Everyone's results will be different. Computer and calculator commands can be found in ES11-p. 284.

c. Everyone's results will be different. Each curve should be approximately normally distributed. Computer and calculator commands can be found in ES11-p. 284.

d. With 55 and 65 as the data values, use the computer and/or calculator commands in ES11-p. 285.
Using computer: P(55 < x < 65) = 0.8944 - 0.6615 = <u>0.2329</u>
Using Table 3:   P(55 < x < 65) = P[(55 - 50)/12 < z < (65 - 50)/12]
= P(0.42 < z < 1.25) = 0.8944 - 0.6628 = <u>0.2316</u>
The difference is due to round-off error in the calculation of z = 0.42.

Additional information on computer and calculator commands regarding the normal distribution, can be found in ES11-pp. 290-292.

6.71 Everyone's generated values will be different, but should have a mean and standard deviation close to 100 and 16, respectively, and be approximately normally distributed.

6.73 Everyone's generated values will be different.

Exercises 6.83 and 6.84 are good exercises to assign. The students will be finding the most common z values that will later be used in hypothesis testing. Relate the results of these exercises to Table 4 (Appendix B, ES11-p718).

SECTION 6.4 EXERCISES

Z - NOTATION

$z(\alpha) = z_\alpha$ = the z value that has α area to the right of it

1. Draw a picture.
2. Shade in desired α area, starting from the far right tail.
3. Based on the diagram and location, determine the z value using Table 3 (Appendix B, ES11-pp716-717).

6.75 a. z(0.03)   b. z(0.14)   c. z(0.75)
      d. z(0.22)   e. z(0.87)   f. z(0.98)

6.77 a. z(0.01)   b. z(0.13)   c. z(0.975)
      d. z(0.90)

6.79 a. z(0.15)   b. z(0.82)

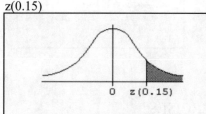
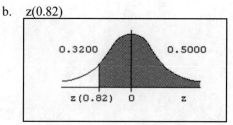

6.81 a. 1.96   b. 1.65   c. 2.33

6.83 a. z(0.05) = 1.65   b. z(0.01) = 2.33   c. z(0.025) = 1.96

      d. z(0.975) = -1.96   e. z(0.98) = -2.05

Note the z values in Exercise 6.84. They are the most common occurring z values. For that reason, Table 4, Part A, has been included in Appendix B (ES11-p718). Note Table 4, Part B, for later use.

6.85 The middle 0.80 of the distribution leaves 0.20 in the 2 tails as Table 4B shows; or 0.10 in one tail as Table 4 A shows. In both cases the z-value is 1.28.

Therefore, the middle 0.80 is bounded on the left by z = -1.28 and on the right by z = +1.28

6.87 Using Table 4B, complete the following chart of z-scores that bound a middle area of a normal distribution.

Middle	0.75	0.90	0.95	0.99
± z	±1.15	±1.65	±1.96	±2.58

6.89 a. A is an area. z is 0.10.
The area to the right of z = 0.10;
A = 1.0000 - 0.5398 = <u>0.4602.</u>

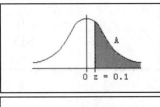

b. B is a z-score.
0.10 is the area to the right of z = B.
Use 0.9000 [1.0000 - 0.1000] to find z-score
on Table 3.
z = B = <u>1.28.</u>

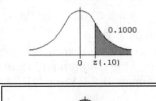

c. C is an area and the area to the right of z.
z = -0.05.
C = 1.0000 - 04801 = <u>0.5199.</u>

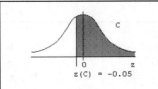

d. D is a z-score.
0.05 is typically the area to the right, but because
of the negative sign, it now is the area to the left.
Using Table 3; z = D = <u>-1.65.</u>

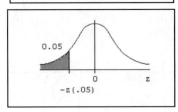

SECTION 6.5 EXERCISES

> **Criteria for the Normal Approximation of the Binomial**
> 1. np and $n(1-p)$ must both be ≥ 5.
> 2. 0.5 must be added to and/or subtracted from the x values to allow for an interval.
>   ex.: x = 5 $\Rightarrow$ 4.5 < x < 5.5
>   discrete   continuous
> 3. $\mu = np$ and $\sigma = \sqrt{npq}$
> 4. $z = (x - \mu)/\sigma$ and Table 3 (Appendix B, ES11-pp. 716-717) are used to find probabilities.

6.91 $np = (100)(0.02) = 2$ $nq = n(1-p) = (100)(0.98) = 98$
No, both np and nq must be greater than or equal to 5.

6.93 Using binomial table (Table 2, Appendix B, ES11-pp. 713-715;
P(x = 0, 1) = 0.463 + 0.366 = <u>0.829</u>

Using normal approximation:
$\mu = np = (15)(0.05) = 0.75$ and $\sigma = \sqrt{npq} = \sqrt{(15)(0.05)(0.95)} = \sqrt{0.7125} = 0.844$
P(x < 1.5) = P[z < (1.5 - 0.75)/0.844]
            = P(z < 0.89) = <u>0.8133</u>

6.95 $P(x = 4,5) = P(3.5 < x < 5.5) = P[(3.5 - 7.0)/\sqrt{3.5} < z < (5.5 - 7.0)/\sqrt{3.5}]$
$= P[-1.87 < z < -0.80] = 0.2119 - 0.0307 = \underline{0.1812}$

$P[x = 4,5 | B(n = 14, p = 0.5)] = 0.061 + 0.122 = \underline{0.183}$ (Table 2, Appendix B, ES11)

6.97 $P(x \geq 9) = P(x > 8.5) = P[z > (8.5 - 9.1)/1.65]$
$= P[z > -0.36] = 1.0000 - 0.3594 = \underline{0.6406}$

$P[x \geq 9 | B(n = 13, p = 0.7)] = P(9) + P(10) + P(11) + P(12) + P(13)$
$= 0.234 + 0.218 + 0.139 + 0.054 + 0.010$
$= \underline{0.655}$ from Table 2 (Appendix B, ES11)

6.99 Let x represent the number of patients in the 250 who will survive melanoma.
x = n(survive)
$\mu = np = (250)(0.98) = \underline{245}$ and $\sigma = \sqrt{npq} = \sqrt{(250)(0.98)(0.02)} = \underline{2.21}$
$P(x > 234.5) = P[z > (234.5 - 245.0)/2.21]$
$= P[z > -4.75] = 1.0000 - 0.000003 = \underline{0.999997}$

6.101 Let x represent the number of licensed female drivers in the 50 sampled.
x = n(licensed female drivers)
$\mu = np = (50)(0.502) = \underline{25.1}$ and $\sigma = \sqrt{npq} = \sqrt{(50)(0.502)(0.498)} = \underline{3.54}$

a. $P(x < 25.5) = P[z < (25.5 - 25.1)/3.54]$
$= P[z < 0.11] = \underline{0.5438}$

b. $P(x > 37.5) = P[z > (37.5 - 25.1)/3.54]$
$= P[z > 3.50] = 1.0000 - 0.9998 = \underline{0.0002}$

6.103 Let x represent the number of games won in the sample of 60 games.
x = n(games won)
$\mu = np = (60)[944/(944 + 1,106)] = (60)(0.4605) = 27.63$
$\sigma = \sqrt{npq} = \sqrt{(60)(0.4605)(0.5395)} = 3.86$

$P(x < 29.5) = P[z < (29.5 - 27.63)/3.86]$
$= P(z < 0.48) = \underline{0.6844}$

6.105 Let x represent the number of parents reporting that their child has been exposed to smoke in their homes.
x = n(parents reporting child exposed to smoke)

$\mu = np = (1200)(0.42) = \underline{504}$ and $\sigma = \sqrt{npq} = \sqrt{(1200)(0.42)(0.58)} = \underline{17.10}$

a. P(449.5 < x < 500.5) = P[(449.5 - 504)/17.1 < z < (500.5 - 504)/17.1]
= P[-3.19 < z < -0.20]
= 0.4207 - 0.0007 = <u>0.4200</u>

b. 0.41819 = 0.4182   c. 0.41866 = 0.4187

6.107 Let x represent the number of students who pick Information Technology as their career choice.
x = n(students picking Information Technology)

$\mu = np = (200)(0.25) = \underline{50}$ and $\sigma = \sqrt{npq} = \sqrt{(200)(0.25)(0.75)} = \underline{6.12}$

a. P(x > 65.5) = P[z > (65.5 – 50)/6.12]
= P[z > 2.53] = 1.0000 - 0.9943 = <u>0.0057</u>

b. P(x < 26.5) = P[z < (26.5 – 50)/6.12]
= P[z < -3.84] = <u>0.00006</u>

c. P(45.5 < x < 59.5) = P[(45.5 - 50)/6.12 < z < (59.5 - 50)/6.12]
= P[-0.74 < z < 1.55]
= 0.9394 - 0.2297 = <u>0.7097</u>

CHAPTER EXERCISES

6.109 -0.84, +0.84

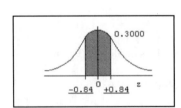

6.111 a. -0.92   b. -2.03   c. -0.74

6.113 a. P(|z| > c) = P(z < -c) + P(z > +c) = 2P(z < -c) = 0.0384
P(z < -c) = 0.0192
-c = -2.07; therefore <u>c = 2.07</u>

b. P(|z| < c) = P(-c < z < +c) = P(-c < z < 0) + P(0 < z < c) = 0.8740
P(-c < z < 0) = 0.4370 ➔ P(z < -c) = 0.0630
-c = -1.53; therefore <u>c = 1.53</u>

6.115 a. $0.9987 - 0.0014 = 0.9973$ b. $0.975 - 0.025 = \underline{0.950}$

  c. $0.10 - 0.01 = \underline{0.09}$

6.117 a. $P[x > 50.0] = P[z > (50.0-39.67)/4.38]$
  $= P[z > 2.36] = 1.0000 - 0.9909 = \underline{0.0091}$

  b. $P[42.0 < x < 48.0] = P[(42.0-39.67)/4.38 < z < (48.0-39.67)/4.38]$
  $= P[0.53 < z < 1.90] = 0.9713 - 0.7019 = \underline{0.2694}$

  c. $P[30.0 < x < 37.5] = P[(30.0-39.67)/4.38 < z < (37.5-39.67)/4.38]$
  $= P[-2.21 < z < -0.50] = 0.3085 - 0.0136 = \underline{0.2949}$

  d. $P[x > 35.0] = P[z > (35.0-39.67)/4.38]$
  $= P[z > -1.07] = 1.0000 - 0.1423 = \underline{0.8577}$

  e. $P[x < 45.0] = P[z < (45.0-39.67)/4.38]$
  $= P[z < 1.22] = \underline{0.8888}$

  f. $P[x < 32.0] = P[z < (32.0-39.67)/4.38]$
  $= P[z < -1.75] = \underline{0.0401}$

6.119 $\mu = 2$ hours, 49 minutes $= 169$ minutes; $\sigma = 21$ minutes

  a. 3 hours $= 180$ minutes
  $P[x > 180] = P[z > (180 - 169)/21]$
  $= P[z > 0.52] = 1.0000 - 0.6985 = \underline{0.3015}$

  b. $P[x < 150] = P[z < (150 - 169)/21]$
  $= P[z < -0.90] = \underline{0.1841}$

  c. Interquartile range is bounded by the first and third quartiles:
  Q_1: $z(0.75) = -0.67$ and Q_3: $z(0.25) = +0.67$
  $z = (x - \mu)/\sigma$: $\pm 0.67 = (x - 169)/21$
  $x - 169 = (\pm 0.67)(21)$
  $x = 169 \pm 14.1$
  Bounds for Interquartile range: from $\underline{154.9 \text{ to } 183.1}$ minutes

  d. Middle 90% is bounded by: $z(0.95) = -1.65$ and $z(0.05) = +1.65$
  $z = (x - \mu)/\sigma$: $\pm 1.65 = (x - 169)/21$
  $x - 169 = (\pm 1.65)(21)$
  $x = 169 \pm 34.7$
  Middle 90% ranges from $\underline{134.3 \text{ to } 203.7}$ minutes

6.121 $z = (x - \mu)/\sigma$
  $-1.65 = (10 - \mu)/0.02$
  $10 - \mu = (-1.65)(0.02)$
  $\mu = 10 + 0.033 = \underline{10.033 \text{ ounces}}$

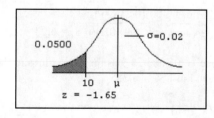

6.123 a. P[x < 350] = P[z < (350 - 525)/80]
                        = P[z < -2.19] = 0.0143

b. z = (x - μ)/σ:
    1.18 = (x – 525)/80
    x – 525 = (1.18)(80)
    x = 525 + 94.4 = 619.4

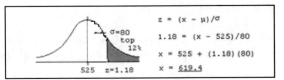

c. Q_1 has a z-score of -0.67 and Q_3 has a z-score of +0.67
    z = (x - μ)/σ:
    -0.67 = (Q_1 - 525)/80   +0.67 = (Q_3 - 525)/80
    Q_1 - 525 = (-0.67)(80) Q_3 – 525 = (+0.67)(80)
    Q_1 = 525 – 53.6 = 471.4 Q_3 = 525 + 53.6 = 578.6
    Interquartile range = Q_3 - Q_1 = 578.6 - 471.4 = 107.2

d. z = (x - μ)/σ:
    2.88 = (x - 525)/80
    x - 525 = (2.88)(80)
    x = 525 + 230.4 = 755.4

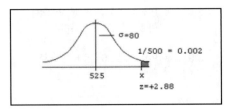

6.125 a. The normal approximation is reasonable since both np = 7.5 and nq = 17.5 are greater than 5.

b. μ = np = (25)(0.3) = 7.5;   σ = $\sqrt{npq}$ = $\sqrt{(25)(0.3)(0.7)}$ = 2.29

6.127 a. Use the binomial probability computer or calculator commands found in ES11-pp. 251-252.

b. P(x ≤ 6) = 0.005154 + 0.028632 + 0.077943 + 0.138565
              + 0.180904 + 0.184925 + 0.154104 = 0.77023

c. μ = np = (50)(0.1) = 5   σ = $\sqrt{npq}$ = $\sqrt{(50)(0.1)(0.9)}$ = 2.12
    Use –0.05 and 6.5 as the data values in the cumulative normal computer or calculator commands found in ES11 p. 285.

    P(x ≤ 6) = 0.760387 - 0.008608 = 0.751779

6.129 a. P[x ≤ 75|B(n = 300, p = 0.2] = P(0) + P(1) + P(2) + P(3) + P(4) + ... + P(75)

b. Use the cumulative binomial probability computer or calculator commands found in ES11-pp. 251-252.   Result: 0.9856

c. μ = np = (300)(0.2) = 60
    σ = $\sqrt{npq}$ = $\sqrt{(300)(0.2)(0.8)}$ = 6.93

    Use –0.05 and 75.5 as the data values in the cumulative normal computer or calculator commands found in ES11- p.285

    P(x ≤ 75) = 0.987346 - 0.0000 = 0.9873

d. (b) and (c) result in answers that are very close in value.

6.131 $\mu = np = (100)(0.80) = \underline{80.0}$
$\sigma = \sqrt{npq} = \sqrt{(100)(0.80)(0.20)} = \underline{4.0}$

$P(x \leq 70) = P(x < 70.5) = P[z < (70.5 - 80.0)/4.0]$
$= P[z < -2.38] = \underline{0.0087}$

6.133 $\mu = np = (50)(0.16) = \underline{8}$
$\sigma = \sqrt{npq} = \sqrt{(50)(0.16)(0.84)} = \underline{2.6}$

a. $P(x > 12.5) = P[z > (12.5 - 8)/2.6]$
$= P[z > 1.73] = 1.0000 - 0.9582 = \underline{0.0418}$

b. $P(x < 7.5) = P[z < (7.5 - 8)/2.6]$
$= P[z < -0.19] = \underline{0.4247}$

c. $P(6.5 < x < 14.5) = P[(6.5 - 8)/2.6 < z < (14.5 - 8)/2.6]$
$= P[-0.58 < z < 2.50] = 0.9938 - 0.2810 = \underline{0.7128}$

6.135 $\mu = np = (100)(0.3333) = \underline{33.3}$
$\sigma = \sqrt{npq} = \sqrt{(100)(0.3333)(0.6667)} = \underline{4.71}$

a. $P(x < 24.5) = P[z < (24.5 - 33.3)/4.71]$
$= P[z < -1.87] = \underline{0.0307}$

b. $P(x > 40.5) = P[z > (40.5 - 33.3)/4.71]$
$= P[z > 1.53] = 1.0000 - 0.9370 = \underline{0.0630}$

6.137 a. mean = -0.00342 std dev. = 0.02089

b.

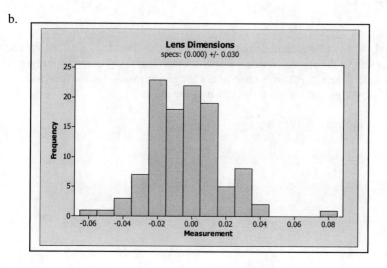

Data is somewhat symmetrical about the center spec of 0.00. The distribution looks approximately normal with the one value 0.08 as questionable.

c.

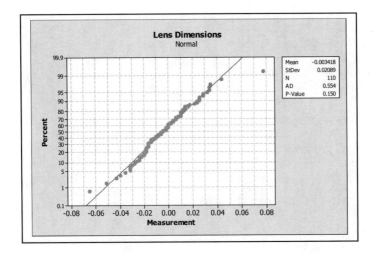

The cumulative relative frequency distribution of the data points follow very closely to a straight line pattern, with one major exception (the point in upper right corner) suggesting an approximately normal distribution.

Empirical rule: $\bar{x} = -0.00342$ $s = 0.02089$

$\bar{x} \pm 1s = -0.00342 \pm 0.02089$
   from -0.02431 to $+0.01747$
   $83/110 = 75.5\%$ versus 68%

$\bar{x} \pm 2s = -0.00342 \pm 2(0.02089)$
   from -0.0452 to $+0.03836$
   $105/110 = 95.5\%$ versus 95%

$\bar{x} \pm 3s = -0.00342 \pm 3(0.02089)$
   from -0.06609 to $+0.05925$
   $109/110 = 99.1\%$ versus 99.7%

The 75.5% for plus and minus one standard deviation is quite high versus the 68%, but the rest of the percents fit very closely to the empirical rule. The one value near 0.08, being so far from the mean, increased the standard deviation and is most likely the major reason for the increased percentage of data falling within one standard deviation of the mean.

d. $P[-0.030 < x < +0.030] = 96/110 = 0.8727 = 87.3\%$

CHAPTER 7 ∇ SAMPLE VARIABILITY

Chapter Preview

Chapters 1 and 2 introduced the concept of a sample and its various measures. Measures of central tendency, measures of dispersion, and the shape of the distribution of the data give a single "snapshot" of the population from which the sample was taken. If repeated samples are taken and statistics noted, a clearer picture of the population from which the sample came will develop. These combined statistics will enable us to better predict the population's parameters. Chapter 7 works with this illustration of repeated sampling in the form of a sampling distribution. A sampling distribution is basically a probability distribution for a sample statistic. Therefore, there can be sampling distributions for the sample mean, for the sample standard deviation or for the sample range, to name a few. The significant results that will surface for the probability distribution of the mean specifically will be contained in the Central Limit Theorem. This theorem justifies the use of the normal distribution in solving a wide range of problems.

Reference to the 2010 U.S. Census is featured in the opening section 'Population Sampling'.

SECTION 7.1 EXERCISES

7.1 a. Histogram

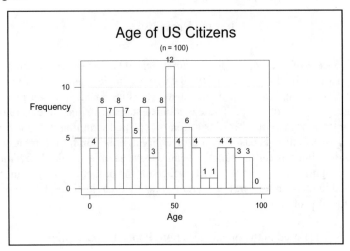

b. Mounded from 0 to 60 with a tail extending from 60 to 100 making the distribution skewed to the right.
c. Not exactly but fairly close.

7.3 a. No, but hope they are fairly close.
b. Variability

7.5 a. A sampling distribution of sample means is the distribution formed by the means from all possible samples of a fixed size that can be taken from a population.
b. It is one element in the sampling distribution of means for samples of size 3.

> Use a tree diagram to find all possible samples, for exercises 7.6a and 7.7a. Each sample will have a probability of 1/n, where n is the number of samples. Remember $\sum P(\text{statistic}) = 1$.

7.7 a.
000 020 040 060 080   600 620 640 660 680
002 022 042 062 082   602 622 642 662 682
004 024 044 064 084   604 624 644 664 684
006 026 046 066 086   606 626 646 666 686
008 028 048 068 088   608 628 648 668 688
200 220 240 260 280   800 820 840 860 880
202 222 242 262 282   802 822 842 862 882
204 224 244 264 284   804 824 844 864 884
206 226 246 266 286   806 826 846 866 886
208 228 248 268 288   808 828 848 868 888
400 420 440 460 480
402 422 442 462 482
404 424 444 464 484
406 426 446 466 486
408 428 448 468 488

b.

$\tilde{x}$	$P(\tilde{x})$
0	13/125
2	31/125
4	37/125
6	31/125
8	13/125
$\sum$	125/125 ck

c.

$\bar{x}$	$P(\bar{x})$
0/3	1/125
2/3	3/125
4/3	6/125
6/3	10/125
8/3	15/125
10/3	18/125
12/3	19/125
14/3	18/125
16/3	15/125
18/3	10/125
20/3	6/125
22/3	3/125
24/3	1/125
$\sum$	125/125 ck

7.9 Every student will have different results, however their graphs should be at least similar to these.

a.

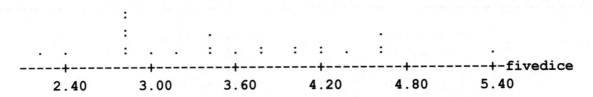

b. The distribution of $\bar{x}$'s is less variable (mostly between 2 and 5) than the distribution of x's (1 to 6) and denser near the middle.

c.

```
                         .
                       :   :   .
                     : . : : . .
              .    : : : : : . .
         .  : .  . : : : : : : :    .
     ------+---------+---------+---------+---------+---------+---------dice
         2.10      2.80      3.50      4.20      4.90      5.60
```

The distribution became more dense and mounded near the middle.

7.11 a. The samples are not all drawn from the same population and they're not the same size - each different type of transit vehicle has a different sample size.

b. The purpose of these repeated samples is for monitoring transit vehicle populations which are continually changing.

7.13 a. Every student will have different results, however the means of the 100 samples, each of size 5, should resemble those listed in (b).

b. sample means

6.0	5.8	3.2	6.2	4.8	3.2	5.0	6.2	6.2	6.4
5.0	2.4	5.8	3.4	2.6	3.6	4.6	4.0	5.0	3.8
5.4	5.2	3.6	5.4	4.4	4.2	4.0	6.0	5.8	3.8
4.8	2.8	4.2	5.6	6.4	5.6	2.8	4.8	3.8	3.0
5.2	4.2	5.0	1.6	4.8	3.2	4.2	5.0	4.8	5.0
4.8	4.6	3.4	4.2	3.8	4.2	4.4	5.6	5.2	5.2
6.0	4.0	4.6	3.4	5.2	6.8	4.4	3.6	5.6	4.4
5.0	6.4	5.2	4.8	5.0	6.0	4.0	5.2	5.6	5.8
6.4	7.0	3.2	2.2	4.6	3.2	7.6	3.0	4.6	4.6
5.4	7.2	5.8	3.6	4.4	2.8	4.8	5.4	4.8	4.0

c.

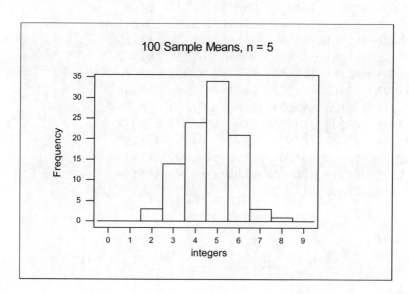

d. The shape is approximately normal, being mounded, approximately symmetrical, and centered near 4.5.

Computer commands for repeated sampling can be found in ES11 p.319.

Computer commands for repeated sampling can be found in ES11 p.319.

7.15 a. Every student will have different results; however, the means of the 200 samples, each of size 10, should resemble those listed in (b).

b. sample means

91.860	109.866	98.004	86.827	102.329	100.573	98.011
114.985	107.411	96.460	86.894	105.660	96.999	93.765
106.034	94.516	105.026	102.436	103.894	90.158	86.329
102.932	98.434	97.764	95.966	104.754	97.277	96.445
86.349	101.668	101.512	93.906	94.665	95.768	108.146
89.229	109.310	100.526	98.969	101.932	108.459	100.079
104.962	101.606	103.315	91.811	93.948	94.520	106.383
107.004	97.663	101.009	99.208	95.563	92.886	107.509
107.101	105.658	103.223	96.658	95.939	108.203	108.573
90.982	101.484	101.294	103.938	101.708	100.146	97.664
96.985	98.901	97.347	95.219	102.195	95.477	92.300
100.841	112.838	102.455	99.470	96.042	107.587	95.336
95.144	92.109	109.939	96.739	91.548	102.796	94.154
109.485	102.767	101.439	88.470	101.822	104.096	87.914
91.202	95.397	94.413	99.975	94.377	89.091	94.101
104.353	90.017	96.030	104.042	94.446	99.678	89.653
96.662	92.773	98.930	102.418	107.959	98.822	101.498
105.041	96.652	105.297	102.878	96.347	104.832	94.467
109.017	105.082	89.613	110.447	115.052	102.291	90.511
92.783	93.481	102.061	100.769	102.865	104.078	87.550
97.086	100.175	89.797	122.981	99.870	104.534	99.702
99.968	98.243	98.681	105.884	96.934	98.235	97.535
103.662	107.472	82.100	97.276	94.818	101.765	99.148
85.795	107.241	104.025	88.025	104.061	97.676	88.778
102.363	110.035	103.318	110.855	97.457	89.955	110.174
95.746	94.286	106.459	101.952	99.420	94.427	103.308
96.455	102.120	95.651	103.804	111.829	102.809	105.131
101.082	89.315	101.396	101.146	101.080	96.246	94.004
103.730	100.669	101.901	97.239			

c.

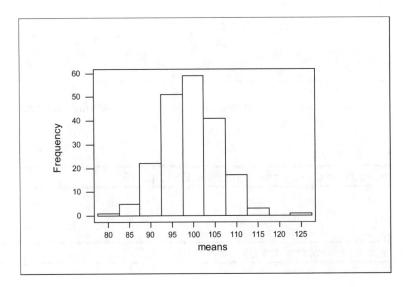

d. The shape is approximately normal, being mounded, approximately symmetrical, and centered near 100.

SECTION 7.2 EXERCISES

> The most important sampling distribution is the sampling distribution of the sample means. It provides the information that makes up the sampling distribution of sample means and the central limit theorem.
>
> **SAMPLING DISTRIBUTION OF SAMPLE MEANS & CENTRAL LIMIT THEOREM**
> If all possible random samples, each of size n, are taken from any population with a mean μ and standard deviation σ, the sampling distribution of sample means ($\bar{x}$'s) will result in the following:
> 1. The mean of the sample means (x bars) will equal the mean of the population, $\mu_{\bar{x}} = \mu$.
> 2. The standard deviation of the sample means (x bars) will be equal to the population standard deviation divided by the square root of the sample size, $\sigma_{\bar{x}} = \sigma / \sqrt{n}$.
> 3. A normal distribution when the parent population is normally distributed or becomes approximately normal distributed as the sample size increases when the parent population is not normally distributed.
>
> In essence, $\bar{x}$ is normally distributed when n is large enough, no matter what shape the population is. The further the population shape is from normal, the larger the sample size needs to be.
> $\sigma_{\bar{x}}$ = the standard deviation of the $\bar{x}$'s is now referred to as the <u>standard error of the mean</u>.

7.17   b. Answers will vary, but should resemble the following results from one simulation:
Mean of xbars = $\bar{\bar{x}}$ = 65.27; very close to μ = 65.15

     c. Standard deviation of xbars = $s_{\bar{x}}$ = 1.383; less than σ
$s_{\bar{x}}$ is approximately $2.754/\sqrt{4}$ = 1.377.

     d. The shape of the histogram is approximately normal.

     e. Took many (1001) samples of size 4 from an approximately normal population and
         1. mean of the xbars ≈ μ
         2. $s_{\bar{x}} \approx \sigma / \sqrt{n}$
         3. approximately normal distribution

7.19   a. 1.0 or one

     b. $\sigma_{\bar{x}} = \sigma / \sqrt{n}$; as n increases the value of this fraction, the standard deviation of sample mean, gets smaller.

7.21   a. <u>500</u>   b. $30/\sqrt{36}$ = <u>5</u>   c. approximately normal

7.23   a. approximately normal   b. 4.58 hours   c. $2.1/\sqrt{250}$ = 0.133

7.25 a. 86.5 pounds/person   b. $29.3/\sqrt{150} = 2.392$ c. approximately normal

7.27 a. Every student will have different results; however, the means of the 100 samples, each of size 6, should resemble those listed in (b).
  b. sample means
  17.0148 21.0960 18.8767 18.4458 23.7957 17.4572 19.9137
  22.9338 21.0164 18.9116 15.8072 23.1245 20.1439 20.8047
  18.7836 19.5104 16.7224 18.1819 19.7173 19.4121 19.7335
  (your 100 sample means should resemble the above)
  c.

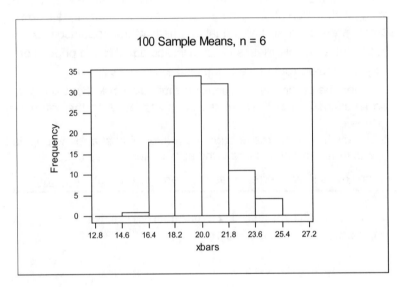

  Mean of xbars = $\bar{\bar{x}}$ = 19.924
  Standard deviation of xbars = $s_{\bar{x}}$ = 1.8023

  d. $\bar{\bar{x}}$ is approximately 20.
  $s_{\bar{x}}$ is approximately $4.5/\sqrt{6} = 1.84$.
  The shape of the histogram is approximately normal.

SECTION 7.3 EXERCISES

It is helpful to draw a normal curve, locating μ and shading in the desired portion for each problem. A new z formula must now be used to determine probabilities about $\bar{x}$.
$$z = \frac{\bar{x} - \mu}{\sigma/\sqrt{n}}$$

7.29 $z = (46.5 - 43)/(5.2/\sqrt{16}) = \underline{2.69}$

7.31 z = +2.00 corresponds to 0.9773 cum prob., z = -2.00 corresponds to 0.0228 cum prob.; need to subtract to find area in between the two z-values

7.33   a. approximately normal

  b. 50

  c. $10/\sqrt{36} = \underline{1.667}$

  d. $P(45 < \bar{x} < 55) = P[(45 - 50)/1.667 < z < (55 - 50)/1.667]$
  $= P[-3.00 < z < +3.00]$
  $= 0.9987 - 0.0014 = \underline{0.9973}$

  e. $P(\bar{x} > 48) = P[z > (48 - 50)/1.667]$
  $= P[z > -1.20]$
  $= 1.0000 - 0.1151 = \underline{0.8849}$

  f. $P(47 < \bar{x} < 53) = P[(47 - 50)/1.667 < z < (53 - 50)/1.667]$
  $= P[-1.80 < z < +1.80]$
  $= 0.9641 - 0.0359 = \underline{0.9282}$

7.35   a. Heights are approximately normally distributed with a $\mu = 69$ and $\sigma = 4$.

  b. $P(x > 70) = P[z > (70 - 69)/4]$
  $= P[z > 0.25]$
  $= 1.0000 - 0.5987 = \underline{0.4013}$

  c. The distribution of $\bar{x}$'s will be approximately normally distributed.

  d. $\mu_{\bar{x}} = \underline{69}$; $\sigma_{\bar{x}} = 4/\sqrt{16} = \underline{1.0}$

  e. $P(\bar{x} > 70) = P[z > (70 - 69)/1.0]$
  $= P[z > +1.00]$
  $= 1.0000 - 0.8413 = \underline{0.1587}$

  f. $P(\bar{x} < 67) = P[z < (67 - 69)/1.0]$
  $= P[z < -2.00]$
  $= \underline{0.0228}$

Watch the wording of the various probability problems.
If the probability for an individual item or person (x) is desired, use $z = (x - \mu)/\sigma$.
If the probability for a sample mean ($\bar{x}$) is desired, use $z = (\bar{x} - \mu) / (\sigma / \sqrt{n})$.

7.37   a. $P(38 < x < 40) = P[(38 - 39)/2 < z < (40 - 39)/2]$
  $= P[-0.50 < z < +0.50]$
  $= 0.6915 - 0.3085 = \underline{0.3830}$

b. $P(38 < \bar{x} < 40) = P[(38 - 39)/(2/\sqrt{30}) < z < (40 - 39)/(2/\sqrt{30})]$
$= P[-2.74 < z < +2.74]$
$= 0.9969 - 0.0031 = \underline{0.9938}$

c. $P(x > 40) = P[z > (40 - 39)/2]$
$= P[z > 0.50]$
$= 1.0000 - 0.6915 = \underline{0.3085}$

d. $P(\bar{x} > 40) = P[z > (40 - 39)/(2/\sqrt{30})]$
$= P[z > 2.74]$
$= 1.0000 - 0.9969 = \underline{0.0031}$

7.39 $\mu = 10.6$ mph, $\sigma = 3.5$ mph

a. $P(x > 13.5) = P[z > (13.5 - 10.6)/3.5] = P(z > 0.83)$
$= 1.0000 - 0.7967 = \underline{0.2033}$

b. $P(\bar{x} > 13.5) = P[z > (13.5 - 10.6)/(3.5/\sqrt{9})]$
$= P[z > 2.49]$
$= 1.0000 - 0.9936 = \underline{0.0064}$

c. No, especially for (a). The wind speed distribution will be skewed to the right, not normal. If they were normally distributed, most of the wind speeds would be between 0 and 22 mph. That does not allow for the high wind speeds that are known to occur. Samples of size 9 are not large enough for the sampling distribution to approach being approximately normal since the population is strongly skewed.

d. The actual probabilities are most likely not nearly as high as those found in (a) and (b).

7.41 a. $P(550 < \bar{x} < 700)$
$= P[(550-638)/(175/\sqrt{36}) < z < (700-638)/(175/\sqrt{36})]$
$= P[-3.02 < z < 2.13]$
$= 0.9834 - 0.0013 = \underline{0.9821}$

b. $P(\bar{x} > 750)$
$= P[z > (750-638)/(175/\sqrt{36})]$
$= P[z > 3.84]$
$= 1.0000 - 0.99994 = \underline{0.00006}$

c. Answers may vary, but if the normality assumption does not hold, the normal distribution still should allow for reasonable estimates for the probabilities since n > 30.

7.43 If z = -0.67, then
$-0.67 = (\bar{x} - 39.0)/(2/\sqrt{25})$
$-0.268 = \bar{x} - 39.0$
$\bar{x} = 38.732 = \underline{38.73 \text{ inches}}$

7.45 a. Using the computer commands in ES11 p.332:
$$P(4 < \bar{x} < 6) = 0.841345 - 0.158655 = \underline{0.68269}$$

Using Table 3: $P(4 < \bar{x} < 6) = P[(4 - 5)/(2/\sqrt{4}) < z < (6 - 5)/(2/\sqrt{4})]$
$$= P[-1.00 < z < +1.00]$$
$$= 0.8413 - 0.1587 = \underline{0.6826}$$

Everybody will get different answers, however, the results should be similar to the results below.

b. sample means

4.45967	4.04628	4.56959	4.74174	5.00541	4.99137	5.27322
4.18889	6.16806	5.10562	4.21561	5.61553	4.73023	3.23896
5.14986	5.83925	3.87701	4.15342	6.95931	6.69219	4.92261
4.00410	5.33392	5.69277	4.04213	6.00010	5.13663	7.04077

(your 100 sample means should resemble the above)

c.

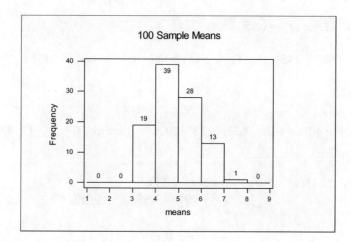

Inspecting the histogram, we find <u>67</u> of the sample means are between 4 and 6; 67/100 = 67%

d. 67% is very close to the expected 0.6826.

CHAPTER EXERCISES

7.47 $\sigma_{\bar{x}} = \sigma / \sqrt{n}$: $18.2/\sqrt{9} = 6.067$ $18.2/\sqrt{25} = 3.64$

$18.2/\sqrt{49} = 2.6$ $18.2/\sqrt{100} = 1.82$

7.49 a. Normally distributed with a mean = $775 and a standard deviation = $115.

b. $P(625 < x < 825) = P[(625-775)/115 < z < (825-775)/115]$
$$= P[-1.30 < z < +0.43]$$
$$= 0.6664 - 0.0968 = \underline{0.5696}$$

c. Approximately normally distributed with a mean = $775 and a standard error = $115/\sqrt{25} = \$23$

d. $P(710 < \bar{x} < 785) = P[(710-775)/(115/\sqrt{25}) < z < (785-775)/(115/\sqrt{25})]$
$$= P[-2.83 < z < +0.43]$$
$$= 0.6664 - 0.0023 = \underline{0.6641}$$

e. z is used in (b) and (d) since the distribution of x is given to be normal, which would also cause the sampling distribution of x-bar to be normal.

f. (b) is a distribution of individual x values; (d) is a sampling distribution of $\bar{x}$ values.

7.51 a. $P(2.63 - e < x < 2.63 + e) = 0.95$

$P(-1.96 < z < +1.96) = 0.95$, using Table 3 (Appendix B, ES11-pp716-717)

$$z = (x - \mu)/\sigma$$

$+1.96 = [(2.63 + e) - 2.63]/0.25$ therefore e = <u>0.49</u>

b. $P(2.63 - E < \bar{x} < 2.63 + E) = 0.95$

$P(-1.96 < z < +1.96) = 0.95$, using Table 3 (Appendix B, ES11-pp716-717)

$$z = (\bar{x} - \mu_{\bar{x}})/\sigma_{\bar{x}}$$

$+1.96 = [(2.63 + E) - 2.63]/(0.25/\sqrt{100})$ therefore E = <u>0.049</u>

7.53 a. $P(x > 1000) = P(z > (1000 - 586)/165) = P(z > 2.51)$
$= 1.0000 - 0.9940 = \underline{0.0060}$

b. $P(\bar{x} < 550) = P[z < (550 - 586)/(165/\sqrt{20})] = P[z < -0.98] = \underline{0.1635}$

c. Daily number of customers can be a very skewed to the right side of the distribution taking into effect holidays, promotions, etc.

7.55 a. $P(245 < x < 255) = P[(245 - 235)/\sqrt{400} < z < (255 - 235)/\sqrt{400}]$
$= P[+0.50 < z < +1.00]$
$= 0.8413 - 0.6915 = \underline{0.1498}$

b. $P(\bar{x} > 250) = P[z > (250 - 235)/(20/\sqrt{10})]$
$= P[z > +2.37]$
$= 1.0000 - 0.9911 = \underline{0.0089}$

7.57 $P(\bar{x} < 680) = P[z < (680 - 700)/(120/\sqrt{144})] = P[z < -2.00] = \underline{0.0228}$

7.59 $P(\Sigma x > 38{,}000) = P[z > (\Sigma x - n\mu)/(\sigma\sqrt{n})]$
$= P[z > (38{,}000 - (50)(750))/(25\sqrt{50})]$
$= P[z > 500/176.777]$
$= P[z > +2.83]$

= 1.0000 - 0.9977 = 0.0023

7.61 a. Let Σx represent the total weight:
$$P(\Sigma x > 4000) = P(\Sigma x/n > 4000/25)$$
$$= P(\overline{x} > 160)$$
$$= P[z > (160 - 300)/(50/\sqrt{25})]$$
$$= P[z > -14.0]$$
$$= \text{approximately } \underline{1.000}$$

b. $P(\Sigma x < 8000) = P(\overline{x} < 320)$
$$= P[z < (320 - 300)/(50/\sqrt{25})]$$
$$= P[z < 2.00]$$
$$= \underline{0.9773}$$

7.63 a. Every student will have different results, however the totals and means of the 50 samples, each of size 10, should resemble those listed below.

sums
1357.86 1344.27 1357.23 1323.87 1335.53 1357.25 1339.91
1365.17 1308.99 1357.99 1346.36 1367.63 1315.55 1354.79
1378.82 1392.20 1320.65 1333.62 1405.80 1373.16 1400.35
1399.88 1414.90 1316.60 1337.11 1309.65 1339.62 1295.95
1350.44 1333.20 1338.42 1389.96 1304.70 1312.90 1312.14
1343.50 1315.46 1390.69 1387.76 1403.76 1339.83 1314.40
1416.16 1355.12 1299.10 1352.39 1367.93 1354.54 1305.35
1325.34

xbars
135.786 134.427 135.723 132.387 133.553 135.725 133.991
136.517 130.899 135.799 134.637 136.763 131.555 135.479
137.882 139.220 132.065 133.362 140.580 137.316 140.035
139.988 141.490 131.660 133.711 130.965 133.962 129.595
135.044 133.320 133.842 138.996 130.470 131.290 131.214
134.350 131.546 139.069 138.776 140.376 133.983 131.440
141.616 135.512 129.910 135.239 136.793 135.454 130.535
132.534

b.

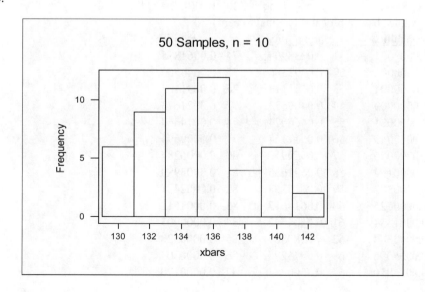

Mean of xbars   =   134.93
Standard deviation of xbars   =   3.2776

c.
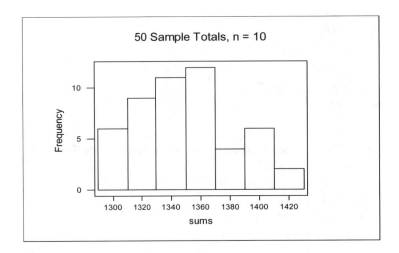

Mean of sums   =   1349.3
Standard deviation of sums   =   32.776

d. The histograms in (b) and (c) are identical in shape. Using the DESCribe command, it is evident that the xbar results are just the sum results divided by 10, the sample size.

Variable	N	Mean	Median	TrMean	StDev	SEMean
sums	50	1349.3	1345.3	1348.3	32.8	4.6
xbars	50	134.93	134.53	134.83	3.28	0.46

Variable	Min	Max	Q1	Q3
sums	1296.0	1416.2	1319.6	1369.2
xbars	129.60	141.62	131.96	136.92

7.65 a. $\mu = np = 200(0.3) = \underline{60}$ $\sigma = \sqrt{npq} = \sqrt{200(0.3)(0.7)} = \sqrt{42} = \underline{6.48}$
  b. (all other x values have a probability equal to zero)

x	probability	x	probability	x	probability
28	0.0000001	51	0.0238909	74	0.0061875
29	0.0000002	52	0.0293386	75	0.0044550
30	0.0000004	53	0.0351114	76	0.0031403
31	0.0000009	54	0.0409633	77	0.0021673
32	0.0000021	55	0.0466024	78	0.0014647
33	0.0000045	56	0.0517144	79	0.0009694
34	0.0000096	57	0.0559916	80	0.0006284
35	0.0000194	58	0.0591635	81	0.0003990
36	0.0000382	59	0.0610258	82	0.0002481
37	0.0000725	60	0.0614617	83	0.0001512
38	0.0001334	61	0.0604542	84	0.0000903
39	0.0002374	62	0.0580861	85	0.0000528
40	0.0004096	63	0.0545298	86	0.0000303
41	0.0006850	64	0.0500263	87	0.0000170

42	0.0011114	65	0.0448587	88	0.0000093
43	0.0017501	66	0.0393242	89	0.0000050
44	0.0026763	67	0.0337065	90	0.0000027
45	0.0039762	68	0.0282539	91	0.0000014
46	0.0057421	69	0.0231647	92	0.0000007
47	0.0080633	70	0.0185791	93	0.0000003
48	0.0110151	71	0.0145791	94	0.0000002
49	0.0146440	72	0.0111947	95	0.0000001
50	0.0189535	73	0.0084125		

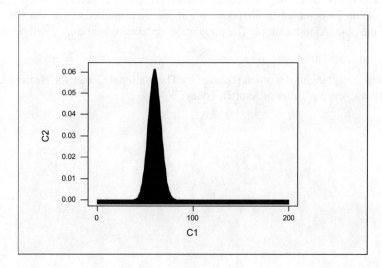

c. Every student will have different results, however the histograms should resemble that in (d)

d.

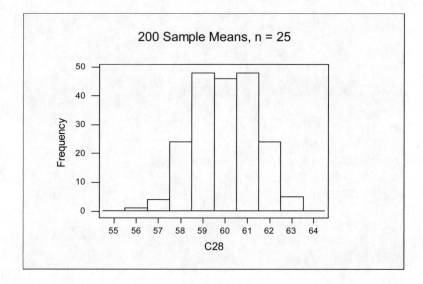

Mean of C28 = $\overline{\overline{x}}$ = 59.979   Standard deviation of C28 = $s_{\overline{x}}$ = 1.368

e. The mean of the sample means, $\overline{\overline{x}}$ = 59.979, is approximately equal to μ = 60. The standard deviation of the sample means, $s_{\overline{x}}$ = 1.368, is approximately equal to $6.48/\sqrt{25}$ = 1.296. The distribution of the sample means is approximately normally distributed.

CHAPTER 8 ∇ INTRODUCTION TO STATISTICAL INFERENCES

Chapter Preview

Chapter 8 introduces inferential statistics. Generalizations about population parameters are made based on sample data in inferential statistics. These generalizations can be made in the form of hypothesis tests or confidence interval estimations. Each is calculated with a degree of uncertainty. The integral elements and procedure for obtaining a confidence interval and for completing a hypothesis test will be presented in this chapter. They will be performed with respect to the population mean, μ. The population standard deviation, σ, will be considered as a known quantity.

Statistical information on the average height of women reported by The National Center for Health Statistics is presented in the opening section, 'Are We Taller or Shorter Today?'.

SECTION 8.1 EXERCISES

8.1 a. American females, or American female health professionals

b. $\bar{x} = 64.78 = 64.8$, s = 3.5

c. Distribution is mounded about center, approximately symmetrical

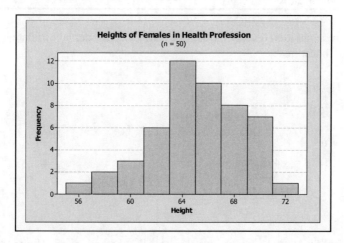

d. around 65 inches, between 62 and 69 inches, the heights most frequent in this data (ans. will vary)

e. A narrower interval would be very desirable and/or a larger sample size

---

ESTIMATION

Point Estimate for a Population Parameter - the value of the corresponding sample statistic
  ex.) $\bar{x}$ is the point estimate of μ
       s is the point estimate of σ
       s^2 is the point estimate of σ^2

Confidence intervals are used to estimate a population parameter on an interval with a degree of certainty. We could begin by taking a sample and finding $\bar{x}$ to just estimate μ. The sample statistic, $\bar{x}$, is a <u>point estimate</u> of the population parameter μ. How good an estimate it is depends not only on the sample size and variability of the data, but also whether or not the sample statistic is unbiased.

(continued)

> **Unbiased Statistic** - a sample statistic whose sampling distribution has a mean value equal to the corresponding population parameter
>
> One would assume that $\bar{x}$ is not exactly equal to μ, but hopefully relatively close. It is by this reasoning that we work with interval estimates of population parameters.
>
> **Level of Confidence** = $1 - \alpha$ = the probability that the interval constructed, based on the sample, contains the true population parameter.

8.3 A point estimate is a single number, usually the sample statistic. An interval estimate is an interval of some width centered at the point estimate

8.5 $n = 15$, $\Sigma x = 271$, $\Sigma x^2 = 5015$

  a. $\bar{x} = \Sigma x/n = 271/15 = 18.0667 = \underline{18.1 \text{ dollars}}$

  b. $s^2 = (\Sigma x^2 - (\Sigma x)^2/n)/(n-1)$
  $= (5015 - (271)^2/15)/14 = 8.4952 = \underline{8.5}$

  c. $s = \sqrt{s^2} = \sqrt{8.4952} = 2.9146 = \underline{2.9 \text{ dollars}}$

8.7
  a. II has the lower variability; both have a mean value equal to the parameter. II would be the better estimator.

  b. II has a mean value equal to the parameter, I does not. II would be the better estimator.

  c. Neither is a good choice; I is negatively biased with less variability, while II is not biased with a larger variability. II would be the better estimator.

8.9 Difficulty and collector fatigue in obtaining and also in evaluating a very large sample; cost of sampling; destruction of product in cases like the rivets illustration.

8.11
  a. US married-couple families; family income

  b. mean; $\bar{x} = \$90{,}835$

  c. Margin of error is the same as the maximum error of estimate

  d. $101   e. 0.90

  f. $90,734 to $90,936; an interval of values that is 0.90 likely to include the population true mean value

8.13 $\sigma_{\bar{x}} = \sigma/\sqrt{n} = 18/\sqrt{36} = 18/6 = 3$

> The level of confidence depends on the number of standard errors from the sample mean.
>
> For $\bar{x} \pm z\sigma_{\bar{x}}$: $\bar{x}$ = mean (given)
>
> $\sigma_{\bar{x}}$ = standard error of the mean (given)
>
> z = number of standard errors

Find the corresponding probability for the z-value using Table 3, Appendix B. Multiply the probability by 2 to cover both parts of the interval. This total probability is equal to the level of confidence.

8.15 a. $1 - 2(0.1003) = 0.7994 = 80.0\%$

b. $1 - 2(0.0749) = 0.8502 = 85.0\%$

c. $1 - 2(0.0250) = 0.9500 = 95.0\%$

d. $1 - 2(0.0099) = 0.9802 = 98.0\%$

8.17 a. An interval of values, 101 to 113, that is 0.95 likely to include the population true number per 1000 showed a prevalence of self-reported hip pain amongst the men.

b. $(113 - 101)/2 = 6;\quad 6 = 1.96 \times$ standard error
$6 = 1.96(\sigma_{\bar{x}})$
$\sigma_{\bar{x}} = 3.06$

c. $(180 - 166)/2 = 7;\quad 7 = 1.96 \times$ standard error
$7 = 1.96(\sigma_{\bar{x}})$
$\sigma_{\bar{x}} = 3.57$

8.19 a. Between 3:09 PM and 3:29 PM, the next eruption should occur.

b. Yes; the snapshot was recorded on 8/14/2009 at 3:25:19 PM, which is within the interval from 3:09 to 3:29 PM.

c. 90% of the eruptions occur within predicted interval

SECTION 8.2 EXERCISES

ESTIMATION OF THE POPULATION MEAN - μ

Point estimate of μ: $\bar{x}$

Interval estimate of μ = confidence interval

$1 - \alpha$ = level of confidence, the probability or degree of certainty desired (ex.: 95%, 99% ...)

A $(1-\alpha)$ confidence interval estimate for μ is: $\bar{x} - z(\alpha/2)\cdot\sigma/\sqrt{n}$ to $\bar{x} + z(\alpha/2)\cdot\sigma/\sqrt{n}$ **

$\bar{x} - z(\alpha/2)\cdot\sigma/\sqrt{n}$ = lower confidence limit
$\bar{x} + z(\alpha/2)\cdot\sigma/\sqrt{n}$ = upper confidence limit
$E = z(\alpha/2)\cdot\sigma/\sqrt{n}$ = maximum error of the estimate

**To find $z(\alpha/2)$:
(suppose for example that a 95% confidence interval is desired)
$95\% = 1 - \alpha$, that is,
$0.95 = 1 - \alpha$

> solving for alpha, α, gives
> $$\alpha = 0.05$$
> dividing both sides by 2 gives
> $$\alpha/2 = 0.025$$
> Now determine the probability associated with z(0.025) using Table 4B (Appendix A, ES11-p718), the Critical Values of Standard Normal Distribution for Two-Tailed Situations. (This table conveniently gives the most popular critical values for z.)

8.21 Either the sampled population is normally distributed or the random sample is sufficiently large for the Central Limit Theorem to hold.

8.23   a. $\alpha = 0.02$; $z(\alpha/2) = z(0.01) = \underline{2.33}$

b. $\alpha = 0.01$; $z(\alpha/2) = z(0.005) = \underline{2.58}$

THE CONFIDENCE INTERVAL: A FIVE-STEP PROCEDURE

Step 1: The Set-Up:
　　Describe the population parameter of concern.
Step 2: The Confidence Interval Criteria:
　　a. Check the assumptions.
　　b. Identify the probability distribution and the formula to be used.
　　c. Determine the level of confidence, $1 - \alpha$.
Step 3: The sample evidence:
　　Collect the sample information.
Step 4: The Confidence Interval:
　　a. Determine the confidence coefficient.
　　b. Find the maximum error of estimate.
　　c. Find the lower and upper confidence limits.
Step 5: The Results:
　　State the confidence interval.

8.25   a.   Step 1: The mean, μ
　　　Step 2: a. normality indicated
　　　　　　　b. z,　$\sigma = 6$　　　　c. $1-\alpha = 0.95$
　　　Step 3: $n = 16$, $\overline{x} = 28.7$
　　　Step 4: a. $\alpha/2 = 0.05/2 = 0.025$; $z(0.025) = 1.96$
　　　　　　　b. $E = z(\alpha/2) \cdot \sigma/\sqrt{n} = (1.96)(6/\sqrt{16}) = (1.96)(1.5) = 2.94$
　　　　　　　c. $\overline{x} \pm E = 28.7 \pm 2.94$
　　　Step 5: $\underline{25.76 \text{ to } 31.64}$, the 0.95 confidence interval for μ

b.   <u>Yes</u>; the sampled population is normally distributed.

8.27 a. Step 1: The mean, μ
Step 2: a. normality assumed because of CLT with n = 86.
b. z , σ = 16.4   c. 1-α = 0.90
Step 3: n = 86, $\bar{x}$ = 128.5
Step 4: a. $\alpha/2$ = 0.10/2 = 0.05; z(0.05) = 1.65
b. E = z($\alpha/2$)·$\sigma/\sqrt{n}$ = (1.65)(16.4/$\sqrt{86}$) = (1.65)(1.76845) = 2.92
c. $\bar{x} \pm E$ = 128.5 ± 2.92
Step 5: 125.58 to 131.42, the 0.90 confidence interval for μ

b. Yes; the sample size is sufficiently large to satisfy the CLT.

8.29 a. point estimate = $\bar{x}$ = 128.5
b. confidence coefficient = z($\alpha/2$) = z(0.05) = 1.65
c. Standard error of the mean = $\sigma/\sqrt{n}$ = (16.4/$\sqrt{86}$) = 1.76845
d. max. error of estimate, E = z($\alpha/2$)·$\sigma/\sqrt{n}$ = (1.65)(1.76845) = 2.92
e. lower confidence limit = $\bar{x}$ - E = 128.5 – 2.92 = 125.58
f. upper confidence limit = $\bar{x}$ + E = 128.5 + 2.92 = 131.42

8.31 Answers will vary.

8.33 a. 15.9; $\approx$ 68%
b. 31.4; $\approx$ 95%
c. 41.2; $\approx$ 99%
d. higher level makes for a wider width; to be more certain the parameter is contained in the interval

8.35 a. 75.92
b. E = z($\alpha/2$)·$\sigma/\sqrt{n}$ = (2.33)(0.5/$\sqrt{10}$) = 0.3684 = 0.368
c. $\bar{x} \pm E$ = 75.92 ± 0.368
    75.552 to 76.288, the 0.98 confidence interval for μ

REMEMBER: variance = σ^2, therefore $\sigma = \sqrt{\sigma^2}$

8.37 a. Step 1: The mean length of fish caught in Cayuga Lake
Step 2: a. normality assumed, CLT with n = 200.
b. z, σ = 2.5   c. 1-α = 0.90
Step 3: n = 200, $\bar{x}$ = 14.3
Step 4: a. $\alpha/2$ = 0.10/2 = 0.05; z(0.05) = 1.65
b. E = z($\alpha/2$)·$\sigma/\sqrt{n}$ = (1.65)(2.5/$\sqrt{200}$)
       = (1.65)(0.17678) = 0.29
c. $\bar{x} \pm E$ = 14.3 ± 0.29

Step 5: 14.01 to 14.59, the 0.90 confidence interval for μ

b. Step 1-3: as shown in (a), except 2c. 1-α = 0.98
Step 4: a. α/2 = 0.02/2 = 0.01; z(0.01) = 2.33
b. E = z(α/2)·σ/√n = (2.33)(2.5/√200)
= (2.33)(0.17678) = 0.41
c. $\bar{x}$ ± E = 14.3 ± 0.41
Step 5: 13.89 to 14.71, the 0.98 confidence interval for μ

8.39 a. Speed readings for the Channel Tunnel train.

b. Step 1: The mean speed of the Channel Tunnel train
Step 2: a. normality indicated
b. z, σ = 19   c. 1-α = 0.90
Step 3: n = 20, $\bar{x}$ = 184
Step 4: a. α/2 = 0.10/2 = 0.05; z(0.05) = 1.65
b. E = z(α/2)·σ/√n = (1.65)(19/√20)
= (1.65)(4.25) = 7.01
c. $\bar{x}$ ± E = 184 ± 7.01
Step 5: 176.99 to 191.01, the 0.90 confidence interval for μ

c. Step 1-3: as shown in (b), except 2c. 1-α = 0.95
Step 4: a. α/2 = 0.05/2 = 0.025; z(0.025) = 1.96
b. E = z(α/2)·σ/√n = (1.96)(19/√20)
= (1.96)(4.25) = 8.33
c. $\bar{x}$ ± E = 184 ± 8.33
Step 5: 175.67 to 192.33, the 0.95 confidence interval for μ

Computer and/or calculator commands to calculate a confidence interval for μ, provided σ is known, can be found in ES11-pp354-355. The output will also contain the sample mean and standard deviation.

8.41 a. The mean length of parts being produced after adjustment.

b. $\bar{x}$ = Σx/n = 759.2/10 = 75.92

c. Step 1: See part 'a'
Step 2: a. normality indicated
b. z, σ = 0.5   c. 1-α = 0.99
Step 3: n = 10, $\bar{x}$ = 75.92
Step 4: a. α/2 = 0.01/2 = 0.005; z(0.005) = 2.58
b. E = z(α/2)·σ/√n = (2.58)(0.5/√10)
= (2.58)(0.158) = 0.408
c. $\bar{x}$ ± E = 75.92 ± 0.408
Step 5: 75.512 to 76.328, the 0.99 confidence interval for μ

8.43 a.

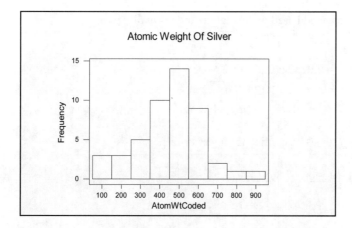

The units are "ten millionths" of the original unit of measure.

b. Mean = 450.6 and standard deviation = 173.4

c. Mean = 107.86814506 and standard deviation = 0.00001734

d. The histogram definitely shows an approximately normal distribution. Also, the Normal Probability Plot shows a straight-line pattern.

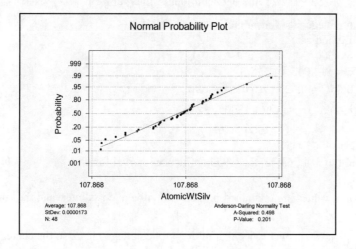

e. The SDSM and CLT both apply. The SDSM defines the mean and standard error for the sampling distribution and says that if the population is normal then the sampling distribution is normal. The histogram for the sample strongly suggests that the population has a normal distribution. The CLT tells us that the sampling distribution approaches a normal distribution for larger samples.

f. Sigma is not known.

g. A value is needed for the standard deviation of the population, and if it is not available then the next best thing is to estimate it with a reliable point estimate. The sample standard deviation is the value that will be used.

h. Step 1: The mean value of all such observations
Step 2: a. normality assumed based on histogram and normality plot
b. z, s = 173.4   c. 1-α = 0.95
Step 3   n = 48, $\bar{x}$ = 450.6
Step 4: a. α/2 = 0.05/2 = 0.025; z(0.025) = 1.96
b. E = z(α/2)·σ/$\sqrt{n}$ = (1.96)(173.4/$\sqrt{48}$)
= (1.96)(25.03) = 49.055
c. $\bar{x}$ ± E = 450.6 ± 49.055
Step 5: 401.5 to 499.7, the 0.95 confidence interval for μ
or

107.86814015 to 107.86814997

8.45 a. Step 1: The mean college cost for private colleges in NY.
Step 2: a. normality indicated.
b. z, σ = $2200   c. 1-α = 0.95
Step 3   n = 32, $\bar{x}$ = $34020
Step 4: a. α/2 = 0.05/2 = 0.025; z(0.025) = 1.96
b. E = z(α/2)·σ/$\sqrt{n}$ = (1.96)(2200/$\sqrt{32}$)
= (1.96)(388.9087) = 762.261
c. $\bar{x}$ ± E = 34020 ± 762.26
Step 5: $33,257.74 to $34,782.26, the 0.95 confidence interval for μ

b. Step 1: The mean college cost for public colleges in NY.
Step 2: a. normality indicated.
b. z, σ = $1500   c. 1-α = 0.95
Step 3   n = 32, $\bar{x}$ = $14,045
Step 4: a. α/2 = 0.05/2 = 0.025; z(0.025) = 1.96
b. E = z(α/2)·σ/$\sqrt{n}$ = (1.96)(1500/$\sqrt{32}$)
= (1.96)(265.1650) = 519.723
c. $\bar{x}$ ± E = 14045 ± 519.72
Step 5: $13,525.28 to $14,564.72, the 0.95 confidence interval for μ

c. Both College Board values are contained in the NYS college costs intervals; the costs are basically the same.

d. The private college mean is over twice that of the public college mean, so the center points for the confidence intervals are quite dramatically different.

e. The decrease in the standard deviation for the public colleges caused the maximum error term to decrease compared to the that of the private colleges, thereby forming a more narrow interval.

To find the <u>sample size n</u> required for a 1-α confidence interval, use the formula: n = [z(α/2)·σ/E]²,
where
   z = standard normal distribution
   α = calculated from the 1-α confidence interval desired
   σ = population standard deviation
   E = maximum error of the estimate

> The maximum error of the estimate, E, is the amount of error that is tolerable or allowed. Quite often, finding the word "within" in an exercise will locate the acceptable value for E.

8.47 $n = [z(\alpha/2) \cdot \sigma/E]^2 = [(2.33)(3)/1]^2 = 48.8601 = \underline{49}$

8.49 $n = [z(\alpha/2) \cdot \sigma/E]^2 = [(2.58)(1.0)/0.5]^2 = 26.6 = \underline{27}$

8.51 $n = [z(\alpha/2) \cdot \sigma/E]^2 = [(1.96)(\sigma)/(0.4\sigma)]^2 = 24.01 = \underline{25}$

SECTION 8.3 EXERCISES

> **DEFINITIONS FOR HYPOTHESIS TESTS**
>
> <u>Hypothesis</u> - a statement that something is true.
>
> <u>Null Hypothesis, H_o</u> - a statement that specifies a value for a population parameter
> ex.: H_o: The mean weight is 40 pounds
>
> <u>Alternative Hypothesis, H_a</u> - opposite of H_o, a statement that specifies an "opposite" value for a population parameter
> ex.: H_a: The mean weight is not 40 pounds
>
> <u>Type I Error</u> - the error resulting from rejecting a true null hypothesis
>
> <u>α (alpha)</u> - the probability of a type I error, that is the probability of rejecting H_o when it is true.
>
> <u>Type II Error</u> - the error resulting from not rejecting a false null hypothesis
>
> <u>β (beta)</u> - the probability of a type II error, that is the probability of not rejecting H_o when it is false.
>
> Keep α and β as small as possible, depending on the severity of the respective error.

8.53 Answers will vary but may contain:
  a. Would survey patients who had received this treatment hoping to find an overwhelming majority who were positively affected by the treatment.
  b. Would survey patients who had received this treatment hoping to find many who were NOT positively affected by the treatment.
  c. It might be difficult to determine how many positive reports are required to show, "Positively The Most Effective …" On the other hand, it would definitely not take as many non-positive reports to conclude that this system is not "Positively The Most Effective …"

8.55 H_o: The system is reliable
  H_a: The system is not reliable

8.57 a. H_o: Special delivery mail does not take too much time
    H_a: Special delivery mail takes too much time

b. H_o: The new design is not more comfortable
H_a: The new design is more comfortable

c. H_o: Cigarette smoke has no effect on the quality of a person's life
H_a: Cigarette smoke has an effect on the quality of a person's life

d. H_o: The hair conditioner is not effective on "split ends"
H_a: The hair conditioner is effective on "split ends"

8.59 Type A correct decision:
Truth of situation: the party will be a dud.
Conclusion: the party will be a dud.
Action: did not go [avoided dud party]

Type B correct decision:
Truth of situation: the party will be a great time.
Conclusion: the party will be a great time.
Action: did go [party was great time]

Type I error:
Truth of situation: the party will be a dud.
Conclusion: the party will be a great time.
Action: did go [party was a dud]

Type II error:
Truth of situation: the party will be a great time.
Conclusion: the party will be a dud.
Action: did not go [missed great party]

Remember; the truth of the situation is not known before the decision is made, the conclusion reached and the resulting actions take place. Only after the party is over can the evaluation be made.

8.61 a. H_o: The victim is alive
H_a: The victim is not alive

b. Type A correct decision: The victim is alive and is treated as though alive.
Type I error: The victim is alive, but is treated as though dead.
Type II error: The victim is dead, but treated as if alive.
Type B correct decision: The victim is dead and treated as dead.

c. The type I error is very serious. The victim may very well be dead shortly without the attention that is not being received.
The type II error is not as serious. The victim is receiving attention that is of no value. This would be serious only if there were other victims that needed this attention.

8.63 You missed a great time.

8.65
a. Type A correct decision: The majority of Americans do favor laws against assault weapons and it is decided that they do favor the laws.
Type B correct decision: The majority of Americans do not favor laws against assault weapons and it is decided that they do not favor the laws.

b. Type A correct decision: The fast food menu is not low salt and it is decided that it is not low salt.
Type B correct decision: The fast food menu is low salt and it is decided that it is low salt.

c. Type A correct decision: The building must not be demolished and it is decided that it should not be demolished.
Type B correct decision: The building must be demolished and it is decided that it should be demolished.

d. Type A correct decision: There is no waste in government spending and it is decided that there is no waste.
Type B correct decision: There is waste in government spending and it is decided that there is waste.

8.67 Type I error: Teaching techniques have no significant effect on students' exam scores and it is decided that teaching techniques do have a significant effect on students' exam scores.
Action: Teaching techniques which have no significant effect will be believed to be effective and used accordingly.

Type II error: Teaching techniques have a significant effect on students' exam scores and it is decided that teaching techniques have no significant effect on students' exam scores.
Action: Teaching techniques which have an significant effect will be believed to be ineffective and not used.

TERMINOLOGY FOR DECISIONS IN HYPOTHESIS TESTS

Reject H_o: use when the evidence disagrees with the null hypothesis.

Fail to reject H_o: use when the evidence does not disagree with the null hypothesis.

Note: The purpose of the hypothesis test is to allow the evidence a chance to discredit the null hypothesis.
Remember: If one believes the null hypothesis to be true, generally there is no test.

8.69   a. Type I   b. Type II   c. Type I   d. Type II

8.71   a. Commercial is not effective.
           b. Commercial is effective.

8.73 a. The type I error is very serious and, therefore, we are willing to allow it to occur with a probability of 0.001; that is, only 1 chance in 1000.

b. The type I error is somewhat serious and, therefore, we are willing to allow it to occur with a probability of 0.05; that is, 1 chance in 20.

c. The type I error is not at all serious and, therefore, we are willing to allow it to occur with a probability of 0.10; that is, 1 chance in 10.

8.75 a. α b. β

8.77 α is the probability of rejecting a TRUE null hypothesis; $1-\beta$ is the probability of rejecting a FALSE null hypothesis; they are two distinctly different acts that both result in rejecting the null hypothesis.

8.79 a. When the test procedure begins, the experimenter is thoroughly convinced the alternative hypothesis can be shown to be true; thus when the decision *reject H_o* is attained, the experimenter will want to say something like "see I told you so." Thus the statement of the conclusion is a fairly strong statement like; "the evidence shows beyond a shadow of a doubt (is significant) that the alternative hypothesis is correct."

b. When the test procedure begins, the experimenter is thoroughly convinced the alternative hypothesis can be shown to be true; thus when the decision *fail to reject H_o* is attained, the experimenter is disappointed and will want to say something like "okay so this evidence was not significant, I'll try again tomorrow." Thus the statement of the conclusion is a fairly mild statement like, "the evidence was not sufficient to show the alternative hypothesis to be correct."

> The **power** of a test is equal to $1 - \beta$. It is the measure of the ability of a hypothesis test to reject a false null hypothesis.

8.81 a. α = P(rejecting H_o when the H_o is true)
= $P(x \geq 86 | \mu=80) = P(z > (86 - 80)/5) = P(z > 1.20)$
= $1.0000 - 0.8849 = \underline{0.1151}$

b. β = P(accepting H_o when the H_o is false)
= $P(x < 86 | \mu=90) = P(z < (86 - 90)/5) = P(z < -0.80)$
= $\underline{0.2119}$

8.83 a. Acceptable diameters: from 23.5 mm to 24.5 mm
(1 unacceptable: Cork 1, 24.51 mm)
Acceptable ovalization: ≤ 0.7 mm
(2 unacceptable: Cork 2, 0.88 mm and cork 21, 0.76 mm)
Acceptable lengths: from 44.3 to 45.7
(1 unacceptable: Cork 21, 44.27 mm)
Therefore 29 corks pass Part 1 inspection.

b. The batch will be refused, there are 3 corks that do not meet the limits of specification.

c. Answers will vary.

SECTION 8.4 EXERCISES

THE PROBABILITY-VALUE HYPOTHESIS TEST: A FIVE-STEP PROCEDURE

Step 1: The Set-Up:
    a. Describe the population parameter of concern.
    b. State the null hypothesis (H_o) and the alternative hypothesis (H_a).

Step 2: The Hypothesis Test Criteria:
    a. Check the assumptions.
    b. Identify the probability distribution and the test statistic formula to be used.
    c. Determine the level of significance, α.

Step 3: The Sample Evidence:
    a. Collect the sample information.
    b. Calculate the value of the test statistic.

Step 4: The Probability Distribution:
    a. Calculate the p-value for the test statistic.
    b. Determine whether or not the p-value is smaller than α.

Step 5: The Results:
    a. State the decision about H_o.
    b. State the conclusion about H_a.

The Null and Alternative Hypotheses, H_o and H_a

H_o: $\mu = 100$ versus H_a: $\mu \neq 100$ (= and $\neq$ form the opposite of each other)
    H_a is a two-sided alternative
        possible wording for this combination:
            a) mean is different from 100 ($\neq$)
            b) mean is not 100 ($\neq$)
            c) mean is 100 (=)

OR ───

H_o: $\mu = 100$ ($\leq$) versus H_a: $\mu > 100$ ($\leq$ and > form the opposite of each other)
    H_a is a one-sided alternative
        possible wording for this combination:
            a) mean is greater than 100 (>)
            b) mean is at most 100 ($\leq$)
            c) mean is no more than 100 ($\leq$)

OR ───

                                                              ...

> H_o: $\mu = 100$ ($\geq$) versus H_a: $\mu < 100$ ($\geq$ and $<$ form the opposite of each other)
>
> H_a is a one-sided alternative
>
> possible wording for this combination:
> a) mean is less than 100 ($<$)
> b) mean is at least 100 ($\geq$)
> c) mean is no less than 100 ($\geq$)
>
> ---
>
> Always show equality (=) in the null hypothesis, since the null hypothesis must specify a single specific value for μ (like $\mu = 100$).
>
> The null hypothesis could be rejected in favor of the alternative hypothesis for three different reasons; 1) $\mu \neq 100$ or 2) $\mu > 100$ or 3) $\mu < 100$. Together, the two opposing statements, H_o and H_a, must contain or account for all numerical values around and including μ. This allows for the addition of $\geq$ or $\leq$ to the null hypothesis. Therefore, if the alternative hypothesis is $<$ or $>$, $\geq$ or $\leq$, respectively, may be added to the null hypothesis. If $\geq$ or $\leq$ is being tested, the appropriate symbol should be written in parentheses after the amount stated for μ.
>
> Sometimes, depending on the wording, it is easier to write the alternative hypothesis first. The alternative hypothesis can only contain $>$, $<$ or $\neq$.
>
> **Hint:** Sometimes it is helpful to either: 1) <u>remove</u> the word "no" or "not" when it is included in the claim, or 2) <u>insert</u> "no" or "not" when it is not in the claim, to form the opposite of the claim.

8.85 H_o: The mean shearing strength is at least 925 lbs.

H_a: The mean shearing strength is less than 925 lbs.

8.87 a. H_o: $\mu = 1.25$ ($\leq$)

H_a: $\mu > 1.25$

b. H_o: $\mu = 335$ ($\geq$)

H_a: $\mu < 335$

c. H_o: $\mu = 230,000$

H_a: $\mu \neq 230,000$

d. H_o: $\mu = 260$ ($\leq$)

H_a: $\mu > 260$

e. $H_o: \mu = 15.00 \ (\leq)$
$H_a: \mu > 15.00$

ERRORS

Type I error - occurs when H_o is rejected and it is a true statement.

Type II error - occurs when H_o is accepted and it is a false statement.

8.89 Type A correct decision: The mean shearing strength is at least 925 lbs and it is decided that it is.
Type I error: The mean shearing strength is at least 925 lbs and it is decided that it is less than 925 lbs.
Type II error: The mean shearing strength is less than 925 lbs and it is decided that it is greater than or equal to 925 lbs.
Type B correct decision: The mean shearing strength is less than 925 lbs and it is decided that it is less than 925 lbs.

Type II error; you buy and use weak rivets.

H_o DECISIONS

Since we work under the assumption that H_o is a true statement,
    all decisions are made <u>based on</u> or <u>pertaining to</u> H_o.
If we are unable to reject H_o, the terminology "fail to reject H_o" is used;
    whereas if we are able to reject H_o, "reject H_o" is used.
After this decision statement, include an additional statement, explaining how the test results support or did not support the experimenter's convictions, to form the conclusion.

8.91 A type I error would be committed if a decision of reject H_o was reached and interpreted as 'mean hourly charge is less than $60 per hour' when in fact the mean hourly charge is at least $60 per hour.

A type II error would be committed if a decision of fail to reject H_o was reached and interpreted as 'mean hourly charge is at least $60 per hour' when in fact the mean hourly charge is less than $60 per hour.

Use the sample information (sample mean and size) and the population parameter in the null hypothesis (μ) to calculate the test statistic z^*.

$$z^* = (\overline{x} - \mu)/(\sigma/\sqrt{n})$$

8.93 a. $z = (\bar{x} - \mu)/(\sigma/\sqrt{n})$
$z^* = (10.6 - 10)/(3/\sqrt{40}) = \underline{1.26}$

b. $z = (\bar{x} - \mu)/(\sigma/\sqrt{n})$
$z^* = (126.2 - 120)/(23/\sqrt{25}) = \underline{1.35}$

c. $z = (\bar{x} - \mu)/(\sigma/\sqrt{n})$
$z^* = (18.93 - 18.2)/(3.7/\sqrt{140}) = \underline{2.33}$

d. $z = (\bar{x} - \mu)/(\sigma/\sqrt{n})$
$z^* = (79.6 - 81)/(13.3/\sqrt{50}) = \underline{-0.74}$

DECISIONS AND CONCLUSIONS

Since the null hypothesis, H_o, is usually thought to be the statement whose truth is being challenged by the experimenter, all decisions are about the null hypothesis. The alternative hypothesis, H_a, however is usually thought to express the experimenter's viewpoint. Thus, the interpretation of the decision or conclusion is expressed from the experimenter and alternative hypothesis point of view.

Decision:
  1) If the p-value is less than or equal to the specified level of significance (α), the null hypothesis will be rejected
  (if **P** $\leq \alpha$, reject H_o).

  2) If the p-value is greater than the specified level of significance (α), fail to reject the null hypothesis
  (if **P** $> \alpha$, fail to reject H_o).

Conclusion:
  1) If the decision is "reject H_o," the conclusion should read "There is sufficient evidence at the α level of significance to show that ... (the meaning of the alternative hypothesis)."

  2) If the decision is "fail to reject H_o," the conclusion should read "There is not sufficient evidence at the α level of significance to show that ... (the meaning of the alternative hypothesis)."

8.95 a. *Reject H_o* or *Fail to reject H_o*

b. When the calculated p-value is smaller than or equal to α, the decision will be *reject H_o*.
When the calculated p-value is larger than α, the decision will be *fail to reject H_o*.

8.97 a. Reject H_o, **P** $< \alpha$

b. Fail to reject H_o, **P** $> \alpha$

c. Reject H_o, **P** $< \alpha$

d. Reject H_o, **P** $< \alpha$

8.99 a. Fail to reject H_o b. Reject H_o

8.101 a. Directions – no answer.
      b. ≈ 0.0000
      c. no means down by 1451 or lower; The probability of taking a sample of 24 and having a mean less than 1451 when the true mean is equal to 1500 is 0.0000.
      d. Reject Ho.

The p-value approach uses the calculated test statistic to find the area under the curve that contains the calculated test statistic and any values "beyond" it, in the direction of the alternative hypothesis. This probability (area under the curve), based on the position of the calculated test statistic, is compared to the level of significance (α) for the test and a decision is made.

8.103 The p-value measures the likeliness of the sample results based on a true null hypothesis.

Rules for calculating the p-value
1) If H_a contains <, then the p-value = $P(z < z^*)$.
2) If H_a contains >, then the p-value = $P(z > z^*)$.
3) If H_a contains $\neq$, then the p-value = $2P(z > |z^*|)$.

The p-value can then be calculated by using the z* value with Table 3, or it can be found directly using Table 5 (Appendix B, in ES11-p718), or it can be found using a computer and/or calculator.

8.105 p-value = $2 \cdot P(z > 1.1) = 2(1.0000 - 0.8643) = 2(0.1357) = \underline{0.2714}$

8.107 a. p-value = $P(z > 1.48) = 1 - 0.9306 = \underline{0.0694}$

      b. p-value = $P(z < -0.85) = \underline{0.1977}$

      c. p-value = $2 \cdot P(z > 1.17) = 2(1.0000 - 0.8790) = \underline{0.2420}$

      d. p-value = $P(z < -2.11) = \underline{0.0174}$

      e. p-value = $2 \cdot P(z < -0.93) = 2(0.1762) = \underline{0.3524}$

8.109 a. **P** = P(z > z*) = 0.0582

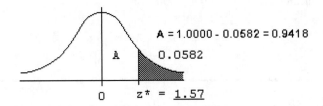

b. **P** = P(z < z*) = 0.0166

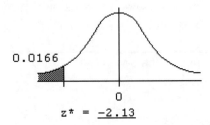

c. **P** = P(z < -z*) + P(z > +z*) = 2·P(z < -z*) = 0.0042

P(z < -z*) = 0.0021

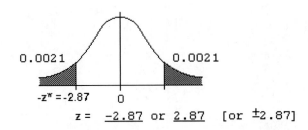

8.111 $P = P(z < z^*) = 0.0170$
   $z^* = -2.12$

   $z = (\bar{x} - \mu)/(\sigma/\sqrt{n})$
   $-2.12 = (14 - 16)/(\sigma/\sqrt{50})$
   $(-2.12)(\sigma/\sqrt{50}) = -2$
   $\sigma = (-2)(\sqrt{50})/(-2.12) = \underline{6.67}$

Computer and calculator commands to complete a hypothesis test for a mean µ with a standard deviation σ can be found in ES11-p382.
Compare the calculated p-value to the given level of significance, α. Using the rules for comparison as stated in IRM-p347, a decision can be made about the null hypothesis.

8.113   a. $H_o: \mu = 525$ vs. $H_a: \mu < 525$

   b.   Fail to reject H_o; the population mean is not significantly less than 525.

   c. $\sigma_{\bar{x}} = \sigma/\sqrt{n} = 60.0/\sqrt{38} = 9.7333 = \underline{9.733}$

8.115   a. $H_o: \mu = 6.25$ vs. $H_a: \mu \neq 6.25$
   b.   Reject H_o; the population mean is significantly different than 6.25.
   c. $\sigma_{\bar{x}} = \sigma/\sqrt{n} = 1.4/\sqrt{78} = \underline{0.1585}$
   d. $\bar{x} = \Sigma x/n$; therefore $\Sigma x = n \cdot \bar{x} = (78)(6.596) = \underline{514.488}$

   $s^2 = [\Sigma x^2 - ((\Sigma x)^2/n)]/(n - 1)$
   $(1.273)^2 = [\Sigma x^2 - ((514.488)^2/78)]/(78 - 1)$
   $1.62053 = [\Sigma x^2 - 3393.5628]/77$
   $\Sigma x^2 = 3393.5628 + (1.62053)(77) = \underline{3518.3437}$

WORD PROBLEMS

1. Look for the key words that indicate the need for a hypothesis test for µ using a z-test. Statements that mention; "testing a <u>mean</u> amount" or "make a decision about the mean value" and the fact that the population standard deviation (σ) or variance (σ^2) are given are examples of these key words.

2. Write down the values needed: μ, σ, $\bar{x}$, n, α.

   If the mean is mentioned in a sentence with no reference to a sample, then it is most likely µ (population mean). If the mean is mentioned in a sentence involving the sample or thereafter, then it is usually $\bar{x}$ (sample mean). Often the sample size (n) is mentioned in the same sentence.

   If the standard deviation or variance is given in a sentence without any reference to a sample, then it also is usually the population σ or σ^2 respectively.

3. Proceed with the hypothesis steps as outlined in: ES11-p371   or   IRM-p343.

8.117 Step 1: a. The mean price for all healthcare laptop replacement costs.
b. H_o: $\mu = \$49,246$ ($\leq$)
H_a: $\mu > \$49,246$

Step 2: a. normality is assumed, CLT with n = 30 indicated
b. z, $\sigma = \$25,000$;   c. $\alpha = 0.001$

Step 3: a. n = 30, $\bar{x} = \$67,783$
b. $z = (\bar{x} - \mu)/(\sigma/\sqrt{n})$
$z^* = (67,783 - 49,246)/(25,000/\sqrt{30}) = 4.08$

Step 4: a. **P** = P(z > 4.08);
Using Table 3, Appendix B, ES11-pp716-717:
P = 1.0000 - 0.99998 = 0.00002
Using Table 5, Appendix B, ES11-p718:
P = 0+
b. p-value is smaller than α

Step 5: a. Reject Ho
b. There is sufficient evidence to conclude that the mean healthcare laptop replacement costs are higher in general, at the 0.001 level of significance.

8.119 Step 1: a. The mean checkout time at a local grocery store
b. H_o: $\mu = 12$ ($\geq$)
H_a: $\mu < 12$

Step 2: a. normality is indicated
b. z, $\sigma = 2.3$;   c. $\alpha = 0.02$

Step 3: a. n = 28, $\bar{x} = 10.9$
b. $z = (\bar{x} - \mu)/(\sigma/\sqrt{n})$
$z^* = (10.9 - 12)/(2.3/\sqrt{28}) = -2.53$

Step 4: a. **P** = P(z < -2.53);
Using Table 3, Appendix B, ES11-pp716-717:
P = 0.0057
Using Table 5, Appendix B, ES11-p718:
0.0054 < **P** < 0.0062
b. p-value is smaller than α

Step 5: a. Reject Ho
b. There is sufficient evidence to conclude that the mean checkout time this week was less than 12 minutes, at the 0.02 level of significance.

8.121 Step 1: a. The average amount spent on Mother's Day.
b. H_o: $\mu = \$123.89$ ($\geq$)
H_a: $\mu < \$123.89$

Step 2: a. normality is assumed, CLT with n = 60
b. z, $\sigma = 39.50$;   c. $\alpha = 0.05$

Step 3: a. $n = 60$, $\bar{x} = 106.27$
b. $z = (\bar{x} - \mu)/(\sigma/\sqrt{n})$
$z^* = (106.27 - 123.89)/(39.50/\sqrt{60}) = -3.46$

Step 4: a. $\mathbf{P} = P(z < -3.46)$;
Using Table 3, Appendix B, ES11-pp716-717:
$\mathbf{P} = 0.0003$
Using Table 5, Appendix B, ES11-p718:
$0.0002 < \mathbf{P} < 0.0003$
b. p-value is smaller than α

Step 5: a. Reject Ho
b. There is sufficient evidence to conclude that the average amount spent on Mother's Day is less than $123.89, at the 0.05 level of significance.

8.123 a. The mean accuracy of quartz watches measured in seconds in error per month.
b. H_0: $\mu = 20$ ($\leq$)
H_a: $\mu > 20$
c. normality is assumed, CLT with $n = 36$
use z with $\sigma = 9.1$; an $\alpha = 0.05$ is given
d. $n = 36$, $\bar{x} = 22.7$
e. $z = (\bar{x} - \mu)/(\sigma/\sqrt{n})$
$z^* = (22.7 - 20)/(9.1/\sqrt{36}) = 1.78$
$\mathbf{P} = P(z > 1.78)$;
Using Table 3, Appendix B, ES11-pp716-717:
$\mathbf{P} = 1.0000 - 0.9625 = 0.0375$
Using Table 5, Appendix B, ES11-p718:
$0.0359 < \mathbf{P} < 0.0401$
f. $\mathbf{P} < \alpha$; Reject H_0

At the 0.05 level of significance, there is sufficient evidence to support the contention that the wrist watches priced under $25 exhibit greater error (less accuracy) than watches in general.

8.125 a. Mean number of students per school nurse in New York.
  b. H_o: $\mu = 750$ ($\leq$)
  H_a: $\mu > 750$
  c. $n = 38$, $\bar{x} = 1007.0$
  $z = (\bar{x} - \mu)/(\sigma/\sqrt{n})$
  $z^* = (1007.0 - 750)/(540/\sqrt{38}) = 2.93$
  $P = P(z > 2.93)$;
   Using Table 3, Appendix B, ES11-pp716-717:
   $P = 1.0000 - 0.9983 = 0.0017$
   Using Table 5, Appendix B, ES11-p718:
   $0.0016 < P < 0.0019$
  d. $P < \alpha$; Reject H_o
   At the 0.01 level of significance, there is sufficient evidence to support the contention that the New York mean number of students per school nurse is significantly higher than the CDC standard of 750.

8.127 a. Step 1: The mean height for female American health professionals.
   Step 2: a. normality assumed, CLT with $n = 50$
       b. z, $\sigma = 2.75$ c. $1-\alpha = 0.95$
   Step 3 $n = 50$, $\bar{x} = 64.78$
   Step 4: a. $\alpha/2 = 0.05/2 = 0.025$; $z(0.025) = 1.96$
       b. $E = z(\alpha/2) \cdot \sigma/\sqrt{n} = (1.96)(2.75/\sqrt{50})$
               $= (1.96)(0.3889) = 0.76$
       c. $\bar{x} \pm E = 64.78 \pm 0.76$
   Step 5: <u>64.02 to 65.54</u>, the 0.95 confidence interval for μ

  b. No

8.129 Results will vary; however, expect your results to be similar to those shown in Table 8.8.

SECTION 8.5 EXERCISES

THE CLASSICAL HYPOTHESIS TEST: A FIVE-STEP PROCEDURE

Step 1: The Set-Up:
  a. Describe the population parameter of concern.
  b. State the null hypothesis (H_o) and the alternative hypothesis (H_a).

Step 2: The Hypothesis Test Criteria:
  a. Check the assumptions.
  b. Identify the probability distribution and the test statistic formula to be used.
  c. Determine the level of significance, α.

Step 3: The Sample Evidence:

> a. Collect the sample information.
> b. Calculate the value of the test statistic.   (continued)

> Step 4: The Probability Distribution:
>   a. Determine the critical region and critical value(s).
>   b. Determine whether or not the calculated test statistic is in the critical region.
>
> Step 5: The Results:
>   a. State the decision about H_o.
>   b. State the conclusion about H_a.

> Review "The Null and Alternative Hypotheses, H_o and H_a" in IRM-pp343-344, if necessary.

8.131 H_o: The mean shearing strength is at least 925 lbs.
H_a: The mean shearing strength is less than 925 lbs.

8.133 a. $H_o: \mu = 1.25$ ($\geq$)   b. $H_o: \mu = 335$ c. $H_o: \mu = 230{,}000$ ($\leq$)
$H_a: \mu < 1.25$ $H_a: \mu \neq 335$ $H_a: \mu > 230{,}000$

> Review "Errors" and "H_o Decisions", if necessary, in: ES11-pp365&366, or IRM-p345.

8.135 $H_o: \mu = 350$ ($\leq$) vs. $H_a: \mu > 350$
a. It is decided that the average salt content is more than 350 mg when in fact, it is not more than 350 mg.
b. It is decided that the average salt content is less than or equal to 350 mg when in fact it is greater than 350 mg.

8.137 $H_o: \mu = 95$ ($\leq$) vs. $H_a: \mu > 95$
a. Type I error: It is decided that the mean minimum plumber's call is greater than $95 when in fact it is not more than $95.
b. Type II error: It is decided that the mean minimum plumber's call is at most $95 when in fact it is greater than $95.

> Critical region - that part under the curve where H_o will be rejected   (size based on α)
>
> Noncritical region - the remaining part under the curve where H_o will not be rejected
>
> Critical value(s) - the $z(\alpha)$ or boundary point values of z, separating the critical and noncritical regions
>
> See ES11-pp393&394 for a visual display of these regions and value(s).

8.139 a. The critical region is the set of all values of the test statistic that will cause us to reject H_o.
b. The critical value(s) is the value(s) of the test statistic that forms the boundary between the critical

region and the non-critical region. The critical value is in the critical region.

8.141 Because alpha and beta are interrelated; if one is reduced, the other one becomes larger. Alpha is area of critical region when H_o is true, beta is related to the noncritical region when H_o is false.

Determining the test criteria

1. Draw a picture of the standard normal (z) curve.   (0 is at the center)

2. Locate the critical region (based on α and H_a)
   a) if H_a contains <, all of the α is placed in the left tail
   b) if H_a contains >, all of the α is placed in the right tail
   c) if H_a contains ≠, place $\alpha/2$ in each tail.

3. Shade in the critical region (the area where you will reject H_o).

4. Find the appropriate critical value(s) using the $z(\alpha)$ concept and the Standard Normal Distribution (Table 3, Appendix B, ES11-pp716-717, or Table 4(a) for one-tail and Table 4(b) for two-tails, Appendix B, ES11-p718).

   If H_a contains >, the critical value is $z(\alpha)$.
   If H_a contains <, the critical value is $z(1-\alpha)$ or $-z(\alpha)$.
   If H_a contains ≠, the critical values are $\pm z(\alpha/2)$ or $z(\alpha/2)$ with $z(1-\alpha/2)$.
   Remember this boundary value divides the area under the curve into critical and noncritical regions and is part of the critical region.

8.143 $z \leq -2.33$

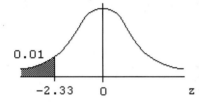

8.145 a.

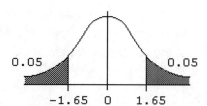

b.

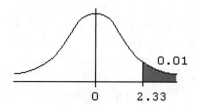

c.

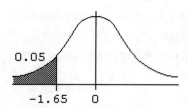

d.

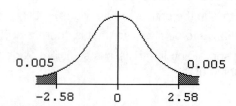

For exercises 8.147 and 8.148, use $z* = (\bar{x} - \mu)/(\sigma/\sqrt{n})$, substituting the given values, then solving for the required unknown.

8.147 $z* = (\bar{x} - \mu)/(\sigma/\sqrt{n})$
  $-1.18 = (\bar{x} - 250)/(22.6/\sqrt{85})$
  $-1.18 = (\bar{x} - 250)/2.451314$
  $-2.89255 = \bar{x} - 250$
  $\bar{x} = 247.107449 = \underline{247.1}$

  $\bar{x} = \Sigma x/n$
  $247.107449 = \Sigma x/85$
  $\Sigma x = \underline{21,004.133}$

NOTE: Standard Error = $\sigma_{\bar{x}} = \sigma/\sqrt{n}$ and $z = (\bar{x} - \mu)/(\sigma/\sqrt{n})$

8.149 a. z = n(standard errors from mean):
  $z = (\bar{x} - \mu)/(\sigma/\sqrt{n})$
  $z = (4.8 - 4.5)/(1.0/\sqrt{100}) = \underline{3.0}$
  $\underline{\bar{x} = 4.8 \text{ is } 3.0 \text{ standard errors } \textbf{above} \text{ the mean } \mu = 4.5}$

  b. If $\alpha = 0.01$, the critical region is $z \geq 2.33$. Since z* is equal to 3.00, it is in the critical region. Therefore, yes, <u>reject H_0</u>.

> **RESULTS, DECISIONS AND CONCLUSIONS**
>
> Since the null hypothesis, H_o, is usually thought to be the statement whose truth is being challenged by the experimenter, all decisions are about the null hypothesis. The alternative hypothesis, H_a, however is usually thought to express the experimenter's viewpoint. Thus, the conclusion (interpretation of the decision) is expressed from the experimenter and alternative hypothesis point of view.
> The two possible outcomes are:
> 1. z^* falls in the critical region or
> 2. z^* falls in the noncritical region.
>
> Decision and Conclusion:
>
> If z^* falls in the critical region, we **reject H_o**. The conclusion is very strong and proclaims the alternative to be the case, that is, there is sufficient evidence to overturn H_o in favor of H_a. It should read something like "There is sufficient evidence at the α level of significance to show that ...(the meaning of the H_a)."
>
> If z^* falls in the noncritical (acceptance) region, we **fail to reject H_o**. The conclusion is much weaker, that is, it suggests that the data does not provide sufficient evidence to overturn H_o. This does not necessarily mean that we have to accept H_o at this point, but only that this sample did not provide sufficient evidence to reject H_o. It should read something like "There is not sufficient evidence at the α level of significance to show that ...(the meaning of the H_a)."

8.151 a. *Reject H_o* or *Fail to reject H_o*

b. When the calculated test statistic falls in the critical region, the decision will be *reject H_o*.
When the calculated test statistic falls in the non-critical region, the decision will be *fail to reject H_o*.

> Computer and calculator commands to complete a hypothesis test using the classical approach can be found in ES11-p382. It is the same command used for the probability approach. Compare the calculated z value (test statistic) with the corresponding critical value(s). The locations of z*, relative to the critical value of z, will determine the decision you must make about the null hypothesis.

8.153 a. $H_o: \mu = 15.0$ vs. $H_a: \mu \neq 15.0$

b. Critical values: $\pm z(0.005) = \pm 2.58$
Decision: reject H_o
Conclusion: There is sufficient evidence to conclude that the mean is different than 15.0.

c. $\sigma_{\bar{x}} = \sigma/\sqrt{n} = 0.5/\sqrt{30} = 0.091287 = \underline{0.0913}$

8.155 a. $H_o: \mu = 72 \ (\leq)$ vs. $H_a: \mu > 72$
 b. Critical values: $z(0.05) = 1.65$
   Decision: Fail to reject H_o
   Conclusion: There is not sufficient evidence to conclude the mean is greater than 72.
 c. $\sigma_{\bar{x}} = \sigma / \sqrt{n} = 12.0 / \sqrt{36} = \underline{2.00}$

See IRM-p352 for information on "Word Problems", if necessary.

Hint for writing the hypotheses for exercise 8.157
Look at the fourth sentence of the exercise, "Is there sufficient evidence to conclude ... scored <u>higher</u> <u>than</u> the state average?". "higher than" indicates <u>greater than</u> (>). Therefore, the alternative hypothesis is <u>greater than</u> (>). Equality (=) is used in the null hypothesis (as usual), but it stands for less than or equal to ($\leq$).

8.157 Step 1: a. The mean score for the Emergency Medical Services Certification Examination
 b. $H_o: \mu = 79.68 \ (\leq)$
   $H_a: \mu > 79.68$
 Step 2: a. normality is assumed, CLT with n = 50
  b. z, $\sigma = 9.06$;   c. $\alpha = 0.05$
 Step 3: a. n = 50, $\bar{x} = 81.05$
  b. $z = (\bar{x} - \mu)/(\sigma/\sqrt{n})$
   $z^* = (81.05 - 79.68)/(9.06/\sqrt{50}) = 1.07$
 Step 4: a. $z(0.05) = 1.65$

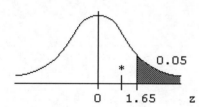

  b. z^* falls in the noncritical region, see Step 4a
 Step 5. a. Fail to reject Ho
  b. There is not sufficient evidence to conclude that the sample average is higher than the state average, at the 0.05 level of significance.

> **Hint for writing the hypotheses for exercise 8.158**
> Look at the fifth sentence in the exercise, "Test the hypothesis that there is no significant difference ...". "No significant difference" indicates <u>equal to</u> (=) (not greater than, not less than). Equality (=) is used in the null hypothesis. The negation is "...NOT equal to...", and ≠ is used in the alternative hypothesis.

> **Hint for writing the hypotheses for exercise 8.159**
> Look at the third sentence in the exercise, "If σ = 6.12, does this sample provide sufficient evidence ... mean number of shoes is greater than the overall ...? 'Greater than' can only go in the alternative hypothesis. Note in the first sentence that the overall average is 15 pairs of shoes.

8.159 Step 1: a. The overall mean number of pairs of shoes for all female adults
b. $H_o: \mu = 15$
$H_a: \mu > 15$

Step 2: a. normality assumed, CLT with n = 35
b. z, σ = 6.12   c. α = 0.10

Step 3: a. n = 35, $\bar{x}$ = 18.37
b. $z = (\bar{x} - \mu)/(\sigma/\sqrt{n})$
$z^* = (18.37 - 15)/(6.12/\sqrt{35}) = 3.26$

Step 4: a. z(0.10) = 1.28
b. z* falls in the critical region, see Step 4a

Step 5: a. Reject H_o

b. At the 0.10 level of significance, the sample does provide sufficient evidence to conclude that the young female college graduates' mean number of shoes is greater than the overall mean number for all female adults.

> **Hint for writing the hypotheses for exercise 8.160**
> Look at the first sentence of the exercise, "A fire insurance company felt that the mean distance ... was at least 4.7 mi." "At least" indicates greater than or equal to (≥). Since the greater than or equal to includes the equality, it belongs in the null hypothesis. You still write the null hypothesis with the equal sign (=), but include the greater than or equal to sign (≥) in parentheses after it. The negation is "..less than", hence the alternative hypothesis must have a less-than sign (<).

> **Hint for writing the hypotheses for exercise 8.161**
> Look at the last sentence of the exercise, "At the 0.05 level of significance, do these data ... mean time of Yankees baseball games is longer than that of other Major League baseball teams?". "longer than" indicates greater than (>). Therefore, the alternative hypothesis is greater than (>). Equality (=) is used in the null hypothesis (as usual), but it stands for less than or equal to (≤).

8.161 Step 1: a. The mean time of Yankee baseball games
 b. H_0: $\mu = 170.1$ (≤)
 H_a: $\mu > 170.1$

Step 2: a. normality indicated
 b. z, $\sigma = 21.0$ c. $\alpha = 0.05$

Step 3: a. $n = 8$, $\bar{x} = 190.375$
 b. $z = (\bar{x} - \mu)/(\sigma/\sqrt{n})$
 $z^* = (190.375 - 170.1)/(21.0/\sqrt{8}) = 2.73$

Step 4: a. $z(0.05) = 1.65$

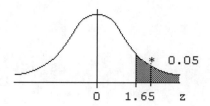

 b. z* falls in the critical region, see Step 4a
Step 5: a. Reject H_0
 b. At the 0.05 level of significance, there is sufficient evidence to support the contention that Yankee games, on the average, last longer than games of Major League Baseball.

> **Hint for writing the hypotheses for exercise 8.162**
> Look at the first sentence in the exercise, "The manager at Air Express feels ... are less than in the past." Since less than (<) does not include the equal to, it must be used in the alternative hypothesis. The negation is "...NOT less than...", which is > or =. Therefore, equality (=) is used in the null hypothesis (as usual), but it stands for greater than or equal to (≥).

8.163 Step 1: a. The average amount of water Americans drink per day
 b. H_0: $\mu = 36.8$ (≤)

H_a: $\mu > 36.8$

Step 2: a. normality assumed, CLT with n = 42
  b. z, $\sigma = 11.2$ c. $\alpha = 0.05$

Step 3: a. n = 42, $\bar{x} = 39.3$
  b. $z = (\bar{x} - \mu)/(\sigma/\sqrt{n})$
    $z^* = (39.3 - 36.8)/(11.2/\sqrt{42}) = 1.45$

Step 4: a. z(0.05) = 1.65

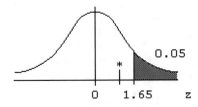

  b. z* falls in the noncritical region, see Step 4a

Step 5: a. Fail to reject H_o
  b. At the 0.05 level of significance, the sample does not provide sufficient evidence to show that education professionals consume, on the average, more water daily than the national average.

8.165 Results will vary, however expect your results to be similar to those shown in Table 8.12.

CHAPTER EXERCISES

8.167 a. $\bar{x} = \underline{32.0}$

  b. $\sigma = \underline{2.4}$

  c. n = $\underline{64}$

  d. $1 - \alpha = \underline{0.90}$

  e. $z(\alpha/2) = z(0.05) = \underline{1.65}$

  f. $\sigma_{\bar{x}} = 2.4/\sqrt{64} = \underline{0.3}$

  g. $E = z(\alpha/2) \cdot \sigma/\sqrt{n} = (1.65)(0.3) = \underline{0.495}$

  h. UCL = $\bar{x} + E = 32.0 + 0.495 = \underline{32.495}$

  i. LCL = $\bar{x} - E = 32.0 - 0.495 = \underline{31.505}$

8.169 Step 1: The mean age of volunteer ambulance members in upstate New York
  Step 2: a. normality assumed, CLT with n = 80
    b. z, $\sigma = 7.8$ c. $1-\alpha = 0.95$

Step 3: n = 80, $\bar{x}$ = 45
Step 4: a. $\alpha/2 = 0.05/2 = 0.025$; $z(0.025) = 1.96$
  b. $E = z(\alpha/2) \cdot \sigma/\sqrt{n} = (1.96)(7.8/\sqrt{80})$
  $= (1.96)(0.872) = 1.709$
  c. $\bar{x} \pm E = 45 \pm 1.7$
Step 5: <u>43.3 to 46.7</u>, the 0.95 confidence interval for μ

Exercise 8.171 shows the effect of the level of confidence (1 - α) on the width of a confidence interval.

8.171 a. Step 1: The mean weights of full boxes of a certain kind of cereal
  Step 2: a. normality indicated
    b. z, $\sigma = 0.27$ c. $1-\alpha = 0.95$
  Step 3: n = 18, $\bar{x}$ = 9.87
  Step 4: a. $\alpha/2 = 0.05/2 = 0.025$; $z(0.025) = 1.96$
    b. $E = z(\alpha/2) \cdot \sigma/\sqrt{n} = (1.96)(0.27/\sqrt{18})$
    $= (1.96)(0.0636) = 0.12$
    c. $\bar{x} \pm E = 9.87 \pm 0.12$
  Step 5: <u>9.75 to 9.99</u>, the 0.95 confidence interval for μ

b. Step 1: The mean weights of full boxes of a certain kind of cereal
  Step 2: a. normality indicated
    b. z, $\sigma = 0.27$ c. $1-\alpha = 0.99$
  Step 3: n = 18, $\bar{x}$ = 9.87
  Step 4: a. $\alpha/2 = 0.01/2 = 0.005$; $z(0.005) = 2.58$
    b. $E = z(\alpha/2) \cdot \sigma/\sqrt{n} = (2.58)(0.27/\sqrt{18})$
    $= (2.58)(0.0636) = 0.16$
    c. $\bar{x} \pm E = 9.87 \pm 0.16$
  Step 5: <u>9.71 to 10.03</u>, the 0.99 confidence interval for μ

c. The increased confidence level widened the interval.

Exercise 8.172 shows the effect of sample size on the width of a confidence interval.

8.173 a. Step 1: The mean score for a clerk-typist position
  Step 2: a. normality assumed, CLT with n = 100
    b. z, $\sigma = 10.5$ c. $1-\alpha = 0.99$
  Step 3: n = 100, $\bar{x}$ = 72.6
  Step 4: a. $\alpha/2 = 0.01/2 = 0.005$; $z(0.005) = 2.58$
    b. $E = z(\alpha/2) \cdot \sigma/\sqrt{n} = (2.58)(10.5/\sqrt{100})$
    $= (2.58)(1.05) = 2.71$
    c. $\bar{x} \pm E = 72.6 \pm 2.71$
  Step 5: <u>69.89 to 75.31</u>, the 0.99 confidence interval for μ

b. <u>Yes.</u> 75.0 falls within the interval.

8.175 a. Yes, all measurements are below 1.0 mm

b. Step 1: The mean ovality for a batch of corks
  Step 2: a. normality assumed, CLT with n = 36
    b. z, $\sigma = 0.10$ c. $1-\alpha = 0.95$

Step 3: n = 36, $\bar{x}$ = 0.254
Step 4: a. α/2 = 0.05/2 = 0.025; z(0.025) = 1.96
  b. E = z(α/2)·σ/√n = (1.96)(0.10/√36)
           = (1.96)(0.016667) = 0.0327
  c. $\bar{x}$ ± E = 0.254 ± 0.0327
Step 5: 0.221 to 0.287, the 0.95 confidence interval for μ

c. With 95% certainty, it can be said that the mean ovality for the batch of corks is between 0.221 and 0.287 millimeters.

8.177 n = [z(α/2)·σ/E]² = [(2.58)(3.7)/1]² = 91.126 = 92

8.179 n = [z(α/2)·σ/E]² = [(2.58)(σ)/(σ/3)]² = 59.9 = 60

8.181 When using probability-value approach:
a. The 0.01 is used as a 'boundary' separating the values of P that will lead to a rejection of the null hypothesis from the values of P that do not lead to a rejection of the null hypothesis.
b. The value of the test statistic would have to be less extreme to lead to a rejection of the null hypothesis.

8.183 a. H$_o$: μ = 100   b. H$_a$: μ ≠ 100

c. α = 0.01   d. μ = 100

e. $\bar{x}$ = 96   f. σ = 12

g. $\sigma_{\bar{x}}$ = 12/√50 = 1.697 = 1.70

h. z* = ($\bar{x}$ - μ)/(σ/√n) = (96-100)/1.7 = -2.35

i. p-value = 2·P(z < -2.35) = 2·P(z > 2.35) = 2(0.5000 - 0.4906)
        = 2(0.0094) = 0.0188

j. Fail to reject H$_o$

k. p-value = 0.0188

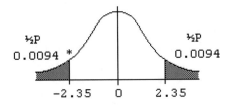

NOTE: When both methods of hypothesis testing are asked for; steps 1, 2, 3 and 5 are the same, and step 4 is shown twice, p-value approach followed by the classical. Dashed lines are used to separate the answers.

8.185 Step 1: a. The mean delay time in a garden sprinkler system
b. H_o: $\mu = 45$ ($\leq$)
H_a: $\mu > 45$
Step 2: a. normality indicated
b. z, $\sigma = 8$ c. $\alpha = 0.02$
Step 3: a. n = 15, $\bar{x} = 50.1$
b. $z = (\bar{x} - \mu)/(\sigma/\sqrt{n})$
$z^* = (50.1 - 45)/(8/\sqrt{15}) = 2.47$
Step 4: --- using p-value approach ----------------------
a. $P = P(z > 2.47)$;
Using Table 3, Appendix B, ES11-pp716-717:
$P = 1.0000 - 0.9932 = 0.0068$
Using Table 5, Appendix B, ES11-p718:
$0.0062 < P < 0.0071$
b. $P < \alpha$
--- using classical approach ----------------------
a.

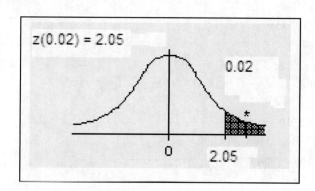

b. z^* falls in the critical region, see Step 4a

Step 5: a. Reject H_o
b. At the 0.02 level of significance, the sample does provide sufficient evidence to conclude that the mean delay time is more than 45 seconds.

8.187 a. H_o: $\mu = 0.50$
H_a: $\mu \neq 0.50$

b. $z^* = (0.51 - 0.50)/(0.04/\sqrt{25}) = 1.25$
$P = 2 \cdot P(z > 1.25) = 2(1.0000 - 0.8944) = \underline{0.2112}$

c. $z = \pm 2.33$

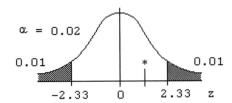

NOTE: When both methods of hypothesis testing are asked for; steps 1, 2, 3 and 5 are the same, and step 4 is shown twice, p-value approach followed by the classical. Dashed lines are used to separate the answers.

8.189 Step 1: a. The mean weight of one load of pollen and nectar being carried by a worker bee to the hive after collecting it
b. H_o: $\mu = 0.0113$ ($\leq$)
H_a: $\mu > 0.0113$

Step 2: a. normality indicated, CLT with n = 200
b. z, $\sigma = 0.0063$ c. $\alpha = 0.01$

Step 3: a. n = 200, $\bar{x} = 0.0124$
b. $z = (\bar{x} - \mu)/(\sigma/\sqrt{n})$
$z^* = (0.0124 - 0.0113)/(0.0063/\sqrt{200}) = 2.47$

Step 4:   --- using p-value approach ----------------------
a. $P = P(z > 2.47)$;
Using Table 3, Appendix B, ES11-pp716-717:
$P = 1.0000 - 0.9932 = 0.0068$
Using Table 5, Appendix B, ES11-p718:
$0.0062 < P < 0.0071$

b. $P < \alpha$
--- using classical approach ---------------------
a.

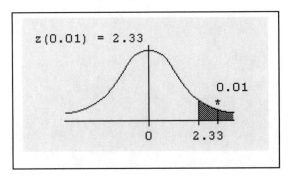

b. z* falls in the critical region, see Step 4a

Step 5: a. Reject H_o

b. At the 0.01 level of significance, the sample does provide sufficient evidence to conclude that the mean load of pollen and nectar carried by Fuzzy's strain of Italian worker bees is greater than the rest of the honey bee population.

8.191 Step 1: a. The mean annual salary for clerk-typists in a large firm
b. H_o: $\mu = 15{,}650$
H_a: $\mu \neq 15{,}650$

Step 2: a. normality assumed, CLT with n = 50
b. z, $\sigma = 1800$ c. $\alpha = 0.05$

Step 3: a. n = 50, $\bar{x} = 16{,}010$
b. $z = (\bar{x} - \mu)/(\sigma/\sqrt{n})$
$z^* = (16{,}010 - 15{,}650)/(1800/\sqrt{50}) = 1.41$

Step 4: -- using p-value approach ----------------------
a. $P = 2P(z > 1.41)$;
Using Table 3, Appendix B, ES11-pp716-717:
$P = 2(1.0000 - 0.9207) = 2(0.0793) = 0.1586$
Using Table 5, Appendix B, ES11-p718:
$2(0.0735 < P < 0.0808)$; $0.1470 < P < 0.1616$

b. $P > \alpha$

-- using classical approach ----------------------
a.

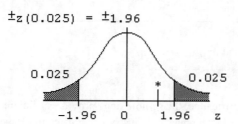

b. z^* falls in the noncritical region, see Step 4a

Step 5: a. Fail to reject H_o

b. At the 0.05 level of significance, the sample does not provide sufficient evidence to conclude the mean salary is different from $15,650.

8.193 Step 1: a. The mean commute driving time for Americans
  b. H_o: $\mu = 24.3$ ($\geq$)
    H_a: $\mu < 24.3$
Step 2: a. normality indicated, CLT with n = 150
  b. z, $\sigma = 10.7$ c. $\alpha = 0.01$
Step 3: a. n = 150, $\bar{x} = 21.7$
  b. $z = (\bar{x} - \mu)/(\sigma/\sqrt{n})$
    $z^* = (21.7 - 24.3)/(10.7/\sqrt{150}) = -2.98$
Step 4: -- using p-value approach ----------------------
  a. $P = P(z < -2.98)$;
    Using Table 3, Appendix B, ES11-pp716-717:
    P = 0.0014
    Using Table 5, Appendix B, ES11-p718:
    $0.0013 < P < 0.0016$
  b. $P < \alpha$
-- using classical approach ---------------------
  a. $-z(0.01) = -2.33$

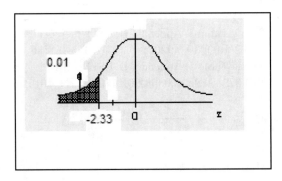

  b. z* falls in the critical region, see Step 4a

Step 5: a. Reject H_o
  b. At the 0.01 level of significance, the sample does provide sufficient evidence to conclude that the mean commute time for the nearby industry is significantly lower than the nationwide average.

8.195 a. Step 1: The mean, μ
  Step 2: a. normality assumed, CLT with n = 100
    b. z, $\sigma = 5.0$ c. $1-\alpha = 0.95$
  Step 3: n = 100, $\bar{x} = 40.6$

Step 4: a. $\alpha/2 = 0.05/2 = 0.025$; $z(0.025) = 1.96$
   b. $E = z(\alpha/2) \cdot \sigma/\sqrt{n} = (1.96)(5/\sqrt{100})$
      $= (1.96)(0.50) = 0.98$
   c. $\bar{x} \pm E = 40.6 \pm 0.98$

Step 5: 39.6 to 41.6, the 0.95 confidence interval for μ

b. Step 1: a. The mean, μ
   b. H_o: $\mu = 40$
      H_a: $\mu \neq 40$

Step 2: a. normality assumed, CLT with n = 100
   b. z, $\sigma = 5.0$ c. $\alpha = 0.05$

Step 3: a. n = 100, $\bar{x} = 40.6$
   b. $z = (\bar{x} - \mu)/(\sigma/\sqrt{n})$
      $z^* = (40.6 - 40)/(5/\sqrt{100}) = 1.20$

Step 4: a. $P = 2P(z > 1.20)$;
   Using Table 3, Appendix B, ES11-pp716-717:
   $P = 2(1.0000 - 0.8849) = 2(0.1151) = 0.2302$
   Using Table 5, Appendix B, ES11-p718:
   $P = 2(0.1151) = 0.2302$
   b. $P > \alpha$

Step 5: a. Fail to reject H_o
   b. At the 0.05 level of significance, there is not sufficient evidence to support the contention that the mean is not equal to 40.

c. Step 1: a. The mean, μ
   b. H_o: $\mu = 40$
      H_a: $\mu \neq 40$

Step 2: a. normality assumed, CLT with n = 100
   b. z, $\sigma = 5.0$ c. $\alpha = 0.05$

Step 3: a. n = 100, $\bar{x} = 40.6$
   b. $z = (\bar{x} - \mu)/(\sigma/\sqrt{n})$
      $z^* = (40.6 - 40)/(5/\sqrt{100}) = 1.20$

Step 4: a. $\pm z(0.025) = \pm 1.96$

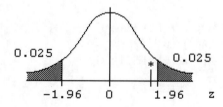

   b. z^* falls in the noncritical region, see * in Step 4a.

Step 5: a. Fail to reject H_o
   b. At the 0.05 level of significance, there is not sufficient evidence to support the contention that the mean is not equal to 40.

d.

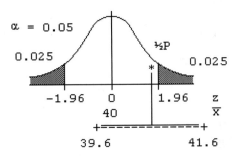

$z^* = 1.20$ is in the noncritical region or $P = 0.2302$ is greater than α, and $\mu = 40$ is within the interval estimate of 39.6 to 41.6.

8.197 a. Step 1: The mean, μ
Step 2: a. normality assumed, CLT with n = 100
  b. z, $\sigma = 5.0$ c. $1-\alpha = 0.95$
Step 3: n = 100, $\bar{x} = 40.9$
Step 4: a. $\alpha/2 = 0.05/2 = 0.025$; $z(0.025) = 1.96$
  b. $E = z(\alpha/2)\cdot\sigma/\sqrt{n} = (1.96)(5/\sqrt{100})$
    $= (1.96)(0.50) = 0.98$
  c. $\bar{x} \pm E = 40.9 \pm 0.98$
Step 5: <u>39.9 to 41.9</u>, the 0.95 confidence interval for μ

b. Step 1: a. The mean, μ
  b. H_o: $\mu = 40$ ($\leq$)
   H_a: $\mu > 40$
Step 2: a. normality assumed, CLT with n = 100
  b. z, $\sigma = 5.0$ c. $\alpha = 0.05$
Step 3: a. n = 100, $\bar{x} = 40.9$
  b. $z = (\bar{x} - \mu)/(\sigma/\sqrt{n})$
    $z^* = (40.9 - 40)/(5/\sqrt{100}) = 1.80$
Step 4: a. **P** = P(z > 1.80);
    Using Table 3, Appendix B, ES11-pp716-717:
    P = (1.0000 - 0.9641) = 0.0359
    Using Table 5, Appendix B, ES11-p718:
    P = 0.0359
  b. **P** < α
Step 5: a. Reject H_o
  b. At the 0.05 level of significance, there is sufficient evidence to support the contention that the mean is greater than 40.

c. Step 1: a. The mean, μ
  b. H_o: $\mu = 40$ ($\leq$)
   H_a: $\mu > 40$

Step 2: a. normality assumed, CLT with n = 100
b. z, $\sigma = 5.0$ c. $\alpha = 0.05$
Step 3: a. n = 100, $\bar{x}$ = 40.9
b. $z = (\bar{x} - \mu)/(\sigma/\sqrt{n})$
$z^* = (40.9 - 40)/(5/\sqrt{100}) = 1.80$
Step 4: a. z(0.05) = 1.65

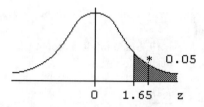

b. z* falls in the critical region, see Step 4a
Step 5: a. Reject H_o
b. At the 0.05 level of significance, there is sufficient evidence to support the contention that the mean is greater than 40.

d.

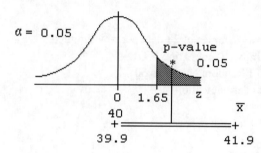

$z^* = 1.80$ is in the critical region and P = 0.0359 is less than α, and $\mu = 40$ is within the interval estimate of 39.9 to 41.9. Since the interval is two-sided and the hypothesis tests are one-sided, it is hard to compare the estimate and the hypothesis tests.

8.199 a. (2)H_a: r > A Failure to reject H_o will result in the drug being marketed. Because of the high current mortality rate, burden of proof is on the old ineffective drug.

b. (1)H_a: r < A Failure to reject H_o will result in the new drug not being marketed. Because of the low current mortality rate, burden of proof is on the new drug.

8.201 a. H_a: $\mu \neq 18$; Fail to reject H_o; The population mean is not significantly different from 18.
b. $\sigma_{\bar{x}} = \sigma / \sqrt{n} = 4.00 / \sqrt{28} = 0.756$

$z = (\bar{x} - \mu)/(\sigma/\sqrt{n})$

$z^* = (17.217 - 18)/(0.756) = -1.04$

p-value = 2·P(z < -1.04) ·
= 2(0.1492) = 0.2984 = <u>0.30</u>

8.203 Every student will have different results, but they should be similar to the following.

    a. <u>Using Minitab</u>:
The commands needed to obtain 50 rows/samples of 28 data per row/sample:
Calc > Random Data > Normal >
    <u>**50**</u> rows in columns <u>**C1-C28**</u>; Mean <u>**19**</u>; Standard deviation <u>**4**</u>
The commands needed to obtain the 50 sample means:
Calc > Row Statistic ... >
    Mean of <u>**C1-C28**</u>; Store results in <u>**C29**</u>
The commands to calculate z*:
Calc > Calculator ... >
    Store result (z*) in <u>**C30**</u>;
    Using expression <u>**(C29-18)/(4/SQRT(28))**</u>
To sort the z* values into ranked order:
Data > Sort ... >
    Sort column <u>**C30**</u>; Store in <u>**C31**</u>; Sort by <u>**C30**</u>

<u>Using Excel</u>:
The commands needed to obtain 50 rows/samples of 28 data per row/sample:
Data > Data Analysis > Random Number Generation > OK
    Number of Variables <u>**28**</u>; Number of Random Numbers <u>**50**</u>;
    Distribution Normal; Mean <u>**19**</u>; Standard deviation <u>**4**</u>;
    Output Range <u>**A1**</u>
The commands needed to obtain the 50 sample means:
Activate cell <u>**AC1**</u>
Insert function f_x > Statistical > AVERAGE > OK
    Number1 <u>**A1:AB1**</u> > OK
    Drag right corner of average value box down to give other averages
The commands to calculate z*:
Activate cell <u>**AD1**</u>
Edit Formula (=) > <u>**(AC1-18)/(4/SQRT(28))**</u>

    b. Examine column C31 or AD1, count the z*'s that are less than –1.04 and greater than 1.04. In one run of the commands, 30/50 or 60% of the values were more extreme than the given z-values; On average, 30% should be more extreme (see 8.201b); an empirical value for the probability that z* is more extreme than the -1.04 in a two-tailed test when the true mean is actually 19.

    c. Critical values = ±1.65; Examine column C31 or AD1, count the z*'s that are between –1.65 and 1.65. For one run of the commands, 33/50 = 66% fell in the noncritical region; an empirical β, probability of type II error.

CHAPTER 9 ▽ INFERENCES INVOLVING ONE POPULATION

Chapter Preview

Chapter 9 continues the work of inferential statistics started in Chapter 8. The concepts of hypothesis tests and confidence intervals will still be presented but on samples where the population standard deviation (σ) is unknown. Also inferences regarding the population binomial probability (p) and the population variance/standard deviation (σ^2/σ) will be introduced.

The chapter's opening section relates a common scenario about a student's Floor to Door time. How long does it take a student to get ready in the morning, from the time your feet hit the floor until you are going out the door?

SECTION 9.1 EXERCISES

9.1 a. American female college students

b.
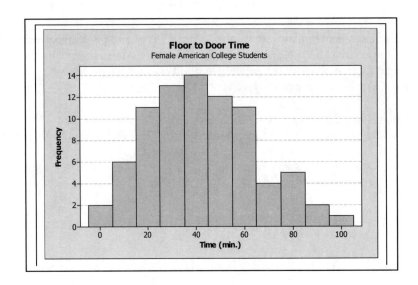

Yes, appears approximately normal with its mounded shape

c.
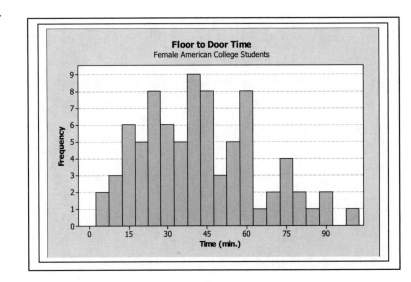

Appears mounded but jagged, most likely caused by time of first class.

d. Yes, approximately., p = 0.193 for normality test

9.3 Pick any 3 numbers; the fourth must be the negative of the sum of the first three. For example; 4, 3, 1, whose sum is 8; the fourth is required to be -8 for the sum to be zero. Three numbers were chosen freely, however there was no choice for the last number.

---

t-Distribution
(used when σ is unknown)

Key facts about the t-distribution:
1. The total area under the t-distribution is 1.
2. It is symmetric about 0.
3. Its shape is a more "spread out" version of the normal shape.
4. A different curve exists for each sample size.   . . .
5. The shape of the distribution approaches the normal distribution shape as n increases [For df > 100, t is approximately normal.].
6. Critical values are determined based on α and degrees of freedom(df)
    - Table 6 (Appendix B, ES11-p719).
7. Degrees of freedom is abbreviated as *df*, where df = n – 1 for this application.

Explore the t-distribution for different degrees of freedom using the Chapter 9 Skillbuilder Applet 'Properties of t-distribution.

Notation: t(df,α) = t(degrees of freedom, area in one tail)
                ↑   ↑   ↑
            Table 6   row id #   column id #

ex.: t(13, 0.025) means df = 13 (row) and α = 0.025 (column), using Table 6, t(13, 0.025) = 2.16
          (df = n-1)

For t(df,α), consider the α given as the amount in one tail and use the top row label - "Area in One Tail". For two-tailed tests, an additional row label is given - "Area in Two-Tails". Note that it is twice the amounts in the one-tail row, therefore α does not have to be divided by two.

For α > 0.5000, use the 1-α amount and negate the t-value.
ex.: t(14,0.90); α = 0.90, 1-α = 0.10,   t(14,0.90) = -t(14,0.10) = -1.35

(Table 6 is in Appendix B, ES11-p719.)

9.5 a. t(12, 0.01) = <u>2.68</u>   b. t(22, 0.025) = <u>2.07</u>
        c. t(50, 0.10) = <u>1.30</u>   d. t(8, 0.005) = <u>3.36</u>

9.7 a. t(18, 0.90) = -t(18, 0.10) = <u>-1.33</u>

     b. t(9, 0.99) = -t(9, 0.01) = <u>-2.82</u>

     c. t(35, 0.975) = -t(35, 0.025) = <u>-2.03</u>

     d. t(14, 0.98) = -t(14, 0.02) = -2.26 (computer calculation) Using Table 6: -2.62 < t < -2.14; Interpolation: -2.30

> For a two-sided test:
> 1. divide α by 2 and use the top row of column labels of Table 6 (Appendix B, ES11-p719) identified as "Area in One Tail" (α/2, in this case)
> or
> 2. use the second row of column labels of Table 6 identified as "Area in Two Tails."

9.9 a. $t(19, 0.05) = 1.73$ b. $\pm t(3, 0.025) = \pm 3.18$

   c. $-t(18, 0.01) = -2.55$ d. $t(17, 0.10) = 1.33$

> Drawing a diagram of a t-curve and labeling the regions with the given information will be helpful in answering exercises 9.11 through 9.16.

9.11 $\alpha/2 = 0.05/2 = 0.025$; $\pm t(12, 0.025) = \pm 2.18$

9.13 a. -2.49 b. 1.71 c. -0.685

9.15 df = 7

9.17 Use the cumulative probability distribution function, Student's t distribution with 18 DF

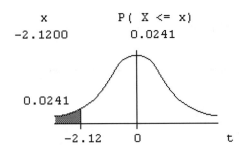

```
    x        P( X <= x)
-2.1200       0.0241
```

9.19 a. Symmetric about mean: mean is 0

   b. Standard deviation of t-distribution is greater than 1; t-distribution is different for each different sample size while there is only one z-distribution, t has df.

> **Estimating μ - the population mean**
> (σ unknown)
>
> 1. point estimate: $\bar{x}$
>
> 2. confidence interval: $\bar{x} \pm t(df, \alpha/2) \cdot (s/\sqrt{n})$, where df = n-1
>
> Review steps for constructing a confidence interval for μ: ES11-pp348-349, IRM-p327. The t-distribution is used when σ is unknown, and sampling is from an approximately normal distribution or the sample size is large.

9.21 Step 1: The mean, μ
  Step 2: a. normality assumed
    b. t c. 1-α = 0.95
  Step 3: n = 24, $\bar{x}$ = 16.7, s = 2.6
  Step 4: a. α/2 = 0.05/2 = 0.025; df = 23; t(23, 0.025) = 2.07
    b. E = t(df,α/2)·(s/$\sqrt{n}$) = (2.07)(2.6/$\sqrt{24}$)
      = 1.0986 = 1.10
    c. $\bar{x} \pm E$ = 16.7 ± 1.10
  Step 5: <u>15.60 to 17.8</u>, the 0.95 confidence interval for μ

9.23 Step 1: The mean notification to arrival time, μ
  Step 2: a. normality assumed
    b. t c. 1-α = 0.98
  Step 3: n = 20, $\bar{x}$ = 5.25, s = 2.78
  Step 4: a. α/2 = 0.02/2 = 0.01; df = 19; t(19, 0.01) = 2.54
    b. E = t(df,α/2)·(s/$\sqrt{n}$) = (2.54)(2.78/$\sqrt{20}$)
      = 1.5789 = 1.58
    c. $\bar{x} \pm E$ = 5.25 ± 1.58
  Step 5: <u>3.67 to 6.83</u>, the 0.98 confidence interval for μ

9.25 Step 1: The mean torque failure strength of #8 square drive wood screws
Step 2: a. Failure torque strength presumed normal
b. t   c. $1-\alpha = 0.95$
Step 3: $n = 22$, $\bar{x} = 45.2$, $s = 5.1$
Step 4: a. $\alpha/2 = 0.05/2 = 0.025$; $df = 21$; $t(21, 0.025) = 2.08$
b. $E = t(df, \alpha/2) \cdot (s/\sqrt{n}) = (2.08)(5.1/\sqrt{22})$
$= 2.26163 = 2.26$
c. $\bar{x} \pm E = 45.2 \pm 2.26$
Step 5: <u>42.94 to 47.46</u>, the 0.95 confidence interval for μ

9.27 Verify - answers given in exercise.

9.29 a. Step 1: The mean drying time of latex paint
Step 2: a. normality indicated
b. t   c. $1-\alpha = 0.95$
Step 3: $n = 8$, $\bar{x} = 4.56$, $s = 1.34$
Step 4: a. $\alpha/2 = 0.05/2 = 0.025$; $df = 7$; $t(7, 0.025) = 2.36$
b. $E = t(df, \alpha/2) \cdot (s/\sqrt{n}) = (2.36)(1.34/\sqrt{8}) = 1.12$
c. $\bar{x} \pm E = 4.56 \pm 1.12$
Step 5: <u>3.44 to 5.68</u>, the 0.95 confidence interval for μ

b. Student answers will differ, depending on how exercise 2.177(c) was answered.

9.31 a.

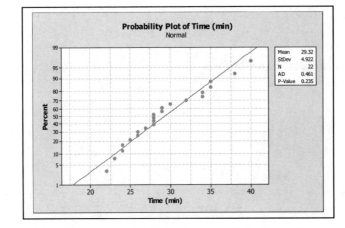

The data appears to follow a straight line, indicating an approximately normal distribution.

b. Step 1: The mean length of lunch breaks at Giant Mart
Step 2: a. Normality indicated in step a
b. t   c. $1-\alpha = 0.95$
Step 3: $n = 22$, $\bar{x} = 29.32$, $s = 4.92$
Step 4: a. $\alpha/2 = 0.05/2 = 0.025$; $df = 21$; $t(21, 0.025) = 2.08$

b. $E = t(df,\alpha/2)\cdot(s/\sqrt{n}) = (2.08)(4.92/\sqrt{22}) = 2.182$
c. $\bar{x} \pm E = 29.32 \pm 2.182$
Step 5: 27.138 to 31.502, the 0.95 confidence interval for μ

Computer and/or calculator commands to calculate a confidence interval for the population mean, μ, when the population standard deviation, σ, is unknown can be found in ES11-p419.

9.33 a.

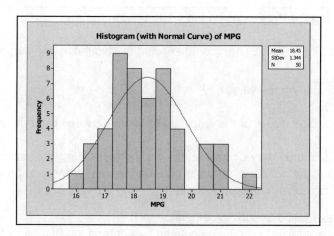

Distribution is mounded in center, approximately symmetrical. With n = 50, the assumptions are satisfied.

b. Step 1: The mean mileage per gallon for SUV's.
Step 2: a. The sampled population appears to be normally distributed.
   b. t   c. $1-\alpha = 0.95$
Step 3: n = 50, $\bar{x}$ = 18.45, s = 1.3440
Step 4: a. $\alpha/2 = 0.05/2 = 0.025$; df = 49; t(49, 0.025) = 2.02
   b. $E = t(df,\alpha/2)\cdot(s/\sqrt{n}) = (2.02)(1.3440/\sqrt{50}) = 0.3839$
   c. $\bar{x} \pm E = 18.45 \pm 0.38$
Step 5: 18.07 to 18.83, the 0.95 confidence interval for μ

c. One can expect at least 18 miles per gallon.

Hypotheses are written the same way as before. Sample size and standard deviation have no effect on the stating of hypotheses.

9.35 a. $H_o: \mu = 11 \, (\geq)$ vs. $H_a: \mu < 11$

b. $H_o: \mu = 54 \, (\leq)$ vs. $H_a: \mu > 54$

c. $H_o: \mu = 75$ vs. $H_a: \mu \neq 75$

9.37 $\mu = 32$, $n = 16$, $\bar{x} = 32.93$, $s = 3.1$
$t = (\bar{x} - \mu)/(s/\sqrt{n})$
$t^* = (32.93 - 32)/(3.1/\sqrt{16}) = \underline{1.20}$

Calculating the **P**-value using the t-distribution

Table 6 or Table 7 (Appendix B, ES11-pp719&720) or a computer and/or calculator, can be used to <u>estimate</u> the p-value

1. Using Table 6 to place bounds on the value of **P**
   a) locate df row
   b) locate the absolute value of the calculated t-value between two critical values in the df row
   c) the p-value is in the interval between the two corresponding probabilities at the top of the columns; read the bounds from either the *one-tail* or *two-tailed* column headings as per H_a.

2. Using Table 7 to estimate or place bounds on the value of **P**
   a) locate the absolute value of the calculated t-value and the df directly for the corresponding probability value

   OR

   b) locate the absolute value of the calculated t-value and its df between appropriate bounds. From the box formed, use the upper left and lower right values for the interval. (see ES11-p422) ...

---

3. Using a computer and/or calculator
   a) the p-value is calculated directly and given in the output when completing the hypothesis test (see ES11-p425)

   OR

   b) the p-value is calculated using the cumulative probability commands: Subtract the probability value from 1 or multiply it by 2, depending on the exercise. The cumulative probability given is $P(t \leq \text{t-value})$.
   (see ES11-pp417)

9.39 a. $\mathbf{P} = P(t < -2.01 | df=10) = P(t > 2.01 | df=10)$
   using Table 6: $0.025 < \mathbf{P} < 0.05$
   using Table 7: $0.031 < \mathbf{P} < 0.037$
   using computer: $\mathbf{P} = 0.036$

   b. $\mathbf{P} = P(t > +2.01 | df=10)$;
   using Table 6: $0.025 < \mathbf{P} < 0.05$
   using Table 7: $0.031 < \mathbf{P} < 0.037$
   using computer: $\mathbf{P} = 0.036$

c. **P** = P(t < -2.01|df=10) + P(t > +2.01|df=10) = 2P(t > 2.01|df=10)
   using Table 6: 0.05 < **P** < 0.10
   using Table 7: 0.062 < **P** < 0.074
   using computer: **P** = 0.072

d. **P** = P(t < -2.01|df=10) + P(t > +2.01|df=10) = 2P(t > 2.01|df=10)
   using Table 6: 0.05 < **P** < 0.10
   using Table 7: 0.062 < **P** < 0.074
   using computer: **P** = 0.072

Draw a picture as before, of an "approximately" normal distribution curve. Shade in the critical regions based on the alternative hypothesis (H_a). Using α and df, find the critical value(s) using Table 6 (Appendix B, ES11-p719).

9.41 a. **P** = P(t > 1.20|df = 15);
   Using Table 6, Appendix B, ES11-p719: 0.10 < **P** < 0.25
   Using Table 7, Appendix B, ES11-p720: **P** = 0.124
   Using computer: **P** = 0.124
   $\alpha = 0.05$; **P** > α; <u>Fail to reject H_o</u>

b. t(15, 0.05) = 1.75; t* = 1.20

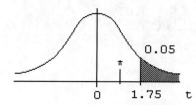

t* falls in the noncritical region; <u>Fail to reject H_o</u>

9.43 Test of mu = 32 vs mu > 32

N   Mean   StDev   T   P
16 32.9300 3.1000 1.20 <u>0.124</u>

9.45   a. $P = 2P(t > 1.60 | df = 14)$
using Table 6: $0.10 < P < 0.20$
using Table 7: $0.065 < \tfrac{1}{2}P < 0.068$; $0.130 < P < 0.136$
using computer: $P = 0.132$

$\alpha = 0.05$; $P > \alpha$; Fail to reject H_o

```
α = 0.05
  α/2 = 0.025
±t(14, 0.025) = ±2.14

    t* = 1.60
    Fail to reject H_o
```

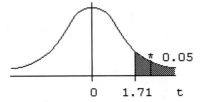

b. $P = P(t > 2.16 | df = 24)$
using Table 6: $0.01 < P < 0.025$
using Table 7: $0.019 < P < 0.024$
using computer: $P = 0.0205$

$\alpha = 0.05$; $P < \alpha$; Reject H_o

```
α = 0.05
t(24, 0.05) = 1.71


t* = 2.16
Reject H_o
```

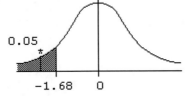

c. $P = P(t < -1.73 | df = 44) = P(t > 1.73 | df = 44)$
using Table 6: $0.025 < P < 0.05$
using Table 7: $0.039 < P < 0.049$
using computer: $P = 0.045$

$\alpha = 0.05$; $P < \alpha$; Reject H_o

```
α = 0.05
-t(44, 0.05) = -1.68



t* = -1.73
  Reject H_o
```

d.   Results for each of the decisions are identical.

> Hypothesis tests, (p-value approach), will be completed in the same format as before. You may want to review: ES11-p371, IRM-p343. The only differences are in:
>
> 1. the calculated test statistic, which is t, where $t = (\bar{x} - \mu)/(s/\sqrt{n})$
>
> 2. using Table 6 or Table 7 (Appendix B, ES11-pp719&720) to <u>estimate</u> the p-value
>
> Remember: σ (population standard deviation) is unknown,
> therefore s (sample standard deviation) is used.

> Hint for writing the hypotheses for exercise 9.47
>
> Look at the first sentence of the exercise, "A student group maintains that the average student must travel for <u>at least</u> 25 minutes ...". "At least" indicates <u>greater than or equal to</u> (≥). Since the <u>greater than or equal to</u> includes the equality, it belongs in the null hypothesis. Continue to write the null hypothesis with the equal sign (=), but include the greater-than or equal-to sign (≥) in parentheses after it. The negation is "..less than", hence the alternative hypothesis must have a less-than sign (<).

9.47   a.   P-value approach:

       Step 1:   a. The mean travel time to college
                  b. H_o: μ = 25 (at least) (≥)
                     H_a: μ < 25 (less than)

       Step 2:   a. Travel times are mounded; CLT is satisfied with n = 31
                  b. t   c. α = 0.01

       Step 3:   a. n = 31, $\bar{x}$ = 19.4, s = 9.6
                  b. $t = (\bar{x} - \mu)/(s/\sqrt{n})$
                    $t^* = (19.4 - 25.0)/(9.6/\sqrt{31}) = -3.25$

       Step 4:   a. **P** = P(t < -3.25|df = 30) = P(t > 3.25|df = 30);
                         Using Table 6, Appendix B, ES11-p719: **P** < 0.005
                         Using Table 7, Appendix B, ES11-p720: 0.001 < **P** < 0.002
                         Using computer: **P** = 0.0014
                  b. **P** < α

       Step 5:   a. Reject H_o
                  b. At the 0.01 level of significance, the sample does provide sufficient evidence to justify the contention that mean travel time is less than 25 minutes.

> Hypothesis tests (classical approach) will be completed using the same format as before. You may want to review: ES11-pp388, IRM-p357. The only differences are in:
>
> 1. finding the critical value(s) of t
>   remember you need α (column) and degrees of freedom (df = n - 1)(row) for the t-distribution
>
> 2. the calculated test statistic, which is t, where $t = (\bar{x} - \mu)/(s/\sqrt{n})$.

       b.   Classical approach:

Step 1: a. The mean travel time to college
        b. H_o: $\mu = 25$ (at least) ($\geq$)
           H_a: $\mu < 25$ (less than)
Step 2: a. Travel times are mounded; assume normality, CLT with n = 31
        b. t   c. $\alpha = 0.01$
Step 3: a. n = 31, $\bar{x} = 19.4$, s = 9.6
        b. $t = (\bar{x} - \mu)/(s/\sqrt{n})$
           $t^* = (19.4 - 25.0)/(9.6/\sqrt{31}) = -3.25$
Step 4: a. $-t(30, 0.01) = -2.46$

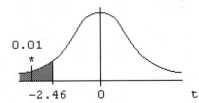

        b. t^* falls in the critical region, see Step 4a
Step 5: a. Reject H_o
        b. At the 0.01 level of significance, the sample does provide sufficient evidence to justify the contention that mean travel time is less than 25 minutes.

Hint for writing the hypotheses for exercise 9.48
Look at the second sentence in the exercise, "It is assumed ... have a <u>higher</u> value". "Higher" indicates greater than. Since greater than (>) does not include the equal to, it must be used in the alternative hypothesis. The negation is " ... NOT greater than", which is < or =. Therefore, equality (=) is used in the null hypothesis (as usual), but it stands for less than or equal to ($\leq$).

Hint for writing the hypotheses for exercise 9.49
Look at the second to last sentence of the exercise, "Is there sufficient evidence that the ...has significantly <u>lower</u> average daily garbage amounts ...?". Since the word "<u>lower</u>" indicate (<) which does not include the equal to, it must be used in the alternative hypothesis. The negation becomes "not lower than", which is > or =. The null hypothesis is written with an equality sign (=), but it stands for greater than or equal to ($\geq$).

9.49 Step 1: a. The mean daily garbage amounts in a Vermont town
              b. H_o: $\mu = 4.6$ ($\geq$)
                 H_a: $\mu < 4.6$ (lower)

Step 2: a. normality indicated
b. t   c. $\alpha = 0.05$
Step 3: a. $n = 18$, $\bar{x} = 3.89$, $s = 1.322$
b. $t = (\bar{x} - \mu)/(s/\sqrt{n})$
$t^* = (3.89 - 4.6)/(1.322/\sqrt{18}) = -2.28$
Step 4: -- using p-value approach --------------------
a. $P = P(t < -2.28 | df = 17)$;
Using Table 6, Appendix B, ES11-p719: $0.01 < P < 0.025$
Using Table 7, Appendix B, ES11-p720: $0.017 < P < 0.022$
Using computer: $P = 0.018$
b. $P < \alpha$
-- using classical approach ------------------
a. $-t(17, 0.05) = -1.74$

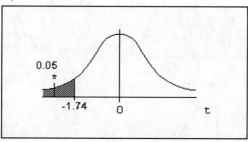

b. t^* falls in the critical region, see * Step 4a

Step 5: a. Reject H_o
b. The sample does provide sufficient evidence to justify the contention that the mean daily garbage amount for a Vermont town is lower than the average American household, at the 0.05 level of significance.

9.51 Verify - answers given in exercise.

> Computer and/or calculator commands to perform a hypothesis test for a population mean if the population standard deviation (σ) is unknown can be found in ES11-p425.
> The alternative command works the same as in the z-distribution (σ known) command. The output will also look the same except a t-value will be calculated in place of the z-value.

9.53 Test of mu = 52.00 vs mu < 52.00

Variable	N	Mean	StDev	SE Mean	T	P
C1	12	49.75	5.48	1.58	-1.42	0.091

$\alpha = 0.01$; $P > \alpha$; Fail to reject H_o
OR
$-t(11, 0.01) = -2.72$; $t^* = -1.42$ falls in the noncritical region; Fail to reject H_o

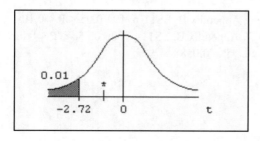

9.55 Sample statistics: $n = 6$, $\Sigma x = 222$, $\Sigma x^2 = 8330$, $\bar{x} = 37.0$, $s = 4.817$

Step 1: a. The mean test score at a certain university
b. H_o: $\mu = 35$ (reasonable)
H_a: $\mu \neq 35$ (not reasonable)

Step 2: a. normality indicated
b. t c. $\alpha = 0.05$

Step 3: a. $n = 6$, $\bar{x} = 37.0$, $s = 4.817$
b. $t = (\bar{x} - \mu)/(s/\sqrt{n})$
$t^* = (37.0 - 35.0)/(4.817/\sqrt{6}) = 1.02$

Step 4: -- using p-value approach --------------------
a. $P = 2P(t > 1.02 | df = 5)$;
Using Table 6, Appendix B, ES11-p719: $0.20 < P < 0.50$
Using Table 7, Appendix B, ES11-p720: $0.161 < \frac{1}{2}P < 0.182$] $0.322 < P < 0.364$
Using computer: $P = 0.3545$

b. $P > \alpha$

-- using classical approach ------------------
a. $\pm t(5, 0.025) = \pm 2.57$

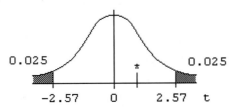

b. t^* falls in the noncritical region, see Step 4a

Step 5: a. Fail to reject H_o
b. The sample does not provide sufficient evidence to reject the claim that the mean score is 35, at the 0.05 level of significance.

9.57 a. Step 1: a. The mean amount of exercise time per week
b. H_o: $\mu = 60$ ($\geq$)
H_a: $\mu < 60$

Step 2: a. Sample appears approximately normal, CLT satisfied
b. t c. $\alpha = 0.05$

Step 3: a. $n = 40$, $\bar{x} = 53.50$, $s = 25.68$
b. $t = (\bar{x} - \mu)/(s/\sqrt{n})$
$t^* = (53.50 - 60)/(25.68/\sqrt{40}) = -1.60$

Step 4: -- using p-value approach --------------------
a. $P = P(t < -1.60 | df = 39)$;
Using Table 6, Appendix B, ES11-p719: $0.05 < P < 0.10$
Using Table 7, Appendix B, ES11-p720: $0.058 < P < 0.059$
Using computer: $P = 0.0588$

b. $P > \alpha$

-- using classical approach ------------------
a. -t(39, 0.05) = -1.69; t ≤ -1.69 critical region
b. t* falls in the noncritical region

Step 5: a. Fail to reject H_o

b. There is sufficient evidence to conclude that the technicians exercise at least 60 minutes a week, at the 0.05 level of significance.

b. Yes, the confidence interval contained the 'at least 60' minutes.

9.59   a.

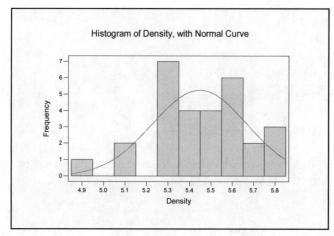

The distribution is mounded and appears to have an approximately normal distribution.

b.   Step 1:   a. The mean density of the earth to the density of water
b. $H_o: \mu = 5.517$
$H_a: \mu < 5.517$ (less than)

Step 2:   a. normality indicated above
b. t   c. $\alpha = 0.05$

Step 3:   a. $n = 29$, $\bar{x} = 5.4479$, $s = 0.2209$
b. $t = (\bar{x} - \mu)/(s/\sqrt{n})$
$t^* = (5.4479 - 5.517)/(0.2209/\sqrt{29}) = -1.68$

Step 4:   -- using p-value approach --------------------
a. $P = P(t < -1.68 | df = 28)$;
   Using Table 6, Appendix B, ES11-p719: $0.05 < P < 0.10$
   Using Table 7, Appendix B, ES11-p720: $0.05 < P < 0.061$
   Using computer: $P = 0.052$
b. $P > \alpha$

-- using classical approach ------------------
a. -t(28, 0.05) = -1.70

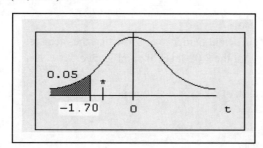

    b. t* falls in the noncritical region, see Step 4a

Step 5:   a. Fail to reject H_o

    b. There is not sufficient evidence to show, the mean of Cavendish's data is significantly less than today's recognized standard of 5.517 g/cm^3, at the 0.05 level of significance.

9.61 a. Yes, the assumption of normality seems reasonable. Manufacturers are required to keep manufacturing processes within specifications.

  b. Step 1: The mean amount of acetaminophen per tablet.
      Step 2: a. Normality assumed, CLT with n = 30.
              b. t   c. 1-α = 0.99
      Step 3: n = 30, $\overline{x}$ = 596.3, s = 4.7
      Step 4: a. α/2 = 0.01/2 = 0.005; df = 29; t(29, 0.005) = 2.76
              b. E = t(df,α/2)·(s/$\sqrt{n}$) = (2.76)(4.7/$\sqrt{30}$) = 2.368
              c. $\overline{x}$ ± E = 596.3 ± 2.37
      Step 5: <u>593.93 to 598.67</u>, the 0.99 confidence interval for μ

  c. The confidence interval in part b suggests that the mean amount of acetaminophen in one pill is less than 600 mg. The interval did not contain the 600 mg.

9.63 a. Step 1: a. The mean length of corks in a shipment
                      b. H_o: μ = 45.0
                         H_a: μ ≠ 45.0
              Step 2: a. normality indicated
                      b. t   c. α = 0.02
              Step 3: a. n = 12, $\overline{x}$ = 44.8775, s = 0.2176
                      b. t = ($\overline{x}$ - μ)/(s/$\sqrt{n}$)
                         t* = (44.8775 – 45.0)/(0.2176/$\sqrt{12}$) = -1.95
              Step 4: -- using p-value approach --------------------
                      a. **P** = 2P(t < -1.95|df = 11) = 2P(t > 1.95|df = 11);
                              Using Table 6, Appendix B, ES11-p719: 0.05 < **P** < 0.10
                              Using Table 7, Appendix B, ES11-p720: (0.034 < **P** < 0.043) =
                                                                    (0.068 < P < 0.086)
                              Using computer: **P** = 0.0771
                      b. **P** > α
                      -- using classical approach ------------------
                      a. ±t(11, 0.01) = ±2.72
                      b. t* falls in the noncritical region, see Step 4a
                      ----
              Step 5: a. Fail to reject H_o

                      b. There is not sufficient evidence to show that the mean length of the corks is different from 45.0mm, at the 0.02 level of significance.

      b. Step 1: a. The mean length of corks in a shipment
                  b. H_o: μ = 45.0
                     H_a: μ ≠ 45.0

Step 2: a. normality indicated
b. t c. α = 0.02
Step 3: a. n = 18, $\bar{x}$ = 45.0183, s = 0.3199
b. t = ($\bar{x}$ - μ)/(s/√n)
t* = (45.0183 – 45.0)/(0.3199/$\sqrt{18}$) = 0.24
Step 4: -- using p-value approach --------------------
a. **P** = 2P(t > 0.24|df = 17);
Using Table 6, Appendix B, ES11-p719: **P** > 0.500
Using Table 7, Appendix B, ES11-p720: 2(0.384 < **P** < 0.422) =
(0.768 < P < 0.844)
Using computer: **P** = 0.813
b. **P** > α
-- using classical approach ------------------
a. ±t(17, 0.01) = ±2.57
b. t* falls in the noncritical region, see Step 4a

Step 5: a. Fail to reject H_o
b. There is not sufficient evidence to show that the second sample mean length of the corks is different from 45.0mm, at the 0.02 level of significance.

c. The test statistic in part d was much smaller since the sample mean was closer to the hypothesized mean.
d. The increase in the sample size made up for the difference in standard deviations between the two samples. The larger standard deviation was with the larger sample size.
e. The standard deviation in part d was smaller that that in part c. In combination with a sample mean closer to the claimed mean produced a very small test statistic, resulting in a Fail to Reject Ho decision.

Computer commands to perform the simulation from the various distributions can be found on pages 385-386 of this manual. Slight variations need to be made for t* versus the z* that was used in Chapter 8.

SECTION 9.2 EXERCISES

p' = sample proportion	p' = x/n
x = number of successes	n = sample size (number of independent trials)

9.65 a. x = number of successes = 45 (with only two outcomes, "success' and "failure")
n = sample size = number of independent trials = 150
b. p' = 45/150 = # of successes/# of trials = 0.30
p' = sample proportion of success
c. p' = x/n = 24/250 = 0.096
d. p' = x/n = 640/2050 = 0.312195 = 0.312
e. p' = x/n = 892/1280 = 0.696875 = 0.697

9.67 a. Yes, it seems likely that the mean of the observed proportions would be the true proportion for the population.
b. Unbiased because the mean of the p' distribution is p, the parameter being estimated.

9.69 a. $\alpha = 0.10$, $\alpha/2 = 0.05$; $z(\alpha/2) = 1.65$ or $-z(\alpha/2) = -1.65$
b. $\alpha = 0.05$, $\alpha/2 = 0.025$; $z(\alpha/2) = 1.96$ or $-z(\alpha/2) = -1.96$
c. $\alpha = 0.01$, $\alpha/2 = 0.005$; $z(\alpha/2) = 2.58$ or $-z(\alpha/2) = -2.58$

Estimating p - the population proportion

1. point estimate: $p' = x/n$

2. confidence interval: $p' \pm z(\alpha/2) \cdot \sqrt{p'q'/n}$
   ↑   ⌣⌣⌣⌣⌣⌣⌣
   point   maximum error
   estimate   of estimate

Computer and/or calculator commands to calculate a confidence interval for the population proportion, p, can be found in ES11-p437.

Review: ES11-pp348-349, IRM-p327, "The Confidence Interval: A Five-Step Procedure" if necessary.
Explore the relationship between a z* value and its corresponding confidence interval using the Skillbuilder Applet 'z* & Confidence Level'.

9.71 a. $\sqrt{p'q'/n} = \sqrt{(0.23)(0.77)/400} = \underline{0.02104}$

b. Step 1: The proportion of convertibles driven by students
Step 2: a. The sample was randomly selected and each subject's response was independent of those of the others surveyed.
  b. $n = 400$; $n > 20$, $np = (400)(92/400) = 92$,
     $nq = (400)(308/400) = 308$, np and nq both > 5
  c. $1 - \alpha = 0.95$
Step 3: $n = 400$, $x = 92$, $p' = x/n = 92/400 = 0.23$
Step 4: a. $z(\alpha/2) = z(0.025) = 1.96$
  b. $E = z(\alpha/2) \cdot \sqrt{p'q'/n} = 1.96 \cdot \sqrt{(0.23)(0.77)/400} = (1.96)(0.02104) = 0.041$
  c. $p' \pm E = 0.23 \pm 0.041$
Step 5: $\underline{0.189 \text{ to } 0.271}$ is the 0.95 interval for $p = P(\text{drives convertible})$

9.73 a. $n = 3,003$ people surveyed, a trial is the surveying of each person, success is when they "do not know that caffeine dehydrates", $p = P(\text{did not know})$, x = number of surveyed 3,003 people who say they did not know that caffeine dehydrates
b. 0.20 (20%). It is a statistic. It is being used to estimate the parameter.
c. $1.96 \cdot \sqrt{(0.20)(0.80)/3003} = 1.96(0.0073) = 0.0143$
d. Maximum error of 1.4% is smaller than the quoted margin of error = 1.8%.
e. 0.20 ± 0.014 or 0.186 to 0.214

9.75 Step 1: The proportion of students that support the proposed budget amount
Step 2: a. The sample was randomly selected and each subject's response was independent of

those of the others surveyed.
　　　　　b. n = 60; n > 20, np = (60)(22/60) = 22, nq = (60)(38/60) = 38, np and nq both > 5
　　　　　c. 1 - α = 0.99
Step 3: n = 60, x = 22, p' = x/n = 22/60 = 0.367
Step 4: a. z(α/2) = z(0.005) = 2.58
　　　　　b. E = z(α/2)· $\sqrt{p'q'/n}$ = 2.58$\sqrt{(0.367)(0.633)/60}$ = (2.58)(0.0622) = 0.161
　　　　　c. p' ± E = 0.367 ± 0.161
Step 5: <u>0.206 to 0.528</u>, the 0.99 interval for p = P(favor budget)

9.77 Step 1: The proportion of video-gaming youngsters that may go on to have an addition
Step 2: a. The sample was randomly selected and each subject was selected independent of the others in the sample.
　　　　　b. n = 1179; n > 20, np = (1179)(0.085) = 100.215, nq = (1179)(0.915) = 1078.785, np and nq both > 5
　　　　　c. 1 - α = 0.99
Step 3: n = 1179, p' = 0.085
Step 4: a. z(α/2) = z(0.005) = 2.58
　　　　　b. E = z(α/2)· $\sqrt{p'q'/n}$ = 2.58· $\sqrt{(0.085)(0.915)/1179}$ = 0.021
　　　　　c. p' ± E = 0.085 ± 0.021
Step 5: <u>0.064 to 0.106</u>, the 0.99 interval for p = P(youngster having an addiction)

9.79 Verify – answers given in exercise

9.81 There are many possible reasons why the results could be biased. Here are a few:
1. Many people will not reveal information about credit cards to telephone callers.
2. Many people will not ever admit to being "easy prey" to offers like this.
3. Many people answer questions with answers they think the caller wants to hear hoping to end the call quickly.

9.83 a. E = 1.96 $\sqrt{(0.39)(0.61)/1000}$ = 1.96(0.01542) = 0.030
　　　E = 1.96 $\sqrt{(0.70)(0.30)/1010}$ = 1.96(0.014419) = 0.028
　　　E = 1.96 $\sqrt{(0.16)(0.84)/1006}$ = 1.96(0.011558) = 0.023

b. The variation was caused by the differing values of p.
　　More explicitly, the differing product of pq: 0.2379, 0.2100, and 0.1344.

c. Yes.

d. The "margin of error" (MoE) is typically reported as the maximum error of estimate calculated using p = 0.5 because this yields the maximum value. Rounding up yields a slightly larger error, in turn a wider interval. "Conservative" equates with a less restrictive (narrower) interval.

e. p = 0.5

9.85 a. p = P(head) = 12,012/24,000 = <u>0.5005</u>

b. $\sqrt{p'q'/n} = \sqrt{(0.5005)(0.4995)/24000} = \underline{0.003227}$

c. $E = z(\alpha/2) \cdot \sqrt{p'q'/n} = (1.96)(0.003227) = 0.006325 = 0.0063$
 $p \pm E = 0.5005 \pm 0.0063$
 $\underline{0.4942 \text{ to } 0.5068}$, the 0.95 interval for $p = P(\text{head})$

d. - f. Each student will have different results. Each set of results will yield an empirical probability whose value is very close to 0.50; in fact, you should expect 95% of such results to be within 0.0063 of 0.50.

9.87 a. – e. The distributions do not look normal, they are skewed right. The gaps are caused by working with a discrete distribution, the binomial distribution. Both histograms look the same because they are both showing the same set of data, one as a proportion and the other as a standardized proportion.

f. The distribution is not normal and the normal distribution should not be used to calculate probabilities.

Sample Size Determination Formula for a Population Proportion

$$n = \frac{[z(\alpha/2)]^2 \cdot p^* \cdot q^*}{E^2}$$

<u>Sample Size - A Four-Step Procedure</u>

Step 1: Use the level of confidence, 1-α, to find $z(\alpha/2)$
Step 2: Find the maximum error of estimate
Step 3: Determine p^* and $q^* = 1 - p^*$ (if not given, use $p^* = 0.5$)
Step 4: Use formula to find n

9.89 Step 1: $1 - \alpha = 0.95$; $z(\alpha/2) = z(0.025) = 1.96$
 Step 2: $E = 0.02$
 Step 3: no estimate given, $p^* = 0.5$ and $q^* = 0.5$
 Step 4: $n = \{[z(\alpha/2)]^2 \cdot p^* \cdot q^*\}/E^2$
  $= (1.96^2)(0.5)(0.5)/(0.02^2) = \underline{2401}$

9.91 a. Step 1: $1 - \alpha = 0.90$; $z(\alpha/2) = z(0.05) = 1.65$
  Step 2: $E = 0.02$
  Step 3: $p^* = 0.72$ and $q^* = 0.28$
  Step 4: $n = \{[z(\alpha/2)]^2 \cdot p^* \cdot q^*\}/E^2$
   $= (1.65^2)(0.72)(0.28)/(0.02^2) = 1372.14 = 1373$

b. Step 1: $1 - \alpha = 0.90$; $z(\alpha/2) = z(0.05) = 1.65$
  Step 2: $E = 0.04$
  Step 3: $p^* = 0.72$ and $q^* = 0.28$
  Step 4: $n = \{[z(\alpha/2)]^2 \cdot p^* \cdot q^*\}/E^2$

$\quad\quad\quad\quad = (1.65^2)(0.72)(0.28)/(0.04^2) = 343.035 = \underline{344}$

c. Step 1: $1 - \alpha = 0.98$; $z(\alpha/2) = z(0.01) = 2.33$
Step 2: $E = 0.02$
Step 3: $p^* = 0.72$ and $q^* = 0.28$
Step 4: $n = \{[z(\alpha/2)]^2 \cdot p^* \cdot q^*\}/E^2$
$\quad\quad = (2.33^2)(0.72)(0.28)/(0.02^2) = 2736.1656 = \underline{2737}$

d. Increasing the maximum error decreases the required sample size. The maximum error is located in the denominator of formula 9.8 and therefore an increase will reduce the resulting value for n.

e. Increasing the level of confidence increases the required sample size. The level of confidence determines the value of z used, and it is located in the numerator of formula 9.8 and therefore an increase in 1-α will increase z and will increase the resulting value for n.

Hypotheses are written with the same rules as before. Now replace μ, the population mean, with p, the population proportion.
(ex.: H_o: p = P(driving a convertible) = 0.45 vs.

H_a: p = P(driving a convertible) $\neq$ 0.45,   if driving a convertible is considered the success)
Review: ES11-pp361-363, IRM-pp343-344, if necessary.

9.93 a. H_o: p = P(work) = 0.60 ($\leq$) vs. H_a: p > 0.60

b. H_o: p = P(interested in quitting) = 1/3 ($\leq$)   vs. H_a: p > 1/3

c. H_o: p = P(vote for) = 0.50 ($\leq$)   vs. H_a: p > 0.50

d. H_o: p = P(seriously damaged) = 3/4 ($\geq$)   vs. H_a: p < 3/4

e. H_o: p = P(H|tossed fairly) = 0.50   vs. H_a: p $\neq$ 0.50

9.95 a. p = 0.70, n = 300, x = 224, p' = x/n = 224/300 = 0.747
$z = (p' - p)/\sqrt{pq/n}$
$z\star = (0.747 - 0.70)/\sqrt{(0.7)(0.3)/300} = 1.78$

b. p = 0.50, n = 450, x = 207, p' = x/n = 207/450 = 0.46
$z = (p' - p)/\sqrt{pq/n}$
$z\star = (0.46 - 0.50)/\sqrt{(0.50)(0.50)/450} = -1.70$

c. p = 0.35, n = 280, x = 94, p' = x/n = 94/280 = 0.336
$z = (p' - p)/\sqrt{pq/n}$
$z\star = (0.336 - 0.35)/\sqrt{(0.35)(0.65)/280} = -0.49$

d. p = 0.90, n = 550, x = 508, p' = x/n = 508/550 = 0.924
$z = (p' - p)/\sqrt{pq/n}$
$z\star = (0.924 - 0.90)/\sqrt{(0.90)(0.10)/550} = 1.88$

> Determining the p-value will follow the same procedures as before. We are again working with the normal distribution. Review: ES11-p377, Table 8.7; IRM-pp348, if necessary.

9.97
a. $P = 2(P(z > 1.48)) = 2(1 - 0.9306) = 2(0.0694) = 0.1388$
b. $P = 2(P < -2.26) = 2(0.0119) = 0.0238$
c. $P = P(z > 0.98) = (1 - 0.8365) = 0.1635$
d. $P = P(z < -1.59) = 0.0559$

> Determining the test criteria will also follow the same procedures as before. We are again working with the normal distribution. Review: ES11-pp392-394, IRM-pp359-360, if necessary.

9.99 a. b.

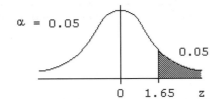 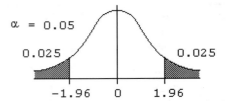

c. d.

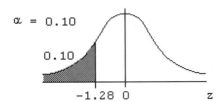 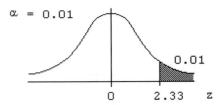

> Since $n \leq 15$ and x is discrete, Table 2 (Appendix B, ES11-pp713-715) will be used to determine the level of significance, α.
>
> x can be any value, 0 through n, for each experiment.
> List all values in numerical sequence.
> Draw a vertical line separating the set of values that belong in the critical region and the set of values that belong in the noncritical region.
> Add all of the probabilities associated with those numbers in the critical region to find α.

9.101 a. $\alpha = P[x = 12, 13, 14, 15 | B(n=15, p=0.5)]$
$= 0.014 + 0.003 + 2(0+) = \underline{0.017}$

b. $\alpha = P[x = 0, 1 | B(n=12, p=0.3)]$
$= 0.014 + 0.071 = \underline{0.085}$

c. $\alpha = P[x = 0, 1, 2, 3, 9, 10|B(n=10,p=0.6)]$
   $= (0+) + 0.002 + 0.011 + 0.042 + 0.040 + 0.006 = \underline{0.101}$

d. $\alpha = P[x = 4, 5, 6,..., 14|B(n=14,p=0.05)]$
   $= 0.004 + 10(0+) = \underline{0.004}$

List all x values in numerical order, as before. Based on H_a#, add consecutive probabilities until the sum is as close to the given α as possible, without exceeding it. Draw a vertical line at this point, separating the critical and noncritical regions.

If H_a contains <, begin adding from x = 0 towards x = n.
If H_a contains >, begin adding from x = n towards x = 0.
If H_a contains ≠, begin adding simultaneously from x = 0 and x = n toward the center.

9.103 a. (1) Correctly fail to reject H_o

b. $\alpha = P[x = 14,15|B(n = 15, p = 0.7)] = 0.031 + 0.005 = \underline{0.036}$

c. (4) Commit a type II error

d. $P = P[x = 13,14,15|B(n = 15, p = 0.7)] = 0.092 + 0.031 + 0.005 = \underline{0.128}$

Review: "The Probability-Value Hypothesis Test: A Five-Step Procedure"; ES11-p371, IRM-p343 and/or "The Classical Hypothesis Test: A Five-Step Procedure"; ES11-p388, IRM-p357, if necessary.

Use $z = \dfrac{p' - p}{\sqrt{\dfrac{pq}{n}}}$, for calculating the test statistic. $p' = x/n$, if not given directly.

Computer and/or calculator commands to perform a hypothesis test for a population proportion can be found in ES11-p444.

Hint for writing the hypotheses for exercise 9.105
Look at the last sentence in the exercise, "If the consumer group ... that less than 90% ...?" The words "less than",(<.) do not indicate any equality, therefore the alternative is less than (<). The negation is "NOT less than", which is > or =. Equality (=) is used in the null hypothesis, but stands for greater than or equal to (≥). Remember to use *p* as the population parameter in the hypotheses.

9.105 Step 1: a. The proportion of claims settled within 30 days
b. H_o: p = P(claim is settled within 30 days) = 0.90 (≥)
H_a: p < 0.90

Step 2: a. independence assumed
b. z; n = 75; n > 20, np = (75)(0.90) = 67.5, nq = (75)(0.10) = 7.5, both np and nq > 5
c. $\alpha = 0.05$

Step 3: a. n = 75, x = 55, p' = x/n = 55/75 = 0.733
b. $z = (p' - p)/\sqrt{pq/n}$

$z* = (0.733 - 0.900)/\sqrt{(0.9)(0.1)/75} = -4.82$

Step 4: -- using p-value approach --------------------
a. **P** = P(z < -4.82) =
Using Table 3, Appendix B, ES11-pp716-717: **P** = 0.000003
Using Table 5, Appendix B, ES11-p718: **P** > 0+
b. **P** < α
-- using classical approach ------------------
a. $-z(0.05) = -1.65$

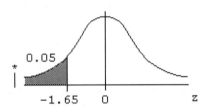

b. z* falls in the critical region, see Step 4a

Step 5: a. Reject H_o
b. The sample provides sufficient evidence that p is significantly less than 0.90; it appears that less than 90% are settled within 30 days as claimed, at the 0.05 level of significance.

Hint for writing the hypothesis for exercise 9.106
Look at the second sentence of the exercise, "Does a random sample ... the proportion of male and of female students who take this course is the same ...." Since "same" indicates equality (=), it must be used in the null hypothesis. The negation is "NOT the same", which is not equal, hence the alternative hypothesis must have a not equal sign (≠).

Hint for writing the hypothesis for exercise 9.107
Look at the first sentence of the exercise, "A politician claims she will receive 60% of the vote ..." The words "will receive" indicate equality (=). Since a politician would be interested in a majority, anything equal to or greater than the stated percentage would be acceptable. Greater than or equal to (≥) includes the equality, therefore it belongs in the null hypothesis. Continue to write the null hypothesis with the equal sign (=), but include the greater-than or equal-to sign(≥) in parentheses after it. The negation is "less than," hence the alternative hypothesis must have a < sign.

9.107 Step 1: a. The proportion of vote for a politician in an upcoming election
b. H_o: p = P(vote for) = 0.60 [will receive 60% of vote]
H_a: p < 0.60 [will receive less than 60%]

Step 2: a. independence assumed
b. z; n = 100; n > 20, np = (100)(0.60) = 60, nq = (100)(0.40) = 40, both np and nq > 5
c. α = 0.05

Step 3: a. n = 100, x = 50, p' = x/n = 50/100 = 0.500
b. $z = (p' - p)/\sqrt{pq/n}$
$z* = (0.500 - 0.600)/\sqrt{(0.6)(0.4)/100} = -2.04$

Step 4: -- using p-value approach --------------------
a. **P** = P(z < -2.04) ;
Using Table 3, Appendix B, ES11-pp716-717: **P** = 0.0207
Using Table 5, Appendix B, ES11-p718: 0.0202 < **P** < 0.0228
b. **P** < α
-- using classical approach ------------------
a. -z(0.05) = -1.65

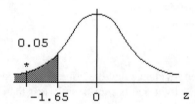

b. z* falls in the critical region, see Step 4a

Step 5: a. Reject H$_o$

b. The sample provides sufficient evidence that the proportion is significantly less than 0.60, at the 0.05 level; it appears that less than 60% support her.

Hint for writing the hypotheses for exercise 9.108

Look at the next to last sentence of the exercise, "Does the watercraft accident rate for PWC's in the state of Nebraska <u>exceed</u> the nation as a whole?" Since exceed, which indicates <u>greater than</u> (>), does not include equality, it must be used in the alternative hypothesis. The negation is "not exceed" which is < or =. Equality (=) is used in the null hypothesis, but stands for ≤. The null hypothesis will indicate that the watercraft accident rate for PWC's in Nebraska is less than or equal to 78%, whereby Nebraska does not exceed the national average proportion. The alternative hypothesis indicates that the watercraft accident rate for PWC's in Nebraska is greater than 78%, whereby Nebraska does exceed the national average proportion.

9.109 Step 1: a. The proportion of adults who speed up to beat a yellow light
b. H$_o$: p = P(speed up) = 0.58
H$_a$: p < 0.58 [lower rate than 58%]

Step 2: a. independence assumed
b. z; n = 150; n > 20, np = (150)(0.58) = 87, nq = (150)(0.42) = 63, both np and nq > 5
c. α = 0.05

Step 3: a. n = 150, x = 71, p' = x/n = 71/150 = 0.473
b. z = (p' - p)/$\sqrt{pq/n}$
z* = (0.473 - 0.580)/$\sqrt{(0.58)(0.42)/150}$ = -2.66

Step 4: -- using p-value approach --------------------
a. **P** = P(z < -2.66) ;
Using Table 3, Appendix B, ES11-pp716-717: **P** = 0.0039
Using Table 5, Appendix B, ES11-p718: 0.0035 < **P** < 0.0040
b. **P** < α
-- using classical approach ------------------

a. -z(0.05) = -1.65

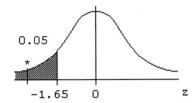

b. z* falls in the critical region, see Step 4a

Step 5: a. Reject H_o

b. The sample provides sufficient evidence that the hometown has a proportion of adults who speed up to beat a yellow light that is significantly less than 0.58, at the 0.05 level of significance. (less than the nation as a whole)

9.111 a. p' = 651/1000 = 0.651

z = (p' - p)/√pq/n

z* = (0.651 − 0.68)/ $\sqrt{(0.68)(0.32)/1000}$ = -1.97

b & c.
Step 1: a. The proportion of American adults who own a library card
b. H_o: p = P(library card) = 0.68
H_a: p < 0.68 [less than 68%]
Step 2: a. independence assumed
b. z; n = 1000; n > 20, np = (1000)(0.68) = 680, nq = (1000)(0.32) = 320, both np and nq > 5
c. α = 0.01
Step 3: a. n = 1000, x = 651, p' = 0.651
b. z = (p' - p)/√pq/n
z* = -1.97
Step 4: -- using p-value approach --------------------
a. **P** = P(z < -1.97) ;
Using Table 3, Appendix B, ES11-pp716-717: **P** = 0.0244
Using Table 5, Appendix B, ES11-p718: 0.0228 < **P** < 0.0256
b. **P** > α
-- using classical approach ------------------
a. -z(0.01) = -2.33
b. z* falls in the noncritical region

Step 5: a. Fail to reject H_o

b. The sample does not provide sufficient evidence that the proportion of American adults that own a library card is significantly less than 0.26, at the 0.01 level of significance.

9.113 a. $H_o: p = 0.225 (\leq)$
$H_a: p > 0.225$
b. $P < 0.05$
$z(0.05) = 1.65; z = 2.71 > 1.65$
Reject Ho   Support Ha at the 0.05 level of significance
c. $p' = x/n = 61/200 = 0.305$

9.115 Step 1: a. The proportion of heads when a coin is flipped by The Flipper
b. $H_o: p = 0.88 (\geq)$
$H_a: p < 0.88$
Step 2: a. independence assumed
b. z; n = 200; n > 20, np = (200)(0.88) = 176, nq = (200)(0.12) = 24, both np and nq > 5
c. $\alpha = 0.05$
Step 3: a. n = 200,
a. x = 181, p' = x/n = 181/200 = 0.905
b. x = 172, p' = 172/200 = 0.86
c. x = 168, p' = 168/200 = 0.84
d. x = 153, p' = 153/200 = 0.765
b. a. $z^* = (0.905 - 0.88)/\sqrt{(0.88)(0.12)/200} = 1.09$
b. $z^* = (0.86 - 0.88)/\sqrt{(0.88)(0.12)/200} = -0.87$
c. $z^* = (0.84 - 0.88)/\sqrt{(0.88)(0.12)/200} = -1.74$
d. $z^* = (0.765 - 0.88)/\sqrt{(0.88)(0.12)/200} = -5.00$
Step 4: -- using p-value approach --------------------
a. **P** = P(z < 1.09) = 0.8621; **P** > α
b. **P** = P(z < -0.87) = 0.1922; **P** > α
c. **P** = P(z < -1.74) = 0.0409; **P** < α
d. **P** = P(z < -5.00) = 0.0000003;
P > α for a. and b.   **P** $\leq \alpha$ for c. and d.
-- using classical approach ------------------
$-z(0.05) = -1.65$
a. and b. are in the noncritical region
c. and d. are in the critical region

Step 5: a. and b. Fail to reject Ho   Support Flipper claim
c. and d. Reject Ho   Do not support Flipper claim

SECTION 9.3 EXERCISES

9.117 a. A: n = 6, $\Sigma x = 43$, $\Sigma x^2 = 323$

$s^2 = (\Sigma x^2 - (\Sigma x)^2/n)/(n - 1) = (323 - 43^2/6)/5 = 2.96667$

$s = \sqrt{s^2} = \sqrt{2.96667} = 1.72$

B: $n = 6$, $\Sigma x = 48$, $\Sigma x^2 = 448$

$s^2 = (\Sigma x^2 - (\Sigma x)^2/n)/(n-1) = (448 - 48^2/6)/5 = 12.8$

$s = \sqrt{s^2} = \sqrt{12.8} = 3.58$

b. Increased standard deviation

c. The 15 is quite different than the rest of the data in this sample. It had a big effect on the standard deviation; it approximately doubled the standard deviation.

χ^2 **Distribution**
(used for inferences concerning σ and σ^2)

Key facts about the χ^2 curve:
1) the total area under a χ^2 curve is 1
2) it is skewed to the right (stretched out to the right side, not symmetrical)
3) it is nonnegative, starts at zero and continues out towards $+\infty$
4) a different curve exists for each sample size
5) uses α and degrees of freedom, df, to determine table values
6) degrees of freedom is abbreviated as 'df', where, df = n-1
7) for df > 2, the mean of the distribution is df.

Notation: $\chi^2(df, \alpha) = \chi^2(\text{"degrees of freedom", "area to the right"})$
↑ ↑ ↑
Table 8   row id #   column id #

ex.) Right tail: $\chi^2(13, 0.025) = 24.7$ (n must have been 14)
Left tail: $\chi^2(13, 0.975) = 5.01$
Note: For left tail, "area to right" includes both the area in the "middle" and the area of the "right" tail.

Explore the chi-square distribution for various degrees of freedom using Chapter 9's Skillbuilder Applet "Chi-Square Probabilities".

9.119 a. 23.2   b. 23.3   c. 3.94   d. 8.64

9.121 a. $\chi^2(19, 0.05) = \underline{30.1}$ b. $\chi^2(4, 0.01) = \underline{13.3}$
  c. $\chi^2(17, 0.975) = \underline{7.56}$ d. $\chi^2(60, 0.95) = \underline{43.2}$

Do not forget to divide α by 2 for a two-sided test.

  e. $\chi^2(21, 0.95) = \underline{11.6}$ and $\chi^2(21, 0.05) = \underline{32.7}$
  f. $\chi^2(6, 0.975) = \underline{1.24}$ and $\chi^2(6, 0.025) = \underline{14.4}$

> Draw a diagram of a chi square distribution, labeling the given information and shading the specified region(s).
> Percentiles: P_k = k^{th} percentile $\Rightarrow$ k% of the data lies <u>below</u> (to left of) this value

9.123 a. $\chi^2(5, 0.05)$ = <u>11.1</u>
       b. $\chi^2(5, 0.05)$ = <u>11.1</u>
       c. $\chi^2(5, 0.10)$ = <u>9.24</u>

9.125 1 - (0.01 + 0.05) = <u>0.94</u>

9.127 Chi-Square with 15 DF

x	P(X <= x)
20.2000	0.8356

       a. $P(\chi^2 < 20.2 | df = 15) = 0.8356$
       b. $P(\chi^2 > 20.2 | df = 15) = 1 - 0.8356 = 0.1644$

> Hypotheses for variability are written with the same rules as before. Now, in place of μ or p, the population standard deviation, σ, or the population variance, σ^2, will be used.
> (ex. H_o: σ = 3.7 vs. H_a: $\sigma \neq 3.7$) Review: ES11-p372-374, IRM-pp343-344, if necessary.

9.129 a. H_o: $\sigma = 24$ ($\leq$) vs. H_a: $\sigma > 24$

       b. H_o: $\sigma = 0.5$ ($\leq$) vs. H_a: $\sigma > 0.5$

       c. H_o: $\sigma = 10$ vs. H_a: $\sigma \neq 10$

       d. H_o: $\sigma^2 = 18$ ($\geq$) vs. H_a: $\sigma^2 < 18$

       e. H_o: $\sigma^2 = 0.025$ vs. H_a: $\sigma^2 \neq 0.025$

> Chi-square test statistic - $\chi^{2*} = (n-1)s^2/\sigma^2$

9.131 a. $\chi^{2*} = (n-1)s^2/\sigma^2 = (17)(785)/(532) = $ <u>25.08</u>

       b. $\chi^{2*} = (n-1)s^2/\sigma^2 = (40)(78.2)/(52) = $ <u>60.15</u>

> Hypothesis tests (p-value approach) will be completed in the same format as before. You may want to review: ES11-p371, IRM-p343. The only differences are in:
>
> 1. the calculated test statistic, which is χ^2, where $\chi^2 = (n-1)s^2/\sigma^2$
>
> 2. using Table 8 to <u>estimate</u> the p-value
>   a) locate df row
>   b) locate the calculated χ^2-value between two critical values in the df row, the p-value or ½p-value is in the interval between the two corresponding probabilities of the critical values
>
> Computer and/or calculator probability and cumulative probability commands for values of χ^2 can be found in ES11-p455-456.

9.133 a. $P = 2P(\chi^{2*} > 27.8 | df = 14)$
Using Table 8: $0.01 < ½P < 0.025$; $\underline{0.02 < P < 0.05}$
Using computer: $P = 0.0302$

b. $P = P(\chi^{2*} > 33.4 | df = 17) = \underline{0.01}$ both are same

c. $P = 2P(\chi^{2*} > 37.9 | df = 25)$
Using Table 8: $0.025 < ½P < 0.05$; $\underline{0.05 < P < 0.10}$
Using computer: $P = 0.0946$

d. $P = P(\chi^{2*} < 26.3 | df = 40)$
Using Table 8: $\underline{0.025 < P < 0.05}$
Using computer: $P = 0.0469$

> Hypothesis tests (classical approach) will be completed using the same format as before. You may want to review: ES11-p388, IRM-p357. The only differences are:
>
> 1. the χ^2 distribution
>   a) draw a skewed right distribution (starting at zero)
>   b) locate df value as middle value
>   c) shade in the critical region(s) based on the alternative hypothesis (H_a).
> 2. finding critical value(s) from Table 8
>   a) degrees of freedom (n-1) is the row id #
>   b) α, area to the right, is the column id #
> 3. the left tail uses $1-\alpha$ or $1-\alpha/2$ for its probability
> 4. the calculated test statistic is χ^{2*}, where $\chi^{2*} = (n-1)s^2/\sigma^2$

9.135 a. $P = P(\chi^2 > 25.08 | df = 17)$
Using Table 8: $0.05 < P < 0.10$
Using computer: $P = 0.0929$
$\alpha = 0.01$; $P > \alpha$; Fail to reject H_o

b. $\chi^2(17, 0.01) = 33.4$ $\chi^2 \geq 33.4$

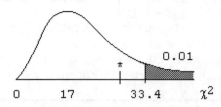

χ^2* falls in the noncritical region; Fail to reject H_o

Hint for writing hypotheses for Exercise 9.137
Look at the second to last sentence of the exercise, "Is the apparent increase in variability ...?". "Increase" indicates greater than. Since greater than (>) does not include the equality (=), it belongs in the alternative hypothesis. The negation is "NOT greater than", which is "< or =". Equality (=) is used in the null hypothesis (as usual), but stands for less than or equal to ($\leq$). The words "in variability" remind you to use the population parameter, σ, in the hypotheses.

9.137 Step 1: a. The standard deviation of weights of 32.0 oz. packages
  b. H_o: $\sigma = 0.25$ [no increase] ($\leq$)
  H_a: $\sigma > 0.25$ [an increase]

Step 2: a. normality indicated
  b. χ^2 c. $\alpha = 0.10$

Step 3: a. $n = 20$, $s = 0.35$
  b. $\chi^{2*} = (n-1)s^2/\sigma^2 = (19)(0.35^2)/(0.25^2) = 37.24$

Step 4: -- using p-value approach --------------------
  a. $P = P(\chi^2 > 37.24 | df = 19)$
    Using Table 8: $0.005 < P < 0.010$
    Using computer: $P = 0.0074$
  b. $P < \alpha$
  -- using classical approach ------------------
  a. $\chi^2(19, 0.10) = 27.2$

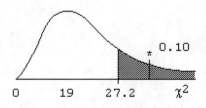

  b. χ^2* falls in the critical region, see Step 4a

Step 5: a. Reject H_o

b. There is sufficient reason to conclude that the apparent increase is significant, at the 0.10 level.

Hint for writing the hypotheses for Exercise 9.138
Look at the second to last sentence in the exercise, "If a standard deviation ... this brand of battery has greater variation than ...?". "Greater than" (>) belongs in the alternative hypothesis. The negation becomes "not greater than" and the null hypothesis would be written with a less than or equal to sign in parentheses (≤). Remember to use the population standard deviation, σ, as the parameter in the hypotheses.

Hint for writing the hypotheses for Exercise 9.139
Look at the last sentence in the exercise, "Does this sample ... the population standard deviation is not equal to 8 ...?". "Not equal to" (≠) belongs in the alternative hypothesis. The negation becomes "equal to" and the null hypothesis would be written with an equality sign (=). Remember to use the population standard deviation, σ, as the parameter in the hypotheses.

9.139 Step 1: a. The standard deviation, σ
b. H_o: σ = 8
H_a: σ ≠ 8

Step 2: a. normality assumed
b. χ^2 c. α = 0.05

Step 3: a. n = 51, $\bar{x}$ = 98.2, s^2 = 37.5
b. χ^{2*} = (n-1)s²/σ² = (50)(37.5)/8² = 29.3

Step 4: -- using p-value approach --------------------
a. **P** = 2P(χ^2 < 29.3|df = 50);
Using Table 8: 0.005 < ½**P** < 0.010; 0.01 < **P** < 0.02
Using computer: **P** = 0.0172
b. **P** < α
-- using classical approach ------------------
a. χ^2(50, 0.975) = 32.4, χ^2(50, 0.025) = 71.4

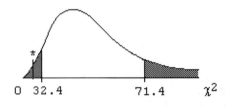

b. χ^{2*} falls in the critical region, see Step 4a

Step 5: a. Reject H_o
b. There is sufficient reason to conclude that the population standard deviation is not equal to 8, at the 0.05 level of significance.

9.141 Step 1: a. The standard deviation of ranchland value in Missouri
b. H_o: $\sigma = 85$ ($\leq$)
H_a: $\sigma > 85$

Step 2: a. normality indicated
b. χ^2 c. $\alpha = 0.05$

Step 3: a. n = 31, s = 125
b. $\chi^{2*} = (n-1)s^2/\sigma^2 = (30)(125^2)/(85^2) = 64.88$

Step 4: -- using p-value approach --------------------
a. **P** = P($\chi^2 > 64.88$|df = 30)
Using Table 8: **P** < 0.005
Using computer: **P** = 0.0002
b. **P** < α

-- using classical approach ------------------
a. $\chi^2(30, 0.05) = 43.8$

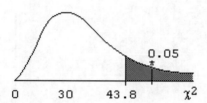

b. χ^{2*} falls in the critical region, see Step 4a

Step 5: a. Reject H_o

b. There is sufficient reason to conclude that the variability in ranchland value in Missouri is greater than the variability for the region, at the 0.05 level of significance.

9.143 a. Allows the use of the chi-square distribution to calculate probabilities.
b.

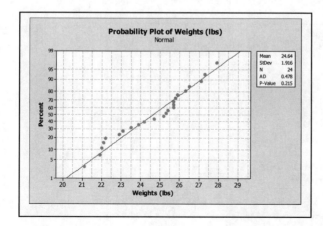

The data values follow very closely to the straight line, indicating an approximately normal distribution. P-value = 0.215 also indicates an approximately normal distribution.

c & d.
Step 1: a. The standard deviation of plates used in weight lifting
b. H_o: $\sigma = 1.0$ ($\leq$)
H_a: $\sigma > 1.0$
Step 2: a. normality indicated
b. χ^2 c. $\alpha = 0.01$
Step 3: a. n = 24, s = 1.916
b. $\chi^{2*} = (n-1)s^2/\sigma^2 = (23)(1.916^2)/(1.0^2) = 84.43$

Step 4: -- using p-value approach --------------------
a. $P = P(\chi^2 > 84.43 | df = 23)$;
Using Table 8: **P** < 0.005
Using computer: **P** = 0.0+
b. **P** < α
-- using classical approach ------------------
a. $\chi^2(23, 0.01) = 41.6$

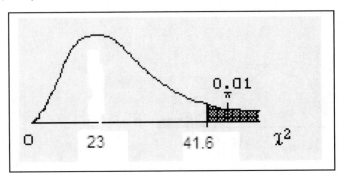

b. χ^{2*} falls in the critical region, see Step 4a

Step 5: a. Reject H_o
b. There is sufficient reason to conclude that the variability in the weights is greater than the acceptable one-pound standard deviation, at the 0.01 level of significance.

9.145   a.   mounded, slightly skewed left, $\overline{x} = 0.01981$, s = 0.01070

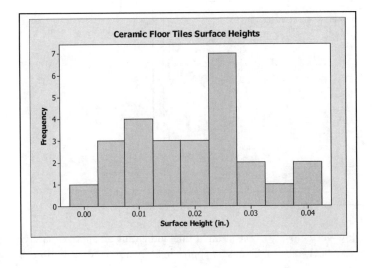

b. Does not appear normal but can be considered approx normal based on normality test: P = 0.620

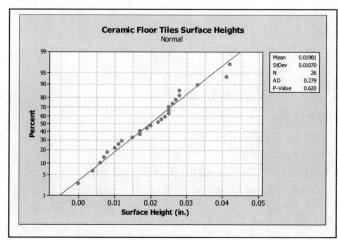

c & d.
Step 1: a. The mean surface height of textured floor tiles
   b. $H_o: \mu = 0.025$
      $H_a: \mu > 0.025$
Step 2: a. normality indicated
   b. t   c. $\alpha = 0.01$
Step 3: a. n = 26, $\bar{x} = 0.01981$, s = 0.01070
   b. $t = (\bar{x} - \mu)/(s/\sqrt{n})$
      $t^* = (0.01981 - 0.025)/(0.01070/\sqrt{26}) = -2.47$
Step 4: -- using p-value approach --------------------
   a. **P** = P(t > -2.47|df = 25);
      Using Table 6, Appendix B, ES11-p719: **P** > 0.50
      Using Table 7, Appendix B, ES11-p720: 0.988 < **P** < 0.99
      Using computer: **P** = 0.990
   b. **P** > α
   -- using classical approach ------------------
   a. t(25, 0.01) = 2.49   t ≥ 2.49
   b. t* falls in the noncritical region

Step 5: a. Fail to reject H_o
   b. The sample does provide sufficient evidence that the mean surface height is no greater than 0.025 inches, at the 0.01 level of significance.

9.147 **a.**
Step 1: a. The standard deviation of dry weights of corks
b. H_o: $\sigma = 0.3275$ [not different]
H_a: $\sigma \ne 0.3275$ [differs]
Step 2: a. normality indicated
b. χ^2 c. $\alpha = 0.02$
Step 3: a. n = 10, s = 0.2920, $s^2 = 0.2920^2 = 0.085264$
b. $\chi^{2*} = (n-1)s^2/\sigma^2 = (9)(0.085264)/0.3275^2 = 7.15$
Step 4: -- using p-value approach --------------------
a. $P = 2 \cdot P(\chi^2 < 7.15 | df = 9)$;
Using Table 8: 2(0.25 < P < 0.50); 0.50 < P < 1.00
Using computer: $P = 0.7570$
b. $P > \alpha$
-- using classical approach ------------------
a. $\chi^2(9, 0.99) = 2.09$, $\chi^2(9, 0.01) = 21.7$

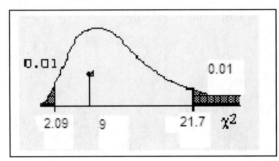

b. χ^{2*} falls in the noncritical region, see Step 4a

Step 5: a. Fail to reject H_o
b. There is not sufficient reason to show that the standard deviation is different from 0.3275 grams, at the 0.02 level of significance.

b.
Step 1: a. The standard deviation of dry weights of corks
b. H_o: $\sigma = 0.3275$ [not different]
H_a: $\sigma \ne 0.3275$ [differs]
Step 2: a. normality indicated
b. χ^2 c. $\alpha = 0.02$
Step 3: a. n = 20, s = 0.2808, $s^2 = 0.2808^2 = 0.078849$
b. $\chi^{2*} = (n-1)s^2/\sigma^2 = (19)(0.078849)/0.3275^2 = 13.97$
Step 4: -- using p-value approach --------------------
a. $P = 2 \cdot P(\chi^2 < 13.97 | df = 19)$;
Using Table 8: 2(0.10 < P < 0.25); 0.20 < P < 0.50
Using computer: $P = 0.4292$
b. $P > \alpha$

-- using classical approach ------------------

a. $\chi^2(19, 0.99) = 7.63$, $\chi^2(19, 0.01) = 36.19 = 36.2$

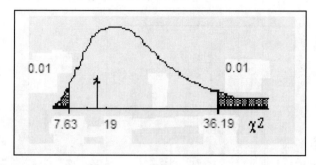

b. χ^{2*} falls in the noncritical region, see Step 4a

Step 5: a. Fail to reject H_o

b. There is not sufficient reason to show that the standard deviation is different from 0.3275 gram, at the 0.02 level of significance.

c. Larger sample standard deviations increase the calculated chi-square value.

d. Larger sample sizes increase the number of degree of freedom and in turn increase the calculated chi-square value. With regards to the p-value and critical-value, since each df has a different distribution, it is not possible to determine, in general, how their values are effected.

9.149 Chi-Square with 23 DF

```
  x        P( X <= x )
36.59      0.964126
```

$P = P(\chi^{2*} > 36.59 | df = 23) = (1 - 0.9641) = \underline{0.0359}$

9.151 Results will vary.
Minitab commands:
  Choose: Calc > Random Data > Normal
  Generate 2000 rows into C1 with mean = 100 and st.dev. = 50.
  Choose: Graph > Histogram
  Use cutpoints with positions –100:300/25.
A histogram of 2000 random data presents a good picture of the population. Notice, the data ranges from -75 to 275.

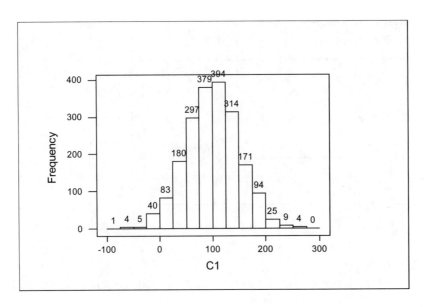

a. Minitab commands:

    Choose: Calc > Random Data > Normal
    Generate 200 rows into C2-C11 with mean = 100 & stdev. = 50.
    Choose: Calc > Row Statistics
    Select Standard deviation for C2-C11 and store in C12.
    Choose: Graph > Histogram
    Use cutpoints with positions 10:100/10.

The histogram below shows the 200 sample standard deviations from samples of n = 10 from N(100,50). Notice, the s-values range from 20 to 90, and appear to be slightly skewed to the right. Remember, the population standard deviation is 50.

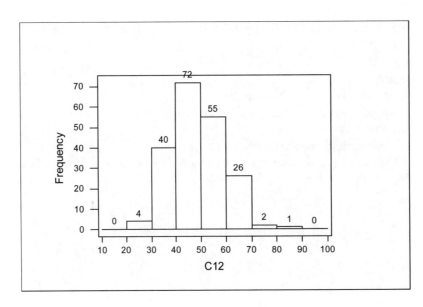

b. Minitab commands:

    Choose: Calc > Calculator
    Store into C13 the expression: (9*C12**2)/(50**2)

Choose: Graph > Histogram
Use midpoints (automatic)

The histogram below shows the 200 calculated chi-square values corresponding to the samples of n = 10 above.

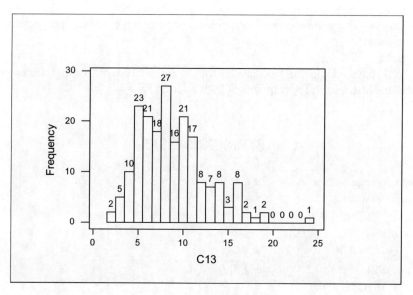

By redrawing the histogram using key critical values from Table 8 (df = 9), we will be able to compare the observed percentage of the chi-square values with the expected percentages.

Minitab commands:
    Enter cutpoints of 0, 2.09, 3.33, 4.17, 5.90, 8.34, 11.4, 14.7, 16.9, 21.7, and 25 into C14.
    Choose: Graph > Histogram
    Use cutpoints with positions in C14.
Table 8 indicates that 1%, 4%, 5%, 15%, 25%, 25%, 15%, 5%, 4%, 1% should occur for the intervals.

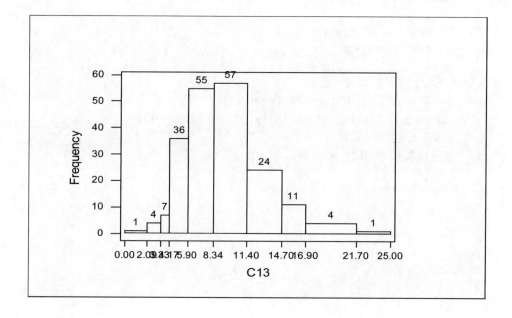

The histogram above shows 1/200 = 0.5%, 4/200 = 2%, 7/200 = 3.5%, 36/200 = 18%, 55/200 = 27.5%, 57/200 = 28.5%, 24/200 = 12%, 11/200 = 5.5%, 4/200 = 2%, and 1/200 = 0.5%. As a set of 10 percentages, they seem very close to what is expected.

c-e.
The results obtained using the exponential distribution are not expected to follow the chi-square distribution as closely. You will need to calculate observed percentages for several intervals in order to detect the true picture.
Minitab commands;
Use the commands above substituting Exponential with a mean of 100 for Normal. Use cutpoints of 0:700/25 for the first histogram. Try cutpoints of 0:40/1 for C13.

CHAPTER EXERCISES

9.153 Step 1: The mean mileage on certain tires for a utility company
Step 2: a. normality assumed
　　　　　b. t　　　　c. $1-\alpha = 0.98$
Step 3: $n = 100$, $\bar{x} = 36{,}000$, $s = 2000$
Step 4: a. $\alpha/2 = 0.02/2 = 0.01$; df = 99; t(99, 0.01) = 2.38
　　　　　b. $E = t(df,\alpha/2)\cdot(s/\sqrt{n}) = (2.38)(2000/\sqrt{100}) = 476$
　　　　　c. $\bar{x} \pm E = 36{,}000 \pm 476$
Step 5: <u>35,524 to 36,476</u>, the 0.98 estimate for μ

9.155 a. $\bar{x} = \Sigma x/n = 878.2/100 = \underline{8.782}$

$s^2 = \Sigma(x - \bar{x})^2/(n-1) = 49.91/99 = 0.5041$

$s = \sqrt{s^2} = \sqrt{0.5041} = \underline{0.710}$

b. Point estimate for μ is $\bar{x} = \underline{8.78}$

c. Step 1: The mean size of oranges
　　Step 2: a. normality assumed
　　　　　　b. t　　　　c. $1-\alpha = 0.95$
　　Step 3: $n = 100$, $\bar{x} = 8.78$, $s = 0.710$
　　Step 4: a. $\alpha/2 = 0.05/2 = 0.025$; df = 99; t(99, 0.025) = 1.99
　　　　　　b. $E = t(df,\alpha/2)\cdot(s/\sqrt{n}) = (1.99)(0.71/\sqrt{100}) = 0.14129 = 0.14$
　　　　　　c. $\bar{x} \pm E = 8.78 \pm 0.14$
　　Step 5: <u>8.64 to 8.92</u>, the 0.95 estimate for μ

9.157 Since df = 71, therefore n = df + 1 = 72

9.159 a. $\bar{X} = \$908.30$, $s = \$118.50$

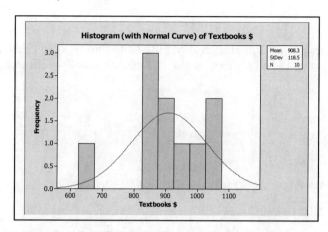

b. The data values are somewhat mounded towards the center.

c. Step 1: The mean cost of required textbooks at NY private colleges
Step 2: a. normality assumed, see part b above.
   b. t c. $1-\alpha = 0.95$
Step 3: $n = 15$, $\bar{X} = 908.30$, $s = 118.50$
Step 4: a. $\alpha/2 = 0.05/2 = 0.025$; $df = 9$; $t(9, 0.025) = 2.26$
   b. $E = t(df, \alpha/2) \cdot (s/\sqrt{n}) = (2.26)(118.50/\sqrt{10}) = 84.689$
   c. $\bar{X} \pm E = 908.30 \pm 84.689$
Step 5: $823.61 to $992.99, the 0.95 estimate for μ

d. 95% confident the mean cost of required textbooks at NY private colleges is between $823.61 and $992.99.
e. Public colleges have a higher mean cost than private colleges, $935.86 versus $908.30.
f. The confidence interval for the public colleges is much wider than that of the private colleges due to the larger standard deviation.

9.161 Step 1: a. The mean amount of trash per year per college student
   b. H_o: $\mu = 640$ ($\geq$)
      H_a: $\mu < 640$ [lower]
Step 2: a. normality indicated
   b. t c. $\alpha = 0.05$
Step 3: a. $n = 18$, $\bar{X} = 559.9$, $s = 158.6$
   b. $t = (\bar{X} - \mu)/(s/\sqrt{n})$
      $t^* = (559.9 - 640)/(158.6/\sqrt{18}) = -2.14$
Step 4: -- using p-value approach --------------------
   a. $P = P(t < -2.14 | df = 17)$;
      Using Table 6, Appendix B, ES11-p719: $0.01 < P < 0.025$
      Using Table 7, Appendix B, ES11-p720: $0.021 < P < 0.027$
      Using computer: $P = 0.023$
   b. $P < \alpha$
   -- using classical approach ------------------
   a. $-t(17, 0.05) = -1.74$

b. t* falls in the critical region

Step 5: a. Reject H_o

b. The sample does provide sufficient evidence that State University students' mean amount of trash is lower than all colleges' mean, at the 0.05 level of significance.

9.163 a. Summary of data: n = 20, $\Sigma x = 629$, $\Sigma x^2 = 21{,}013$

$\bar{x} = \Sigma x/n = 629/20 = \underline{31.45}$

$s^2 = [\Sigma x^2 - (\Sigma x)^2/n]/(n-1) = [21{,}013 - (629^2/20)]/19 = 64.7868$

$s = \sqrt{s^2} = \sqrt{64.7868} = \underline{8.049}$

b. Step 1: a. The mean age at which mothers give birth to abnormal males
b. H_o: $\mu = 28.0$ ($\leq$)
H_a: $\mu > 28.0$ (older)

Step 2: a. normality indicated
b. t   c. $\alpha = 0.05$

Step 3: a. n = 20, $\bar{x} = 31.45$, s = 8.049
b. $t = (\bar{x} - \mu)/(s/\sqrt{n})$
$t^* = (31.45 - 28.0)/(8.049/\sqrt{20}) = 1.92$

Step 4: -- using p-value approach --------------------
a. $P = P(t > 1.92 | df = 19)$;
Using Table 6, Appendix B, ES11-p719: $0.025 < P < 0.05$
Using Table 7, Appendix B, ES11-p720: $0.029 < P < 0.037$
Using computer: $P = 0.035$
b. $P < \alpha$
-- using classical approach -----------------
a. $t(19, 0.05) = 1.73$

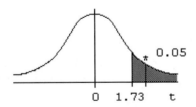

b. t* falls in the critical region, see * Step 4a

Step 5: a. Reject H_o

b. The sample does provide sufficient evidence to justify the contention that the mean age of mothers of abnormal male children is significantly greater than the mean age of mothers with normal male children, at the 0.05 level.

9.165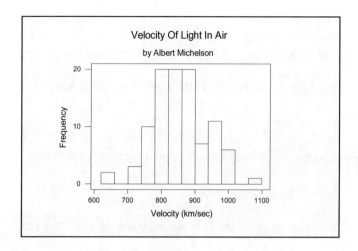

Step 1: a. The mean value for the velocity of light in air
b. H_o: $\mu = 734.5$
H_a: $\mu \neq 734.5$

Step 2: a. normality assumed, note graph above; CLT
b. t   c. $\alpha = 0.01$

Step 3: a. n = 100, $\bar{x}$ = 852.40, s = 79.01
b. $t = (\bar{x} - \mu)/(s/\sqrt{n})$
$t^* = (852.40 - 734.5)/(79.01/\sqrt{100}) = 14.92$

Step 4: -- using p-value approach --------------------
a. **P** = 2P(t > 14.92|df = 99);
Using Table 6, Appendix B, ES11-p719: **P** < 0.005
Using Table 7, Appendix B, ES11-p720: **P** = 0.00+
Using computer: **P** = 0.00+

b. **P** < α

-- using classical approach ------------------
a. ±t(99, 0.005) = ±2.65; critical region: t ≤ -2.65, t ≥ 2.65
b. t* falls in the critical region

Step 5: a. Reject H_o
b. At the 0.01 level of significance there is sufficient evidence to show that the true constant value for the velocity of light in air is different from 299,734.5 km/sec.

9.167 Step 1: The proportion of married men who prefer a brand of instant coffee
Step 2: a. The sample was randomly selected and each subject's response was independent of those of the others surveyed.
b. n = 100; n > 20, np = (100)(20/100) = 20, nq = (100)(80/100) = 80, np and nq both > 5
c. 1 - α = 0.95

Step 3: n = 100, x = 20
Step 4: a. $z(\alpha/2) = z(0.025) = 1.96$
b. $E = z(\alpha/2) \cdot \sqrt{p'q'/n} = 1.96\sqrt{(0.20)(0.80)/100} = (1.96)(0.04) = 0.078$
c. p' ± E = 0.200 ± 0.078

Step 5: 0.122 to 0.278, the 0.95 interval for p = P(preferred this company's brand of instant coffee)

9.169 a.
Step 1: The proportion of dissatisfied customers at a local auto dealership
Step 2: a. The sample was randomly selected and each subject's response was independent of those of the others surveyed.
b. n = 60; n > 20, np = (60)(14/60) = 14, nq = (60)(46/60) = 46, np and nq both > 5
c. $1 - \alpha = 0.95$
Step 3: a. n = 60, x = 14, p' = x/n = 14/60 = 0.233
Step 4: a. $z(\alpha/2) = z(0.025) = 1.96$
b. $E = z(\alpha/2)\cdot\sqrt{p'q'/n} = 1.96\sqrt{(0.233)(0.767)/60} = (1.96)(0.054575) = 0.107$
c. p' ± E = 0.233 ± 0.107
Step 5: 0.126 to 0.340, the 0.95 interval for p = P(dissatisfied)

b. The dealer has overestimated his percent of satisfied customers; it appears to be between 66% and 87%, which is less than 90%.

9.171 Step 1: $1 - \alpha = 0.98$; $z(\alpha/2) = z(0.01) = 2.33$
Step 2: E = 0.02
Step 3: p* = 0.48 and q* = 0.52
Step 4: $n = \{[z(\alpha/2)]^2 \cdot p^* \cdot q^*\}/E^2 = ((2.33^2)(0.48)(0.52))/(0.02^2) = 3387.6336 = \underline{3388}$

9.173
p = 0.1 0.2 0.3 0.4 0.5 0.6 0.7 0.8 0.9
q = 0.9 0.8 0.7 0.6 0.5 0.4 0.3 0.2 0.1
----
pq = 0.09 0.16 0.21 0.24 0.25 0.24 0.21 0.16 0.09

9.175 p' = x/n = 73/100 = 0.73
$z = (p' - p)/\sqrt{pq/n}$
$z^* = (0.73 - 0.80)/\sqrt{(0.8)(0.2)/100} = -1.75$
P = P(z < -1.75) = <u>0.0401</u>

9.177 a. Type A correct decision: The proportion who prefer the new crust is no more than 0.50 and it is decided that no more than 0.50 prefer the crust. Action – stay with current recipe
Type I error: The proportion who prefer the new crust is no more than 0.50 and it is decided that more than 0.50 prefer the crust. Action – change recipes and the majority of customers do not prefer the new recipe.
Type B correct decision: The proportion who prefer the new crust is more than 0.50 and it is decided that more than 0.50 prefer the crust. Action – change recipes and majority of customers are happy
Type II error - The proportion who prefer the new crust is more than 0.50 and it is decided that no more than 0.50 prefer the crust. Action – stay with current recipe when majority of customers prefer the new recipe

b. Type A correct decision: The proportion who prefer the new crust is at least 0.50 and it is decided that at least 0.50 prefer the crust. Action – change recipes and majority of customers are happy
Type I error: The proportion who prefer the new crust is at least 0.50 and it is decided that less than 0.50 prefer the crust. Action – stay with current recipe and the majority of customers are not happy.

Type B correct decision: The proportion who prefer the new crust is less than 0.50 and it is decided that less than 0.50 prefer the crust. Action – stay with current recipe and majority of customers are happy

Type II error - The proportion who prefer the new crust is less than 0.50 and it is decided that at least 0.50 prefer the crust. Action – change recipes and majority of customers will not be happy

    c. If the position is "change only if p is significantly greater than 0.5" then use the alternative hypothesis in part (a).

9.179 a. P = P[x = 9,10,11,...,15|B(n=15,p=0.5)]
          = [0.153 + 0.092 + 0.042 + 0.014 + 0.003 + 2(0+)] = <u>0.304</u>

      P > α; Fail to reject Ho
      There is not significant evidence to conclude that there is a preference for the new crust.

    b. P = P[x = 120,121,122,...,200|B(n=200,p=0.5)]

        Binomial with n = 200 and p = 0.5

          x   P(X <= x)
          120 0.998183

      P = 1.0 − 0.998183 = 0.001817

      P < α; Reject Ho
      There is significant evidence to conclude that there is a preference for the new crust.

    c. Results using the binomial are the same as using the z-distribution.

9.181 a. The percentage of all people in the population who have a specific characteristic is a parameter; it is the binomial parameter p, P(success).

    b. Step 1:   The proportion of cell phone users ages 18-27 who have used text messaging within the past month
        Step 2:   a. The sample was randomly selected and each subject's response was independent of the others surveyed.
                  b. n = 1460, n > 20; np = (1460)(0.63) = 919.8, nq = (1460)(0.37) = 540.2, np and nq both > 5
                  c. 1 − α = 0.95

Step 3: n = 1460, p' = 0.63 (given)
Step 4: a. $z(\alpha/2) = z(0.025) = 1.96$
b. $E = z(\alpha/2) \cdot \sqrt{p'q'/n} = 1.96 \cdot \sqrt{(0.63)(0.37)/1460} = (1.96)(0.013) = 0.025 = 0.03$
c. $p' \pm E = 0.63 \pm 0.03$
Step 5: 0.60 to 0.66, the 0.95 interval for p = P(use text messaging)

c. The 63% is the point estimate, the ±3 percent is the maximum error, 63 ± 3 (60% to 66%) is the confidence interval and as shown, has the typical level of confidence, 95%.

9.183 Step 1: a. The standard deviation for the length of life for 60-watt light bulbs
b. $H_o: \sigma = 81$ [not larger] ($\leq$)
$H_a: \sigma > 81$ [larger]
Step 2: a. normality indicated
b. χ^2 c. $\alpha = 0.05$
Step 3: a. n = 101, $s^2 = 8075$
b. $\chi^{2*} = (n-1)s^2/\sigma^2 = (100)(8075)/81^2 = 123.1$

Step 4: -- using p-value approach --------------------
a. $P = P(\chi^2 > 123.1 | df=100)$
Using Table 8: $0.05 < P < 0.10$
Using computer: $P = 0.0584$
b. $P > \alpha$
-- using classical approach -----------------
a. $\chi^2(100, 0.05) = 124.3$

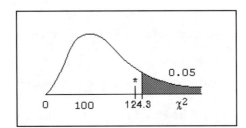

b. χ^{2*} falls in the noncritical region, see Step 4a

Step 5: a. Fail to reject H_o
b. There is not sufficient reason to reject Bright-lite's claim, at the 0.05 level of significance.

9.185 Step 1: a. The standard deviation of sales tickets at Julie's ice cream restaurant franchise
b. $H_o: \sigma = \$2.45$ ($\leq$)
$H_a: \sigma > \$2.45$
Step 2: a. normality indicated
b. χ^2 c. $\alpha = 0.05$
Step 3: a. n = 71, s = 2.95

Step 4:
  b. $\chi^{2*} = (n-1)s^2/\sigma^2 = (70)(2.95^2)/2.45^2 = 101.5$
  -- using p-value approach --------------------
   a. $P = P(\chi^2 > 101.5 | df=70)$
     Using Table 8: $0.005 < P < 0.01$
     Using computer: $P = 0.0082$
   b. $P < \alpha$
  -- using classical approach ------------------
   a. $\chi^2(70, 0.05) = 90.5$ $\chi^2 \geq 90.5$

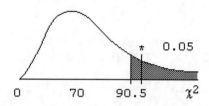

   b. χ^{2*} falls in the critical region, see Step 4a

Step 5: a. Reject H_o

   b. There is sufficient evidence that the variability in sales at Julie's franchise is greater than the variability for the company, at the 0.05 level of significance.

9.187 a. The assumption of normality is reasonable in that the sample size is 35, a size considered large by the CLT.

  b. Step 1: a. The mean length of 2-inch nails
      b. H_o: $\mu = 2.0$
        H_a: $\mu \neq 2.0$
    Step 2: a. Sample assumed approximately normal, CLT satisfied
      b. t   c. $\alpha = 0.05$

Step 3: a. $n = 35$, $\bar{x} = 2.025$, $s = 0.048$
 b. $t = (\bar{x} - \mu)/(s/\sqrt{n})$
  $t^* = (2.025 - 2.0)/(0.048/\sqrt{35}) = 3.08$

Step 4: -- using p-value approach --------------------
 a. $P = 2P(t > 3.08|df = 34)$;
  Using Table 6, Appendix B: $2(P < 0.005) = P < 0.010$
  Using Table 7, Appendix B: $P = 2(0.002 < P < 0.003) =$
   $0.004 < P < 0.006$
  Using computer: $P = 0.0041$
 b. $P < \alpha$
 -- using classical approach ------------------
 a. $\pm t(34, 0.025) = \pm 2.04$; $t \leq -2.04$ & $t \geq 2.04$ critical region
 b. t^* falls in the critical region

Step 5: a. Reject H_o
 b. There is sufficient evidence to reject the idea that the nails have a mean length of 2 inches, at the 0.05 level of significance.

c.
Step 1: a. The standard deviation, σ, of 2-inch nails
 b. H_o: $\sigma = 0.040$ ($\leq$)
  H_a: $\sigma > 0.040$

Step 2: a. normality assumed, based on $n = 35$ and CLT
 b. χ^2 c. $\alpha = 0.05$

Step 3: a. $n = 35$, $\bar{x} = 2.025$, $s = 0.048$
 b. $\chi^{2*} = (n-1)s^2/\sigma^2 = (34)(0.048^2)/0.04^2 = 48.96$

Step 4: -- using p-value approach --------------------
 a. $P = P(\chi^2 > 48.96|df = 34)$
  Using Table 8: $0.01 < P < 0.025$ using $df = 30$
  Using computer: $P = 0.0466$
 b. $P < \alpha$
 -- using classical approach ------------------
 a. $\chi^2(34, 0.05) = 43.8$
 b. χ^{2*} falls in the critical region

Step 5: a. Reject H_o
 b. There is sufficient reason to conclude that the standard deviation is greater than 0.040 inches on this production run, at the 0.05 level of significance.

d. Answers will vary but will include that the length and variability did not meet specifications.

9.189 a.

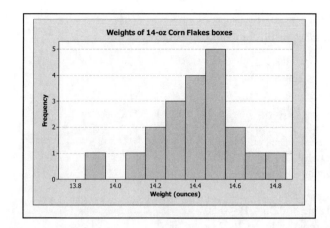

b. $\bar{x} = 14.386$ $s = 0.217$

c. $1/20 = 0.05 = 5\%$ below 14 oz.

d.

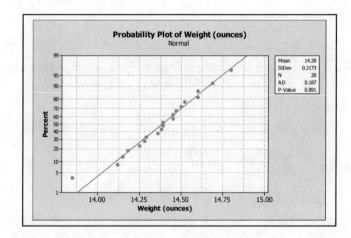

The data values follow closely to a straight line indicating a normal distribution. The P-value also indicates the normality of the data.

e. Step 1: The mean weight of Corn Flakes boxes
Step 2: a. normality assumed, see part d above.
b. t   c. $1-\alpha = 0.95$
Step 3: $n = 20$, $\bar{x} = 14.386$, $s = 0.217$
Step 4: a. $\alpha/2 = 0.05/2 = 0.025$; $df = 19$; $t(19, 0.025) = 2.09$
b. $E = t(df,\alpha/2)\cdot(s/\sqrt{n}) = (2.09)(0.217/\sqrt{20}) = 0.101$
c. $\bar{x} \pm E = 14.386 \pm 0.101$
Step 5: 14.285 to 14.487, the 0.95 estimate for μ

f.
Step 1: a. The standard deviation, σ, of fill for Corn Flakes boxes
   b. H_o: $\sigma = 0.2$ ($\leq$)
      H_a: $\sigma > 0.2$

Step 2: a. normality indicated
   b. χ^2 c. $\alpha = 0.01$

Step 3: a. $n = 20$, $\overline{x} = 14.386$, $s = 0.217$
   b. $\chi^{2*} = (n-1)s^2/\sigma^2 = (19)(0.217^2)/0.2^2 = 22.37$

Step 4: -- using p-value approach --------------------
   a. $P = P(\chi^2 > 22.37 | df = 19)$
      Using Table 8: $0.25 < P < 0.50$
      Using computer: $P = 0.2662$
   b. $P > \alpha$
   -- using classical approach ------------------
   a. $\chi^2(19, 0.01) = 36.2$
   b. χ^{2*} falls in the noncritical region

Step 5: a. Fail to reject H_o
   b. There is not sufficient reason to reject that the filling process is running with a standard deviation of no more than 0.2 oz, at the 0.01 level of significance.

9.191 a. $P(x > 14.2) = P(z > (14.2 - 14.386)/0.217) = P(z > -0.86)$
   $= 1.0000 - 0.1949 = 0.8051$

b. $P(x > 14.2) = P(z > (14.2 - 14.153)/0.0414) = P(z > 1.14)$
   $= 1.0000 - 0.8729 = 0.1271$

c. $1000(14.386) = 14386$ oz; $14386/14.153 = 1016.46$ boxes

d. For every 1000 boxes at current fill-rate, there would be 1016.5 boxes with the new machine. That is an extra 16.5 boxes to sell, which is an increase of 1.65% in revenue at no extra cost of product, except for the initial cost of the machine.

CHAPTER 10 ∇ INFERENCES INVOLVING TWO POPULATIONS

Chapter Preview

In Chapters 8 and 9, the concepts of confidence intervals and hypothesis tests were introduced. Each of these was demonstrated with respect to a single mean, standard deviation, variance or proportion. In Chapter 10, these concepts will be extended to include two means, two proportions, two standard deviations, or two variances, thereby enabling us to compare two populations. Distinctions will have to be made with respect to dependent and independent samples, in order to select the appropriate testing procedure and test statistic.

A look at driving habits by gender is presented in the chapter's opening section, "Battle of the Sexes – Commute Time".

SECTION 10.1 EXERCISES

10.1 a. male commute time to college, female commute time to college

b. male: $\bar{x} = 17.97$, $s = 5.42$; female: $\bar{x} = 25.64$, $s = 9.95$

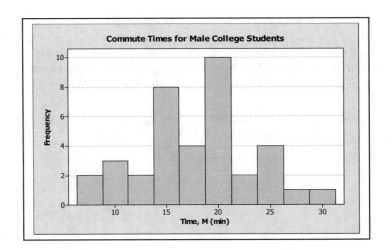

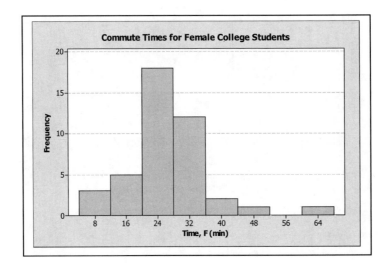

c. Independent, separate, unrelated samples, no pairings

d. Yes, independent; cannot be paired

e. No: if paired, then dependent, if not paired then independent

> **INDEPENDENT SAMPLES** - Two samples are independent if the selection of one sample from a population has no effect on the selection of the other sample from another population. (They do not have to be different populations.)
> ex.: the repair costs for two different brands of VCRs
> **DEPENDENT SAMPLES** (paired samples) - two samples are dependent if the objects or individuals selected for one sample from a population are paired in some meaningful way with the objects or individuals selected for the second sample from the same or another population.
> ex.: "before and after" experiments - change in weight for smokers who became nonsmokers

10.3 Identical twins are so much alike that the information obtained from one would not be independent from the information obtained from the other twin.

10.5 Independent samples. The two samples are from two separate and different sets of students, males and females.

10.7 Dependent samples. The two sets of data were obtained from the same set of 20 people, each person providing one piece of data for each sample.

10.9 Independent samples. The gallon of paint serves as the population of many (probably millions) particles. Each set of 10 specimens forms a separate and independent sample.

10.11 a. Independent samples will result if the two sets are selected in such a way that there is no relationship between the two resulting sets.

b. Dependent samples will result if the 1,000 men and women were husband and wife or if they were brother and sister, or related in some way.

SECTION 10.2 EXERCISES

10.13 a.
Pairs	1	2	3	4	5
d=A-B	1	1	0	2	-1

b. $n = 5$, $\Sigma d = 3$, $\Sigma d^2 = 7$

$\bar{d} = \Sigma d/n = 3/5 = \underline{0.6}$

c. $s_d = \sqrt{(\Sigma d^2 - (\Sigma d)^2/n)/(n-1)} = \sqrt{(7 - (3)^2/5)/(4)} = \sqrt{1.3} = \underline{1.14}$

> Estimating μ_d - the population mean difference

1. point estimate: $\bar{d} = \dfrac{\Sigma d}{n}$

2. confidence interval: $\bar{d} \pm t(df, \alpha/2) \cdot (s_d/\sqrt{n})$

   ↑ point estimate ↑ confidence coefficient ↑ estimated standard error

   $\underbrace{\phantom{\bar{d} \pm t(df, \alpha/2) \cdot (s_d/\sqrt{n})}}_{\text{maximum error of estimate}}$

Follow the steps outlined in "The Confidence Interval: A Five-Step Procedure" in: ES11-pp. 348-349, IRM-p327.
Computer and/or calculator commands to construct a confidence interval for the mean difference can be found in ES11-pp485-486.

10.15 a. Step 1: The mean difference, μ_d
   Step 2: a. normality assumed
   b. t   c. $1-\alpha = 0.95$
   Step 3: $n = 26$, $\bar{d} = 6.3$, $s_d = 5.1$
   Step 4: a. $\alpha/2 = 0.05/2 = 0.025$; $df = 25$; $t(25, 0.025) = 2.06$
   b. $E = t(df,\alpha/2) \cdot (s_d/\sqrt{n}) = (2.06)(5.1/\sqrt{26})$
      $= (2.06)(1.0002) = 2.06$
   c. $\bar{d} \pm E = 6.3 \pm 2.06$
   Step 5: **4.24 to 8.36**, the 0.95 confidence interval for μ_d

b. The same $\bar{d}$ and s_d values were used with a much larger n, resulting in a narrower confidence interval.

10.17 Data Summary: $n = 10$, $\Sigma d = 263$, $\Sigma d^2 = 12319$

$\bar{d} = \Sigma d/n = 263/10 = \underline{26.3}$

$s_d = \sqrt{(\Sigma d^2 - (\Sigma d)^2/n)/(n-1)}$

$= \sqrt{(12319 - (263^2/10))/9} = \sqrt{600.2333} = \underline{24.4997}$

SE Mean $= s_d/\sqrt{n} = 24.4997/\sqrt{10} = \underline{7.747}$

$t(df,\alpha/2) = t(9, 0.025) = 2.26$

$E = t(df,\alpha/2) \cdot (s_d/\sqrt{n}) = (2.26)(7.747) = 17.51$

$\bar{d} \pm E = 26.3 \pm 17.51$
   8.8 to 43.8, the 0.95 confidence interval for μ_d

10.19 Sample statistics: d = A − B   n = 8, $\bar{d}$ = 3.75, s_d = 5.726

Step 1: The mean difference in weight gain for pigs fed ration A as compared to those fed ration B
Step 2: a. normality indicated
b. t   c. 1−α = 0.95
Step 3: n = 8, $\bar{d}$ = 3.75, s_d = 5.726
Step 4: a. α/2 = 0.05/2 = 0.025; df = 7; t(7, 0.025) = 2.36
b. E = t(df,α/2)·($s_d/\sqrt{n}$) = (2.36)(5.726/$\sqrt{8}$)
= (2.36)(2.0244) = 4.78
c. $\bar{d}$ ± E = 3.75 ± 4.78
Step 5: <u>−1.03 to 8.53</u>, the 0.95 interval for μ_d

10.21 a. Male insurance costs appear higher than female insurance costs
b.

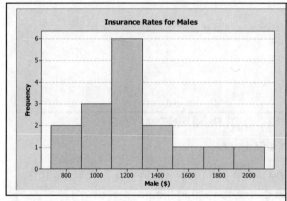
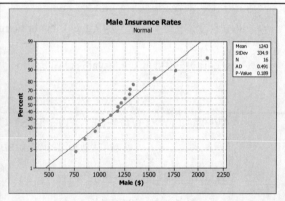
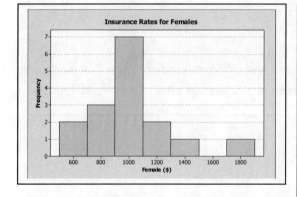
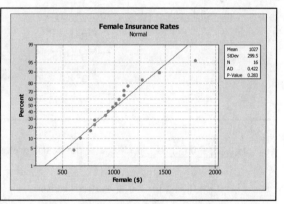
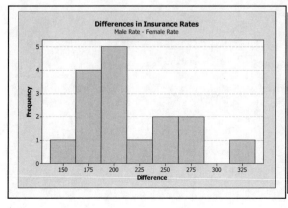
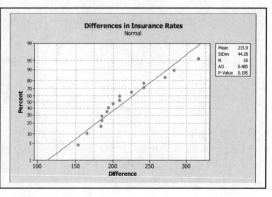

b. males: $\bar{x}$ = $1242.70, s = $334.90; females: $\bar{x}$ = $1026.80, s = $299.50,
    difference: $\bar{x}$ = $215.90, s = $44.30

c. Male and female cost distributions are approximately normal; each pass the normality test (see part b graphs)

e. Step 1: The mean difference, μ_d
    Step 2: a. normality indicated – see step d
             b. t   c. 1-α = 0.95
    Step 3: n = 16, $\bar{d}$ = 215.90, s_d = 44.3
    Step 4: a. $\alpha/2$ = 0.05/2 = 0.025; df = 15; t(15, 0.025) = 2.13
             b. E = t(df,α/2)·($s_d/\sqrt{n}$) = (2.13)(44.3/$\sqrt{16}$)
                = (2.13)(11.075) = 23.58975 = 23.59
             c. $\bar{d}$ ± E = 215.90 ± 23.59
    Step 5: $192.31 to $239.49, the 0.95 confidence interval for μ_d

f. Yes, entire confidence interval is above zero

WRITING HYPOTHESES FOR TEST OF TWO DEPENDENT MEANS

μ_d = population mean difference

null hypothesis - H_o: μ_d = 0   ("the mean difference equals 0, that is, there is no difference within the pairs of data")

possible alternative hypotheses -
H_a: μ_d > 0
H_a: μ_d < 0
H_a: μ_d ≠ 0,   ("the mean difference is significant, that is, there is a difference within the pairs of data")

10.23
a. H_o: μ_d = 0 (≤); H_a: μ_d > 0; d = posttest – pretest

b. H_o: μ_d = 0; H_a: μ_d ≠ 0; d = after - before

c. H_o: μ_d = 0; H_a: μ_d ≠ 0; d = reading1 – reading2

d. H_o: μ_d = 0 (≤); H_a: μ_d > 0; d = post score – pre score

10.25
a. **P** = P(t > 1.86|df = 19)
    Using Table 6, Appendix B, ES11-p719: 0.025 < **P** < 0.05
    Using Table 7, Appendix B, ES11-p720: 0.036 < **P** < 0.044
    Using computer, **P** = 0.0392

b. $P = 2P(t < -1.86|df = 19) = 2P(t > 1.86|df = 19)$
   Using Table 6, Appendix B, ES11-p719: $2(0.025 < \mathbf{P} < 0.05) = 0.05 < \mathbf{P} < 0.10$
   Using Table 7, Appendix B, ES11-p720: $2(0.036 < \mathbf{P} < 0.044) = 0.072 < \mathbf{P} < 0.088$
   Using computer, $\mathbf{P} = 0.0784$

c. $P = P(t < -2.63|df = 28) = P(t > 2.63|df = 28)$
   Using Table 6, Appendix B, ES11-p719: $0.005 < \mathbf{P} < 0.01$
   Using Table 7, Appendix B, ES11-p720: $0.006 < \mathbf{P} < 0.008$
   Using computer, $\mathbf{P} = 0.0069$

d. $P = P(t > 3.57|df = 9)$
   Using Table 6, Appendix B, ES11-p719: $\mathbf{P} < 0.005$
   Using Table 7, Appendix B, ES11-p720: $0.002 < \mathbf{P} < 0.004$
   Using computer, $\mathbf{P} = 0.003$

Reviewing how to determine the test criteria in: ES11-pp392-394, IRM-pp359-360, may be helpful. Remember the t-distribution uses Table 6 (Appendix B, ES11-p719), therefore α and degrees of freedom,
df = n - 1, are needed.

Hypothesis Tests for Two Dependent Means

In this form of hypothesis test, each data value of the first sample is compared to its corresponding (or paired) data value in the second sample. The differences between these paired data values are calculated, thereby forming a sample of differences or *d* values. It is these differences or *d* values that we wish to use to test the difference between two dependent means.

Review the parts to a hypothesis test (p-value & classical) as outlined in: ES11-pp371&388, IRM-pp343&357, if needed. Changes will occur in:
  1) the calculated value of the test statistic, t;

$$t = \frac{\bar{d} - \mu_d}{s_d / \sqrt{n}}, \quad \text{where} \quad \bar{d} = \frac{\sum d}{n}, \; s_d = \sqrt{\frac{\sum d^2 - (\sum d)^2 / n}{n - 1}}$$

and n = # of paired differences

---

2) a. p-value approach
    Use Table 6 (Appendix B, ES11-p719) to <u>estimate</u> the p-value
      1) Locate df row.
      2) Locate the absolute value of the calculated t-value between two critical values in the df row.
      3) The p-value is in the interval between the two corresponding probabilities at the top of the columns; read the bounds from the *one-tailed* heading if H_a is one-tailed, or from the *two-tailed* headings if H_a is two-tailed.
    OR

> Use Table 7 (Appendix B, ES11-p720) to estimate or place bounds on the p-value
> 1) locate the absolute value of the calculated t-value along the left margin and the df along the top, then read the p-value directly from the table where the row and column intersect
>
> OR
>
> 2) locate the absolute value of the calculated t-value and its df between appropriate bounds. From the box formed at the intersection of these row(s) and column(s), use the upper left and lower right values for the bounds on **P**.
>
> b. classical approach
> Use Table 6 (Appendix B, ES11-p719) with df = n - 1 and α to find the critical value
>
> 3) if H_o is rejected, a significant difference as stated in H_a is indicated
> if H_o is not rejected, no significant difference is indicated

---

> Computer and/or calculator commands to perform a hypothesis test for μ_d can be found in ES11-p488.
> **NOTE**: To find $\bar{d}$ and s_d, set up a table of corresponding pairs of data. Calculate d, the difference (be careful to subtract in the same direction each time). Calculate a d^2 for each pair and find summations, Σd and $\Sigma(d^2)$.
> The sample of paired differences are assumed to be selected from an approximately normally distributed population with a mean μ_d and a standard deviation σ_d. Since σ_d is unknown, the calculated t-statistic is found using an estimated standard error of $s_d/\sqrt{n}$.

> **Hint for writing the hypotheses for exercise 10.27**
> Look at the fourth sentence of the exercise; "Does this sample ... sufficient evidence of increased household savings?" For the savings to be increased, there would have to be less current spending. This implies that the differences, "former spending – current spending," would be positive, which in turn implies a greater than zero (> 0). Therefore the alternative hypothesis is a greater than (>). The negation is "NOT increased", which indicates < or =. Therefore the null hypothesis should be written with the equal sign (=), but include the less-than or equal-to sign ($\leq$) in parentheses after it.

10.27 Sample statistics: n = 15, $\bar{d}$ = 75.50, s_d = 66.20

Step 1: a. The mean difference between former spending and current spending
b. H_o: $\mu_d = 0$
H_a: $\mu_d > 0$ (increased savings)

Step 2: a. normality indicated.
b. t   c. $\alpha = 0.05$

Step 3: a. n = 15, $\bar{d}$ = 75.50, s_d = 66.20
b. $t^* = (\bar{d} - \mu_d)/(s_d/\sqrt{n})$
$= (75.50 - 0.0)/(66.20/\sqrt{15}) = 4.42$

Step 4: -- using p-value approach --------------------
a. $P = P(t > 4.42 | df = 14)$;
Using Table 6, Appendix B, ES11-p719: **P** < 0.005
Using Table 7, Appendix B, ES11-p720: **P** < 0.001
Using computer, **P** = 0.000

b. **P** < α

-- using classical approach ------------------
a. t(14, 0.05) = 1.76, using Table 6

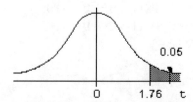

b. t* falls in the critical region, see * Step 4a

Step 5: a. Reject H_o

b. At the 0.05 level of significance, there is a significant increase in household savings.

Hint for writing the hypotheses for exercise 10.28
Look at the third sentence of the exercise; "Does this sample ... sufficient reason that the specialty blade is beneficial in achieving faster times?" For the time to be faster, there would have to be less specialty blade time. This implies that the differences, "current blade time – specialty blade time," would be positive, which in turn implies a greater than zero (> 0). Therefore the alternative hypothesis is a greater than (>). The negation is "NOT beneficial", which indicates < or =. Therefore the null hypothesis should be written with the equal sign (=), but include the less-than or equal-to sign (≤) in parentheses after it.

Hint for writing the hypotheses for exercise 10.29
Look at the second and third sentences of the exercise; "The data ... where d is the amount of corrosion on the coated portion subtracted from the amount of corrosion on the uncoated portion. Does this sample provide sufficient reason to conclude that the coating is beneficial?" For the coating to be beneficial, there would have to be less corrosion on the coated portion. This implies that the differences, "uncoated - coated," if beneficial, would be positive, which in turn implies a greater than zero (> 0). Therefore the alternative hypothesis is a greater than (>). The negation is "NOT beneficial", which indicates < or =. Therefore the null hypothesis should be written with the equal sign (=), but include the less-than or equal-to sign (≤) in parentheses after it.

10.29 Sample statistics: n = 40, $\bar{d}$ = 220/40 = 5.5, s_d = 11.34

Step 1: a. The mean difference between coated and uncoated sections of steel pipe
b. H_o: $\mu_d = 0$
H_a: $\mu_d > 0$ (beneficial)

Step 2: a. normality assumed, CLT with n = 40.
b. t   c. $\alpha = 0.01$

Step 3: a. n = 40, $\bar{d}$ = 5.5, s_d = 11.34
b. $t^* = (\bar{d} - \mu_d)/(s_d/\sqrt{n})$
  $= (5.5 - 0.0)/(11.34/\sqrt{40}) = 3.067$

Step 4: -- using p-value approach --------------------
a. $P = P(t > 3.067 | df = 39)$;
Using Table 6, Appendix B, ES11-p719: $P < 0.005$
Using Table 7, Appendix B, ES11-p720: $P \approx 0.002$
Using computer, $P = 0.002$
b. $P < \alpha$
-- using classical approach ------------------
a. $t(39, 0.01) \approx t(35, 0.01) = 2.44$, using Table 6
$t(39, 0.01) = 2.426$, by computer

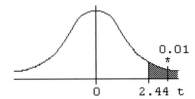

b. t* falls in the critical region, see * Step 4a

Step 5: a. Reject H_o
b. At the 0.01 level of significance, there is a significant benefit to coating the pipe.

Hint for writing the hypotheses for exercise 10.30
Look at the fifth sentences of the exercise; "Comprehension was generally <u>higher</u> on the second reading <u>than</u> on the first by an average of <u>3.2</u> on this scale.". The word "higher" in this problem implies that the "second reading" number is more than the "first reading" number. If the differences are calculated by subtracting the *first* numbers from the *second* numbers, a positive (> 0) difference will coincide with higher. Therefore the alternative hypothesis is written as a <u>greater-than</u> (>) and the 3.2. The negation is "NO higher," which indicates < or =. Therefore the null hypothesis should be written with the equal sign (=) but include the less-than or equal-to sign (≤) in parentheses after it.

Computer and/or calculator commands to perform a hypothesis test for μ_d can be found in ES11-p488. The order of subtraction needs to match the "planned" approach as determined by H_a.

10.31 Data Summary: $n = 5$, $\Sigma d = 90$, $\Sigma d^2 = 2700$

Step 1: a. The mean difference, μ_d
b. $H_o: \mu_d = 0$
$H_a: \mu_d > 0$
Step 2: a. normality indicated
b. t   c. $\alpha = 0.05$
Step 3: a. $n = 5$, $\overline{d} = 18$, $s_d = 16.43$
b. $t^* = (\overline{d} - \mu_d)/(s_d/\sqrt{n}) = (18 - 0)/(16.43/\sqrt{5}) = 2.45$
Step 4: -- using p-value approach --------------------
a. $P = P(t > 2.45 | df = 4)$;

Using Table 6, Appendix B, ES11-p719: $0.025 < P < 0.05$
Using Table 7, Appendix B, ES11-p720: $0.033 < P < 0.037$
Using computer: $P = 0.0352$
b. $P < \alpha$

-- using classical approach ------------------
a. $t(4, 0.05) = 2.13$

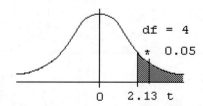

b. t* falls in the critical region, see Step 4a

Step 5: a. Reject H_o
b. At the 0.05 level of significance, there is sufficient evidence that the mean difference is greater than zero.

10.33 Verify - answers given in exercise.

10.35 $d = M - N$, $\alpha = 0.02$
T-Test of the Mean
Test of $\mu_d = 0.00$ vs $\mu_d < 0.00$

Variable	N	Mean	StDev	SEMean	T	P-value
C3	6	-2.33	4.41	1.80	-1.30	0.13

$P > \alpha$; fail to reject H_o

10.37 Step 1: a. The mean difference, μ_d, control group
b. H_o: $\mu_d = 0$ (pre – post)
H_a: $\mu_d > 0$ (improvement)
Step 2: a. normality assumed
b. t   c. $\alpha = 0.05$
Step 3: a. $n = 10$, $\overline{d} = 0.80$, $s_d = 4.492$
b. $t^* = (\overline{d} - \mu_d)/(s_d/\sqrt{n})$
$= (0.80 - 0)/(4.492/\sqrt{10}) = 0.56$
Step 4: -- using p-value approach --------------------
a. $P = P(t > 0.56 | df = 9)$;
Using Table 6, Appendix B, ES11-p719: $P > 0.25$
Using Table 7, Appendix B, ES11-p720: $0.281 < P < 0.315$
Using computer: $P = 0.2946$
b. $P > \alpha$
-- using classical approach ------------------

a. t(9, 0.05) = 1.83

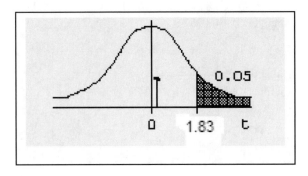

b. t* falls in the noncritical region, see Step 4a

Step 5: a. Fail to reject H_o

b. At the 0.05 level of significance, there is insufficient evidence to show that one's self esteem increases after participation in college courses.

10.39
a. The null hypothesis is, "the average difference is zero."
b. The "t-calculated" and the "t-critical" values are being used to make the decision as in the classical approach.
c. The test is two-tailed and the t-distribution is symmetric, making the number of multiples each statistic is from zero the only information needed. Further, the absence of negative numbers makes the table less confusing to most.
d. The decision was to "fail to reject the null hypothesis" in 12 of them. Actually the calculated t (2.224) for the No. 4 sieve size is less than the critical value (2.228) and therefore leads to a "fail to reject" decision also. But for some reason, they viewed it as too close.
e. The conclusion reached was, "the two methods of sampling are equivalent with respect to Gmb, Gmm, asphalt binder content and gradation."
f. The recommended action was, "the revised Florida method for sampling (FM 1-T 168) be accepted and implemented statewide."

SECTION 10.3 EXERCISES

Estimating ($\mu_1 - \mu_2$) - the difference between two population means, independent samples

1. Point Estimate: $\bar{x}_1 - \bar{x}_2$

2. Confidence Interval

$$(\bar{x}_1 - \bar{x}_2) \pm t(df, \alpha/2) \cdot \sqrt{(s_1^2 / n_1) + (s_2^2 / n_2)}$$

↑ point estimate   ↑ confidence coefficient   ↑ estimated standard error

maximum error of estimate

estimate df by using the smaller value of df_1 or df_2

10.41 $\sqrt{(s_1^2/n_1)+(s_2^2/n_2)} = \sqrt{(190/12)+(150/18)} = \sqrt{24.1667} = \underline{4.92}$

10.43 Case I: df will be between 17 and 40
Case II; df = 17 (smaller df)
Case I will occur when completing the inference on a computer or calculator, where the df are calculated. Case II will occur when completing the test without software, using Table 6 and a conservative approach.

Review "The Confidence Interval: A Five-Step Procedure" in: ES11-pp348-349, IRM-p327, if necessary.
Subtract sample means ($\bar{x}_1 - \bar{x}_2$ or $\bar{x}_2 - \bar{x}_1$) in whichever order results in a positive difference. Also, use appropriate subscripts to designate the source.

10.45 Step 1: The difference between two means, $\mu_1-\mu_2$
Step 2: a. normality indicated
b. t   c. $1-\alpha = 0.90$
Step 3: sample information given in exercise;
$\bar{x}_1 - \bar{x}_2 = 35 - 30 = 5$
Step 4: a. $\alpha/2 = 0.10/2 = 0.05$; df = 14; t(14, 0.05) = 1.76
b. $E = t(df,\alpha/2) \cdot \sqrt{(s_1^2/n_1)+(s_2^2/n_2)}$
$= (1.76)\sqrt{(22^2/20)+(16^2/15)}$
$= (1.76)(6.42) = 11.3$
c. $(\bar{x}_1 - \bar{x}_2) \pm E = 5 \pm 11.3$
Step 5: $\underline{-6.3 \text{ to } 16.3}$, the 0.90 confidence interval for $\mu_1-\mu_2$

10.47 Confidence interval using t from Table 6:
Step 1: The difference between the average daily car rental rates in Boston and NYC, $\mu_{NYC}-\mu_{Bos}$
Step 2: a. normality indicated
b. t   c. $1-\alpha = 0.95$
Step 3: sample information given in exercise;
$\bar{x}_{NYC} - \bar{x}_{Bos} = 127.75 - 95.94 = 31.81$
Step 4: a. $\alpha/2 = 0.05/2 = 0.025$; df = 9; t(9, 0.025) = 2.26
b. $E = t(df,\alpha/2) \cdot \sqrt{(s_{NYC}^2/n_{NYC})+(s_{Bos}^2/n_{Bos})}$
$= (2.26)\sqrt{(15.83^2/16)+(7.50^2/10)}$
$= (2.26)(4.6137627) = 10.427 = 10.43$
c. $(\bar{x}_{NYC} - \bar{x}_{Bos}) \pm E = 31.81 \pm 10.43$
Step 5: $\underline{\$21.38 \text{ to } \$42.24}$, the 0.95 confidence interval for $\mu_{NYS}-\mu_{Bos}$

OR Confidence interval using calculated df:
Difference = mu (NYC) - mu (Bos)
Estimate for difference: 31.81
95% CI for difference: (22.24, 41.38)
DF = 22

Computer and/or calculator commands to construct a confidence interval for the difference between two means can be found in ES11-pp502-503.

10.49 Verify - answer given in exercise

10.51 Sample statistics:
N. Dakota: n = 11, $\bar{x} = 1403$, s = 159
S. Dakota: n = 14, $\bar{x} = 1548$, s = 401

Confidence interval using t from Table 6:
Step 1: The difference between mean sunflower yields for North and South Dakota, $\mu_1 - \mu_2$
Step 2: a. normality assumed.
   b. t   c. $1-\alpha = 0.95$
Step 3: sample information given above;
$\bar{x}_S - \bar{x}_N = 1548 - 1403 = 145$
Step 4: a. $\alpha/2 = 0.05/2 = 0.025$; df = 10; t(10, 0.025) = 2.23
   b. $E = t(df, \alpha/2) \cdot \sqrt{(s_S^2/n_S) + (s_N^2/n_N)}$
   $= (2.23)\sqrt{(401^2/14) + (159^2/11)}$
   $= (2.23)(117.40553) = 261.8143$
   c. $(\bar{x}_S - \bar{x}_N) \pm E = 145 \pm 261.8$
Step 5: <u>-116.8 to 406.8</u>, the 0.95 confidence interval for $\mu_S - \mu_N$

OR Confidence interval using calculated df:
Difference = mu (S. Dakota) - mu (N. Dakota)
Estimate for difference: 145
95% CI for difference: (-102, 393)
DF = 17

WRITING HYPOTHESES FOR THE DIFFERENCE BETWEEN TWO MEANS

null hypothesis:
$H_0: \mu_1 = \mu_2$ **or** $H_0: \mu_1 - \mu_2 = 0$ **or** $H_0: \mu_1 - \mu_2 = \#$

possible alternative hypotheses:
$$H_a: \mu_1 > \mu_2 \quad \text{or} \quad H_a: \mu_1-\mu_2 > 0 \quad \text{or} \quad H_a: \mu_1-\mu_2 > \#$$
$$H_a: \mu_1 < \mu_2 \quad \text{or} \quad H_a: \mu_1-\mu_2 < 0 \quad \text{or} \quad H_a: \mu_1-\mu_2 < \#$$
$$H_a: \mu_1 \neq \mu_2 \quad \text{or} \quad H_a: \mu_1-\mu_2 \neq 0 \quad \text{or} \quad H_a: \mu_1-\mu_2 \neq \#$$

10.53 a. $H_o: \mu_1 - \mu_2 = 0 \quad$ vs. $\quad H_a: \mu_1 - \mu_2 \neq 0$

b. $H_o: \mu_1 - \mu_2 = 0 \ (\leq) \quad$ vs. $\quad H_a: \mu_1 - \mu_2 > 0$

c. $H_o: \mu_N - \mu_S = 0 \ (\geq) \quad$ vs. $\quad H_a: \mu_N - \mu_S < 0$
or equivalently
$H_o: \mu_S - \mu_N = 0 \ (\leq) \quad$ vs. $\quad H_a: \mu_S - \mu_N > 0$

d. $H_o: \mu_M - \mu_F = 0 \quad$ vs. $\quad H_a: \mu_M - \mu_F \neq 0$

10.55 st.err. $= \sqrt{(s_1^2 / n_1) + (s_2^2 / n_2)}$

a. $\sqrt{(12/16) + (15/21)} = \underline{1.21}$

b. $\sqrt{(0.054/8) + (0.087/10)} = \underline{0.1243}$

c. $\sqrt{(2.8^2/16) + (6.4^2/21)} = \sqrt{2.44} = \underline{1.56}$

10.57 $t^* = [(\bar{x}_2 - \bar{x}_1) - (\mu_2 - \mu_1)] / \sqrt{(s_2^2 / n_2) + (s_1^2 / n_1)}$
$= [(1.66 - 1.43) - 0] / \sqrt{(0.29^2 / 21) + (0.18^2 / 9)} = \underline{2.64}$

Review the rules for calculating the p-value in: ES11-p377, IRM-p348, if necessary. Remember to use the t-distribution, therefore either Table 6, Table 7 or the computer/calculator will be used to find probabilities.
Review of the use of the tables can be found in: ES11-pp421-422, IRM-pp397-398.

Use Table 6 (Appendix B, ES11-p719) with the smaller of df_1 or df_2 and the given α to find the critical value(s). Reviewing how to determine the test criteria in: ES11-pp415-416, IRM-pp359-360, as it is applied to the t-distribution may be helpful.

10.59 a. α = 0.05   df=15   b. α = 0.01   df=26

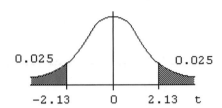
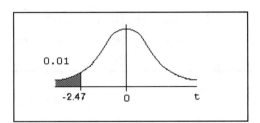

c. α = 0.10   df=7   d. α = 0.05   df=13

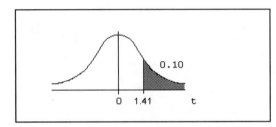
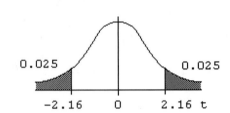

10.61 With the smaller degrees of freedom, df = 9, a higher calculated value is needed making it more difficult to reject H_o. This is due to the lack of reliability with a small sample.

Hypothesis Test for the Difference Between Two Means, Independent Samples

Review the parts to a hypothesis test as outlined in: ES11-pp371& 388, IRM-pp343&357, if needed. Slight changes will occur in:

1. **the hypotheses**:   (see box before exercise 10.53)

2. **the calculated test statistic**
$$t = \frac{(\bar{x}_1 - \bar{x}_2) - (\mu_1 - \mu_2)}{\sqrt{(s_1^2 / n_1) + (s_2^2 / n_2)}}$$, using df = smaller of df_1 or df_2

3. If H_o is rejected, a significant difference between the means is indicated.
   If H_o is not rejected, no significant difference between the means is indicated.

Any subscripts may be used on the hypotheses. Try to use letters that indicate the source. The form $H_o: \mu_2 - \mu_1 = 0$ (versus $H_o: \mu_1 = \mu_2$) is the preferred form since it establishes the order for subtraction that will be needed when calculating the test statistic.

Computer and/or calculator commands to perform a hypothesis test for the difference between two means can be found in ES11-pp502-503.

Hint for writing the hypotheses for exercise 10.62
Look at the last sentence of the exercise; "At the 0.05 level of significance, do the data support ... the mean length of girls' names is longer than the mean length of boys' names?" The word "longer"

indicates a greater than. Therefore, the alternative hypothesis is greater than(>), which can only go in the alternative hypothesis. The negation becomes "not greater than" and the null hypothesis would be written with an equality sign (=) and (≤) after it. The direction of subtraction, ($\mu_G - \mu_B$) will give a positive difference.

10.63 No, the brunettes overall grade average was lower than the overall grade average for the blondes. Lauren could say that there is no difference between blondes and brunettes intelligence and do a hypothesis test and hope the means are not significantly different.

Hint for writing the hypotheses for exercise 10.65

Look at the second to last sentence of the exercise; "Do these results show ... with an ESG to help them is significantly greater than those not using an ESG?". The words "greater than" indicates a (>), which can only go in the alternative hypothesis. The negation becomes "not greater than" and the null hypothesis would be written with an equality sign (=) and (≤) after it. The direction of subtraction, ($\mu_1 - \mu_2$) will give a positive difference.

10.65 Step 1: a. The difference between the mean scores of students using an electronic study guide to help them learn accounting principles and those not using one

b. $H_o: \mu_1 - \mu_2 = 0$

$H_a: \mu_1 - \mu_2 > 0$

Step 2: a. normality assumed, CLT with $n_1 = 38$ and $n_2 = 36$.

b. t   c. $\alpha = 0.01$

Step 3: a. sample information given in exercise

b. $t^* = [(\bar{x}_1 - \bar{x}_2) - (\mu_1 - \mu_2)]/\sqrt{(s_E^2 / n_E) + (s_C^2 / n_C)}$

$= [(79.6 - 72.8)-0]/[\sqrt{(6.9^2 / 38) + (7.6^2 / 36)}]$

$= 4.02$

Step 4: -- using p-value approach --------------------

a. $P = P(t > 4.02 | df = 35)$;

Using computer/calculator, **P** = 0.0001

Using Table 6, Appendix B, ES11-p719: **P** < 0.005

Using Table 7, Appendix B, ES11-p720: **P** = 0+

OR $P = P(t > 4.02 | df = 70)$;

Using computer/calculator, **P** = 0.0001

b. **P** < α

-- using classical approach ------------------

a. critical region: $t \geq 2.44$

b. t^* is in the critical region

Step 5: a. Reject H_o

b. At the 0.01 level of significance, there is sufficient evidence to conclude the ESG students are doing better in accounting principles than those not using an ESG.

10.67 Step 1: a. The difference between the mean weights of two types of cheese

b. $H_o: \mu_G - \mu_B = 0$

$H_a: \mu_G - \mu_B \neq 0$

Step 2: a. normality assumed
b. t   c. α = 0.05

Step 3: a. sample information given in exercise

b. $t^* = [(\bar{x}_G - \bar{x}_B) - (\mu_G - \mu_B)] / \sqrt{(s_G^2/n_G) + (s_B^2/n_B)}$

$= [(1.2 - 1.05) - 0] / [\sqrt{(0.32^2/16) + (0.25^2/14)}]$

$= 1.44$

Step 4: -- using p-value approach --------------------
a. p-value = $2 \cdot P(t > 1.44 | df = 13)$
Using computer/calculator, **P** = 0.1735
Using Table 6, Appendix B, ES11-p719: 2(0.05 < **P** < 0.10) = 0.10 < **P** < 0.20
Using Table 7, Appendix B, ES11-p720: [0.077 < ½**P** < 0.093]; 0.154 < **P** < 0.186
OR **P** = $2P(t > 1.44 | df = 27)$;
Using computer/calculator, **P** = 0.1614
b. **P** > α

-- using classical approach ------------------
a. critical region: t ≤ -2.16 and t ≥ 2.16
b. t* is in the critical region

Step 5: a. Fail to reject H_o
b. At the 0.05 level of significance, there is insufficient evidence to conclude that there is a difference in the mean weights of the two cheeses.

Hint for writing the hypotheses for exercise 10.68

Look at the first sentence of the exercise; "If a random ... , can you conclude that there is a significant difference between the selling prices ...?" The word "difference" indicates a not equal to. Therefore, the alternative hypothesis is not equal to (≠). The negation becomes "equal to" and the null hypothesis would be written with an equality sign (=). Either direction ($\mu_S - \mu_N$) or ($\mu_N - \mu_S$) will work for a not equal to. The effect of the chosen direction will only show in the sign (+/-) of the calculated test statistic.

10.69 a. Verify - answers given in exercise.

b. **P** = $2P(t > 0.59 | df = 20)$;   (Table 7)
2(0.277 < ½**P** < 0.312), 0.554 < **P** < 0.624

c. Using Table 7: **P** = $2P(t > 0.59 | df = 12)$;
2(0.280 < ½**P** < 0.313), 0.560 < **P** < 0.626

10.71 Step 1: a. The difference between the mean selling prices of men's drivers and women's drivers, $\mu_M - \mu_F$
b. $H_o: \mu_M - \mu_F = 0$
   $H_a: \mu_M - \mu_F > 0$

Step 2: a. normality indicated
b. t   c. α = 0.05

Step 3: a. $n_M = 10$, $\bar{x}_M = 232$, $s_M = 141$

$n_F = 6$, $\bar{x}_F = 230$, $s_N = 154$

b. $t^* = [(\bar{x}_M - \bar{x}_F) - (\mu_M - \mu_F)] / \sqrt{(s_M^2/n_M) + (s_F^2/n_F)}$

$= [(232 - 230) - 0] / [\sqrt{(141^2/10) + (154^2/6)}]$

$= 0.02$

Step 4: -- using p-value approach --------------------
a. $\mathbf{P} = P(t > 0.02 | df = 5)$;
Using computer/calculator, $\mathbf{P} = 0.492$
Using Table 6, Appendix B, ES11-p719: $\mathbf{P} > 0.25$
Using Table 7, Appendix B, ES11-p720: $0.462 < \mathbf{P} < 0.500$
OR $\mathbf{P} = P(t > 0.02 | df = 9)$;
Using computer/calculator, $\mathbf{P} = 0.491$
b. $\mathbf{P} > \alpha$
-- using classical approach ------------------
a. critical region: $t \geq 2.02$
b. t^* falls in the noncritical region

Step 5: a. Fail to reject H_o

b. At the 0.05 level of significance, there is not sufficient evidence to show that the mean cost of a men's driver is more expensive than a women's driver.

10.73 Step 1: a. The difference between mean game times for the Yankees and the St. Louis Cardinals, $\mu_Y - \mu_C$

b. H_o: $\mu_Y - \mu_C = 0$
H_a: $\mu_Y - \mu_C > 0$

Step 2: a. normality assumed
b. t   c. $\alpha = 0.05$

Step 3: a. $n_Y = 14$, $\bar{x}_Y = 197.6$, $s_Y = 24.7$
$n_C = 12$, $\bar{x}_C = 170.1$, $s_C = 29.2$

b. $t^* = [(\bar{x}_Y - \bar{x}_C) - (\mu_Y - \mu_C)] / \sqrt{(s_Y^2/n_Y) + (s_C^2/n_C)}$

$= [(197.6 - 170.1) - 0] / [\sqrt{(24.7^2/14) + (29.2^2/12)}]$

$= 2.57$

Step 4: -- using p-value approach --------------------
a. $\mathbf{P} = P(t > 2.57 | df = 11)$;
Using computer/calculator, $\mathbf{P} = 0.013$
Using Table 6, Appendix B, ES11-p719: $0.01 < \mathbf{P} < 0.025$
Using Table 7, Appendix B, ES11-p720: $0.012 < \mathbf{P} < 0.016$
OR $\mathbf{P} = P(t > 2.57 | df = 21)$;
Using computer/calculator, $\mathbf{P} = 0.009$
b. $\mathbf{P} < \alpha$
-- using classical approach ------------------
a. critical region: $t \geq 1.80$
b. t^* is in the critical region

Step 5: a. Reject H_o

b. At the 0.05 level of significance, there is sufficient difference between the mean times for the Yankees' and the Cardinals' games.

10.75 a. Yes, p-values from normality tests > 0.05

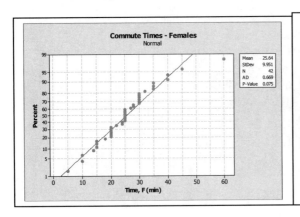
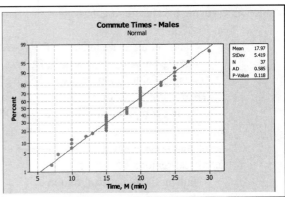

b.
Step 1: a. The difference between mean commute times for males and females, $\mu_F - \mu_M$

b. H_o: $\mu_F - \mu_M = 0$
H_a: $\mu_F - \mu_M \neq 0$

Step 2: a. normality indicated in part a
b. t   c. $\alpha = 0.05$

Step 3: a. $n_F = 42$, $\overline{x}_F = 25.64$, $s_F = 9.95$
$n_M = 37$, $\overline{x}_M = 17.97$, $s_M = 5.42$

b. $t^* = [(\overline{x}_F - \overline{x}_M) - (\mu_F - \mu_M)] / \sqrt{(s_F^2/n_F)+(s_M^2/n_M)}$

$= [(25.64 - 17.97)-0]/[\sqrt{(9.95^2/42)+(5.42^2/37)}\,]$

$= 4.32$

Step 4: -- using p-value approach --------------------
a. $P = 2P(t > 4.32 | df = 35)$;
Using computer/calculator, $P = 0.000122$
Using Table 6, Appendix B, ES11-p719: ½$P < 0.01$; $P < 0.02$
Using Table 7, Appendix B, ES11-p720: $P = 0+$
OR $P = 2P(t > 4.32 | df = 64)$;
Using computer/calculator, $P = 0.000$
b. $P < \alpha$
-- using classical approach -----------------
a. critical region: $t \leq -2.03$ and $t \geq 2.03$
b. t^* is in the critical region

Step 5: a. Reject H_o
b. At the 0.05 level of significance, there is sufficient evidence to show that the mean commute time for females is not the same as males.

c. Factors such as: living at home or not, having children, having a job, method of commuting

10.77 a.

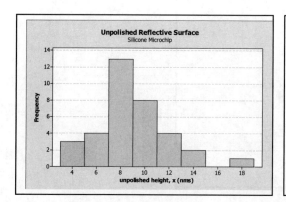 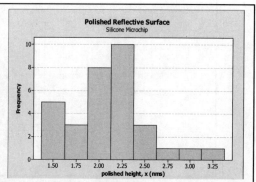

Unpolished: $\bar{x} = 8.98$, s = 3.12,   Polished: $\bar{x} = 2.126$, s = 0.437

b. Histograms show mounded, slightly skewed distributions, normality tests demonstrate normality in both distributions

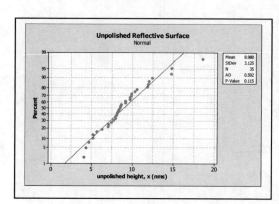 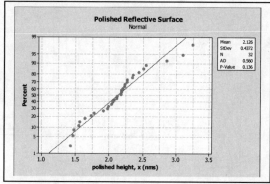

10.79 Everybody will get different results, but they can all be expected to look very similar to the following.
   a. Minitab commands:
   Choose: Calc > Random > Normal to generate both distributions into C1 and C2
   Choose: Stat > Basic Statistics > Display Descriptive Statistics for the means and standard deviations.
   Choose: Graph > Histogram
   Use cutpoints as noted for both distributions.

   Excel commands:
   Choose: Data > Data Analysis > Random Number Generation > Normal to generate both distributions in columns A and B.
   Choose: Data > Data Analysis > Descriptive Statistics for the means and standard deviations.
   Choose: Data > Data Analysis > Histogram
   Use cutpoints as noted for both distributions.

   N(100,20)

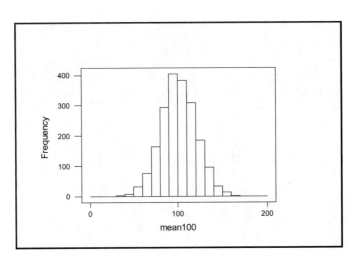

N(120,20)

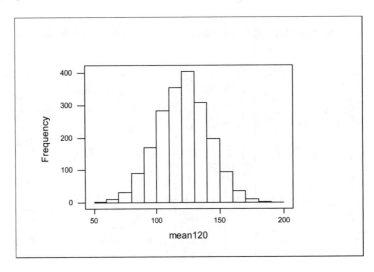

b. The sampling distribution is expected to be normal in shape with a mean of 20 (120-100) and have a standard error of $\sqrt{\frac{20^2}{8} + \frac{20^2}{8}}$ or 10.

c. Minitab commands:
Choose: Calc > Random > Normal
Generate 100 rows into C3-C10 with mean = 100 and standard deviation = 20
Choose: Calc > Row Statistics
Select Mean for C3-C10 and store in C11.
Repeat above for 100 rows into C12-C19 with mean = 120 and standard deviation = 20.
Also select Mean for C12-C19 and store in C20.

Choose: Calc > Calculator
Store in C21 the expression: C20 – C11

Excel commands:
Choose: Data > Data Analysis > Random Number Generation > Normal

Generate 100 rows into columns C through J with mean 100 and standard deviation = 20.
Choose: Insert function > All > Average
Find the average for C1-J1 and store in K1. Drag down for other averages.
Repeat above for 100 rows into L through S with mean 120 and standard deviation = 20.
Also calculate means into column T.
Choose: Edit formula (=)
Find T1 – K1 and store in U1. Drag down for other subtractions.

d. Minitab commands:
Choose: Stat > Basic Statistics > Display Descriptive Statistics for C21
Choose: Graph > Histogram
Use cutpoints –20:60/10 for C21

Excel commands:
Choose: Data > Data Analysis > Descriptive Statistics for column U.
Choose: Data > Data Analysis > Histogram
Use classes from –20 to 60 in increments of 10 for column U.

100 values for the difference between two sample means:

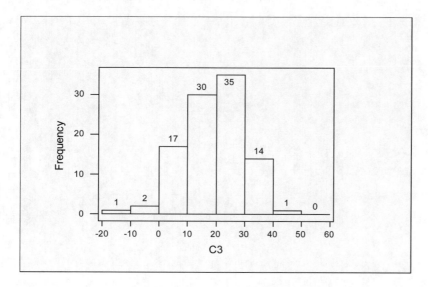

e. For the empirical sampling distribution, the mean is 19.51 and the standard error is 10.71. There is 65%, 96% and 99% of the values within one, two and three standard errors of the expected mean of 20. This seems to agree closely with the empirical rule, thus suggesting a normal distribution occurred.

f. You can expect very similar results to occur on repeated trials.

For Exercises 10.80 – 10.82, adjust the Minitab and/or Excel commands in Exercise 10.79

10.81 Everybody will get different results, but they can all be expected to look very similar to the results found in exercise 10.79. It turns out that the t^* statistic is very "robust", meaning "it works quite well even when the assumptions are not met." This it one of the reasons the t-test for the mean and the t-test for the difference between to means are such important tests.

SECTION 10.4 EXERCISES

Estimating $(p_1 - p_2)$ - the difference between two population proportions - independent samples (large samples)

1. Point estimate: $p'_1 - p'_2$

2. Confidence interval:
$$(p'_1 - p'_2) \pm z(\alpha/2) \cdot \sqrt{(p'_1 \cdot q'_1 / n_1) + (p'_2 \cdot q'_2 / n_2)}$$

↑ point estimate   ↑ confidence coefficient   ↑ estimated standard error

Maximum error of the estimate

Computer and/or calculator commands to construct a confidence interval for the difference between two proportions can be found in ES11-pp513-514.

10.83 $x = \underline{75}$, $n = \underline{250}$,
$p' = x/n = 75/250 = \underline{0.30}$, $q' = 1 - p' = 1 - 0.30 = \underline{0.70}$

10.85 a. $\sqrt{(p'_1 \cdot q'_1 / n_1) + (p'_2 \cdot q'_2 / n_2)} =$
$\sqrt{(0.8 \cdot 0.2 / 40) + (0.8 \cdot 0.2 / 50)} = \sqrt{0.0072} = 0.085$

b. $\sqrt{(p'_1 \cdot q'_1 / n_1) + (p'_2 \cdot q'_2 / n_2)} =$
$\sqrt{(0.6 \cdot 0.4 / 33) + (0.65 \cdot 0.35 / 38)} = \sqrt{0.013259} = 0.115$

10.87 Step 1: The difference in proportions of nurses who experienced a change in position based on their participation in a program, $p_w - p_n$

Step 2: a. n's > 20, np's and nq's all > 5
b. z c. $1 - \alpha = 0.99$

Step 3: $n_w = 341$, $x_w = 87$, $p'_w = 87/341 = 0.255$,
$q'_w = 1 - 0.255 = 0.745$
$n_n = 40$, $x_n = 9$, $p'_n = 9/40 = 0.225$,
$q'_n = 1 - 0.225 = 0.775$
$p'_w - p'_n = 0.255 - 0.225 = 0.03$

Step 4: a. $\alpha/2 = 0.01/2 = 0.005$; $z(0.005) = 2.58$
b. $E = z(\alpha/2) \cdot \sqrt{(p'_w \cdot q'_w / n_w) + (p'_n \cdot q'_n / n_n)}$
$= 2.58 \cdot \sqrt{(0.255)(0.745)/341 + (0.225)(0.775)/40}$
$= (2.58)(0.07) = 0.18$
c. $(p'_w - p'_n) \pm E = 0.03 \pm 0.18$

Step 5: <u>-0.15 to 0.21</u>, the 0.99 interval for $p_w - p_n$

10.89 Step 1: The difference in proportions of defectives parts produced by two machines, $p_1 - p_2$

Step 2: a. n's > 20, np's and nq's all > 5
b. z c. $1 - \alpha = 0.90$

Step 3: $n_1 = 150$, $x_1 = 12$, $p'_1 = 12/150 = 0.08$,
$q'_1 = 1 - 0.08 = 0.92$
$n_2 = 150$, $x_2 = 6$, $p'_2 = 6/150 = 0.04$,
$q'_2 = 1 - 0.04 = 0.96$
$p'_1 - p'_2 = 0.08 - 0.04 = 0.04$

Step 4: a. $\alpha/2 = 0.10/2 = 0.05$; $z(0.05) = 1.65$
b. $E = z(\alpha/2) \cdot \sqrt{(p'_1 \cdot q'_1 / n_1) + (p'_2 \cdot q'_2 / n_2)}$
$= 1.65 \cdot \sqrt{(0.08 \cdot 0.92/150) + (0.04 \cdot 0.96/150)}$
$= (1.65)(0.027) = 0.04$
c. $(p'_1 - p'_2) \pm E = 0.04 \pm 0.04$

Step 5: <u>0.000 to 0.080</u>, the 0.90 interval for $p_1 - p_2$

10.91 Step 1: The difference in proportions of women and men that wash more than 10 times a day, $p_w - p_m$

Step 2: a. n's > 20, np's and nq's all > 5
b. z c. $1 - \alpha = 0.95$

Step 3: $n_w = 442$, $x_w = 274$, $p'_w = 0.62$, $q'_w = 1 - 0.62 = 0.38$
$n_m = 446$, $x_m = 165$, $p'_m = 0.37$, $q'_m = 1 - 0.37 = 0.63$
$p'_w - p'_m = 0.62 - 0.37 = 0.25$

Step 4: a. $\alpha/2 = 0.05/2 = 0.025$; $z(0.025) = 1.96$

b. $E = z(\alpha/2) \cdot \sqrt{(p'_w \cdot q'_w/n_w) + (p'_m \cdot q'_m/n_m)}$
   $= 1.96 \cdot \sqrt{(0.62 \cdot 0.38/442) + (0.37 \cdot 0.63/446)}$
   $= (1.96)(0.0325) = 0.0637$

c. $(p'_w - p'_m) \pm E = 0.25 \pm 0.064$

Step 5: <u>0.186 to 0.314</u>, the 0.95 interval for $p_w - p_m$

WRITING HYPOTHESES FOR THE DIFFERENCE BETWEEN TWO PROPORTIONS

a) null hypothesis:

$H_0: p_1 = p_2$ or $p_1 - p_2 = 0$

b) possible alternative hypotheses:

$H_a: p_1 > p_2$ or $H_a: p_1 - p_2 > 0$

$H_a: p_1 < p_2$ or $H_a: p_1 - p_2 < 0$

$H_a: p_1 \neq p_2$ or $H_a: p_1 - p_2 \neq 0$

10.93 a. $H_0: p_m - p_w = 0$ vs. $H_a: p_m - p_w \neq 0$

b. $H_0: p_b - p_g = 0 \; (\leq)$ vs. $H_a: p_b - p_g > 0$

c. $H_0: p_c - p_{nc} = 0 \; (\leq)$ vs. $H_a: p_c - p_{nc} > 0$

Hypothesis Test for the Difference Between Two Proportions, Independent Samples (Large Samples)

Review parts to a hypothesis test as outlined in: ES11-pp371&388, IRM-pp343&357, if needed. Changes will occur in:

1. **the hypotheses**: (see box before exercise 10.93)

2. **the calculated test statistic**

$$z = \frac{(p'_1 - p'_2) - (p_1 - p_2)}{\sqrt{p'_p q'_p \left(\frac{1}{n_1} + \frac{1}{n_2}\right)}}, \text{ where } p'_1 = \frac{x_1}{n_1}, \; p'_2 = \frac{x_2}{n_2},$$

$$p'_p = \frac{x_1 + x_2}{n_1 + n_2} \text{ and } q'_p = 1 - p'_p$$

3. If H_o is rejected, a significant difference in proportions is indicated.

   If H_o is not rejected, no significant difference is indicated.

NOTE: The sampling distribution of $p_1' - p_2'$ is approximately normally distributed with a mean $(p_1 - p_2)$ and a standard error of $\sqrt{p_1 q_1/n_1 + p_2 q_2/n_2}$, if n_1 and n_2 are sufficiently large. Since H_o is assumed to be true, $p_1 - p_2$ is considered equal to 0. Since p_1 and p_2 are also unknown, the best estimate for $p(p = p_1 = p_2)$ is a pooled estimate p_p'.

Computer and/or calculator commands to perform a hypothesis test for the difference between two proportions can be found in ES11-pp517-518.

10.95 $p_p' = (x_E + x_R)/(n_E + n_R) = (15 + 25)/(250 + 275) = 40/525 = \underline{0.076}$

$q_p' = 1 - p_p' = 1 - 0.076 = \underline{0.924}$

10.97 Rewrite the alternative hypothesis for easier understanding: $H_a: p_R - p_E > 0$

$p_R' = 25/275 = 0.091$, $p_E' = 15/250 = 0.06$

$z^* = (p_R' - p_E')/\sqrt{(p_p')(q_p')[(1/n_R)+(1/n_E)]}$
$= (0.091 - 0.06)/\sqrt{(0.076)(0.924)[(1/275) + (1/250)]}$
$= 0.031/0.0232 = 1.34$

$P = P(z > 1.34) = (1.0000 - 0.9099) = \underline{0.0901}$

Use Table 4 (Appendix B, ES11-p718), (Normal Distribution) to determine the test criteria. If the amount of α is not listed in Table 4, use Table 3 (Appendix B, ES11-pp716-717). A review of determining the test criteria in: ES11-pp393-394, IRM-pp359-360, may be helpful.

10.99 a. $\alpha = 0.05$ b. $\alpha = 0.05$

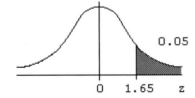
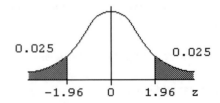

c. $\alpha = 0.04$ d. $\alpha = 0.01$

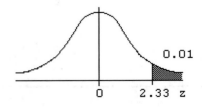

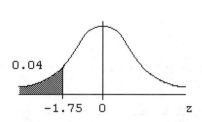
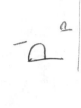

Hint for writing the hypotheses for exercise 10.100

The word "compare" implies 2 populations, namely the home PC owners and the work PC owners. The results are given in the form of number of successes and sample size, therefore a difference between 2 proportions is suggested. Look at the second to last sentence of the exercise; "Did the home PC owners experience more problems...?" The word "more" indicates a greater than. Therefore, the alternative is greater than (>). The negation becomes "not greater than" and the null hypothesis would be written with an inequality sign ($\leq$).

Hint for writing the hypotheses for exercise 10.101

Look at the last sentence of the exercise; "Is there sufficient evidence to show a difference in the effectiveness of the 2 image campaigns...?" The words "2 image campaigns" imply 2 populations, namely the citizens exposed to a conservative campaign and citizens exposed to a moderate campaign. The results are given in the form of proportions, therefore a difference between 2 proportions is suggested. The word "difference" indicates a not equal to. Therefore, the alternative is not equal to ($\neq$). The negation becomes "equal to" and the null hypothesis would be written with an equality sign (=).

10.101 Step 1: a. The difference in the proportions for the effectiveness of two campaign images, $p_m - p_c$

b. H_o: $p_m - p_c = 0$

H_a: $p_m - p_c \neq 0$

Step 2: a. n's > 20, np's and nq's all > 5

b. z c. $\alpha = 0.05$

Step 3: a. $n_m = 100$, $p'_m = 0.50$, $n_c = 100$, $p'_c = 0.40$

$p'_p = (x_m + x_c)/(n_m + n_c) = (50+40)/(100+100) = 0.45$

$q'_p = 1 - p'_p = 1.000 - 0.45 = 0.55$

b. $z = [(p'_m - p'_c)-(p_m - p_c)]/\sqrt{(p'_p)(q'_p)[(1/n_m) + (1/n_c)]}$

$z^* = (0.50 - 0.40)/\sqrt{(0.45)(0.55)[(1/100) + (1/100)]} = 0.10/0.0704 = 1.42$

Step 4: -- using p-value approach --------------------

a. $\mathbf{P} = 2P(z > 1.42)$;

Using Table 3, Appendix B, ES11-pp716-717: $\mathbf{P} = 2(1.0000 - 0.9222)$
$= 2(0.0778) = 0.1556$

Using Table 5, Appendix B, ES11-pp718: $2(0.0735 < ½\mathbf{P} < 0.0808)$;
$0.1470 < \mathbf{P} < 0.1616$

b. $\mathbf{P} > \alpha$

-- using classical approach -----------------

a. Critical region: $z \leq -1.96$ and $z \geq 1.96$

b. z* falls in the noncritical region

Step 5: a. Fail to reject H_0

b. There is not sufficient evidence to show a difference, at the 0.05 level.

Hint for writing the hypotheses for exercise 10.102

Look at the last sentence of the exercise; "Based on this survey, ... is there a <u>difference</u> in the <u>proportion of men and women</u> ...?" The words "proportion of men and women" suggest the parameter proportion for 2 populations, namely men and women, that is, p_M and p_W. The word "difference" indicates a <u>not equal to</u>. Therefore, the alternative is <u>not equal to</u> ($\neq$). The negation becomes "equal to" and the null hypothesis would be written with an equality sign (=).

10.103 Step 1: a. The difference in the proportion of helmet use when there is a child in the home, $p_c - p_{nc}$

b. $H_0: p_c - p_{nc} = 0$
$H_a: p_c - p_{nc} > 0$

Step 2: a. n's > 20, np's and nq's all > 5
b. z   c. $\alpha = 0.01$

Step 3: a. $n_c = 340$, $x_c = 296$, $p'_c = 0.87$

$n_{nc} = 340$, $x_{nc} = 252$, $p'_{nc} = 0.74$

$p'_p = (x_c + x_{nc})/(n_c + n_{nc}) = (296+252)/(340+340) = 0.81$

$q'_p = 1 - p'_p = 1.000 - 0.81 = 0.19$

b. $z = [(p'_c - p'_{nc}) - (p_c - p_{nc})] / \sqrt{(p'_p)(q'_p)[(1/n_c)+(1/n_{nc})]}$

$z^* = (0.87 - 0.74) / \sqrt{(0.81)(0.19)[(1/340)+(1/340)]} = 4.32$

Step 4: -- using p-value approach --------------------
a. $P = P(z > 4.32)$;
Using Table 3, Appendix B, ES11-pp716-717: $P < 1.0000 - 0.99998 = 0.00002$
Using Table 5, Appendix B, ES11-pp718: $P = 0+$
Using Minitab output in exercise: $P = 0.000$

b. $P < \alpha$

-- using classical approach ------------------
a. critical region: $z \geq 2.33$
b. z* falls in the critical region

Step 5: a. Reject H_0

b. There is sufficient evidence to show that the proportion of helmet use is significantly greater when there is a child in the home, at the 0.01 level.

10.105 a. Answers may vary but it there does seem to be a difference in the proportions of men and women who say 'Executives are paid too much'. 60% is a well over a majority and the 50% is not quite a majority.

b. Step 1: a. The difference in the proportion of men and women who say that 'Executives are paid too much', $p_w - p_m$

b. H_o: $p_w - p_m = 0$
H_a: $p_w - p_m \neq 0$

Step 2: a. np's and nq's all > 5
b. z   c. $\alpha = 0.05$

Step 3: a. $n_w = 20$, $p'_w = 0.60$; $x = (20)(0.60) = 12$
$n_m = 20$, $p'_m = 0.50$; $x = (20)(0.50) = 10$
$p'_p = (x_w + x_m)/(n_w + n_m) = (12+10)/(20+20) = 0.55$
$q'_p = 1 - p'_p = 1.000 - 0.55 = 0.45$

b. $z = [(p'_w - p'_m) - (p_w - p_m)] / \sqrt{(p'_p)(q'_p)[(1/n_w) + (1/n_m)]}$

$z^* = (0.60 - 0.50) / \sqrt{(0.55)(0.45)[(1/20) + (1/20)]}$
$= 0.10/0.1573 = 0.64$

Step 4: -- using p-value approach --------------------
a. P = 2P($z^* > 0.64$);
Using Table 3, Appendix B, ES11-pp716-717: P = 2(1.0000 - 0.7389)
= 2(0.2611) = 0.5222
Using Table 5, Appendix B, ES11-pp718: (0.2578 < ½P < 0.2743);
0.5156 < P < 0.5486

b. P > α
-- using classical approach ------------------
a. Critical region: $z \leq -1.96$ and $z \geq 1.96$
b. z^* falls in the noncritical region

Step 5: a. Fail to reject H_o
b. There is not sufficient evidence to show a difference, at the 0.05 level of significance.

c. Step 1: a. The difference in the proportion of men and women who say that 'Executives are paid too much', $p_w - p_m$

b. H_o: $p_w - p_m = 0$
H_a: $p_w - p_m \neq 0$

Step 2: a. n's > 20, np's and nq's all > 5
b. z   c. $\alpha = 0.05$

Step 3: a. $n_w = 500$, $p'_w = 0.60$; $x = (500)(0.60) = 300$
$n_m = 500$, $p'_m = 0.50$; $x = (500)(0.50) = 250$
$p'_p = (x_w + x_m)/(n_w + n_m) = (300+250)/(500+500) = 0.55$
$q'_p = 1 - p'_p = 1.000 - 0.55 = 0.45$

b. $z = [(p'_w - p'_m) - (p_w - p_m)] / \sqrt{(p'_p)(q'_p)[(1/n_w) + (1/n_m)]}$

$$z^* = (0.60 - 0.50)/\sqrt{(0.55)(0.45)[(1/500)+(1/500)]}$$
$$= 0.10/0.03146 = 3.18$$

Step 4: -- using p-value approach --------------------
a. **P** = 2P(z* > 3.18);
Using Table 3, Appendix B, ES11-pp716-717: **P** = 2(1.0000 - 0.9993)
= 2(0.0007) = 0.0014
Using Table 5, Appendix B, ES11-pp718: (0.0007 < ½**P** < 0.0008);
0.0014 < **P** < 0.0016

b. **P** < α
-- using classical approach ------------------
a. Critical region: z ≤ -1.96 and z ≥ 1.96
b. z* falls in the critical region

Step 5: a. Reject H_o
b. There is sufficient evidence to show a difference, at the 0.05 level of significance.

d. Even with a 10% difference, it takes a reasonably large sample size to show significance.

10.107 a.

Step 1: a. The difference in the proportion of men and women who think it is OK for women to make marriage proposals to men, $p_M - p_W$

b. $H_o: p_M - p_W = 0$
$H_a: p_M - p_W \neq 0$

Step 2: a. n's > 20, np's and nq's all > 5
b. z   c. α = 0.05

Step 3: a. $n_M = 250$, $p'_M = 0.63$ $n_W = 250$, $p'_W = 0.55$
$p'_p = (x_M + x_W)/(n_M + n_W) = (158+138)/(250+250) = 0.592$
$q'_p = 1 - p'_p = 1.000 - 0.592 = 0.408$

b. $z = [(p'_M - p'_W) - (p_M - p_W)]/\sqrt{(p'_p)(q'_p)[(1/n_M)+(1/n_W)]}$
$$z^* = (0.63 - 0.55)/\sqrt{(0.592)(0.408)[(1/250)+(1/250)]}$$
$$= 0.08/0.0439578 = 1.82$$

Step 4: -- using p-value approach --------------------
a. **P** = 2P(z* > 1.82);
Using Table 3, Appendix B, ES11-pp716-717: **P** = 2(1.0000 - 0.9656)
= 2(0.0344) = 0.0688
Using Table 5, Appendix B, ES11-pp718: (0.0322 < ½**P** < 0.0359);
0.0644 < **P** < 0.0718

b. **P** > α
-- using classical approach ------------------
a. Critical region: z ≤ -1.96 and z ≥ 1.96
b. z* falls in the noncritical region

Step 5: a. Fail to reject H_o
b. There is not sufficient evidence to show a difference, at the 0.05 level.

b. Steps 1 & 2 are same as in part a.

Step 3:   a. $n_M = 500$, $p'_M = 0.63$ $n_W = 500$, $p'_W = 0.55$

$p'_p = (x_M + x_W)/(n_M + n_W) = (315+275)/(500+500) = 0.59$

$q'_p = 1 - p'_p = 1.000 - 0.59 = 0.41$

b. $z = [(p'_M - p'_W) - (p_M - p_W)]/\sqrt{(p'_p)(q'_p)[(1/n_M) + (1/n_W)]}$

$z^* = (0.63 - 0.55)/\sqrt{(0.59)(0.41)[(1/500) + (1/500)]}$

$= 0.08/0.0311 = 2.57$

Step 4:   -- using p-value approach --------------------
a. $P = 2P(z^* > 2.57)$;

Using Table 3, Appendix B, ES11-pp716-717: $P = 2(1.0000 - 0.9949)$
$= 2(0.0051) = 0.0102$

Using Table 5, Appendix B, ES11-pp718: $(0.0047 < \frac{1}{2}P < 0.0054)$;
$0.0094 < P < 0.0108$

b. $P < \alpha$

-- using classical approach ------------------
a. Critical region: $z \leq -1.96$ and $z \geq 1.96$
b. z^* falls in the critical region

Step 5:   a. Reject H_o
b. There is sufficient evidence to show a difference, at the 0.05 level.

c. $P = 0.05$ for a 2-tailed test – 0.025 on each tail
$z(0.025) = 1.96$

$1.96 = (0.63 - 0.55)/\sqrt{(0.59)(0.41)[(1/n) + (1/n)]}$

$1.96 = (0.08)/\sqrt{(0.4838)/n}$

$1.96(\sqrt{(0.4838)/n}) = 0.08$

$\sqrt{(0.4838)/n} = 0.0408$

$0.4838/n = 0.00166$
$0.4838 = 0.00166n$
$0.4838/0.00166 = n$
$290.4 = n = \underline{291}$

10.109   Step 1:   a. The difference in the proportion of reject product between two methods, $p_2 - p_1$
b. $H_o: p_2 - p_1 = 0$
$H_a: p_2 - p_1 \neq 0$

Step 2:   a. n's > 20, np's and nq's all > 5
b. z   c. $\alpha = 0.05$

Step 3:   a. $n_2 = 992$, $p'_2 = 26/992 = 0.0262$, $n_1 = 320$, $p'_1 = 4/320 = 0.0125$

$p'_p = (x_2 + x_1)/(n_2 + n_1) = (26+4)/(992+320) = 0.0229$

$q'_p = 1 - p'_p = 1.000 - 0.0229 = 0.9771$

b. $z = [(p'_2 - p'_1) - (p_2 - p_1)]/\sqrt{(p'_p)(q'_p)[(1/n_2) + (1/n_1)]}$

$z^* = (0.0262 - 0.0125)/\sqrt{(0.0229)(0.9771)[(1/992) + (1/320)]}$

$= 0.0137/0.0096 = 1.43$

Step 4: -- using p-value approach --------------------
a. $P = 2P(z > 1.43)$;
Using Table 3, Appendix B, ES11-pp716-717: $P = 2(1.0000 - 0.9236)$
$= 2(0.0764) = 0.1528$
Using Table 5, Appendix B, ES11-pp718: $(0.0735 < ½P < 0.0808)$;
$0.1470 < P < 0.1616$
b. $P > \alpha$
-- using classical approach ------------------
a. Critical region: $z \leq -1.96$ and $z \geq 1.96$
b. z^* falls in the noncritical region

Step 5: a. Fail to reject H_o
b. There is not sufficient evidence to show a difference, at the 0.05 level of significance.

```
Variable  N    Mean    Median   TrMean   StDev   SE Mean
C5       200  0.09270  0.10000  0.09189  0.06059  0.00428

Variable    Minimum   Maximum    Q1       Q3
C5          -0.04000  0.26000   0.05000  0.13000
```

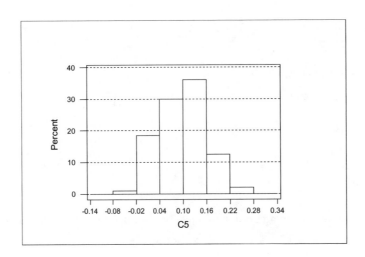

d. Yes

SECTION 10.5 EXERCISES

WRITING HYPOTHESES FOR THE RATIO BETWEEN TWO STANDARD DEVIATIONS OR VARIANCES
null hypothesis: $H_o: \sigma_1 = \sigma_2$ or $H_o: \sigma_1^2 = \sigma_2^2$ or $H_o: \sigma_1^2 / \sigma_2^2 = 1$

> possible alternative hypotheses:
>
> $H_a: \sigma_1 > \sigma_2$ or $H_a: \sigma_1^2 > \sigma_2^2$ or $H_a: \sigma_1^2/\sigma_2^2 > 1$
>
> $H_a: \sigma_1 \neq \sigma_2$ or $H_a: \sigma_1^2 \neq \sigma_2^2$ or $H_a: \sigma_1^2/\sigma_2^2 \neq 1$
>
> $H_a: \sigma_2 > \sigma_1$ or $H_a: \sigma_2^2 > \sigma_1^2$ or $H_a: \sigma_2^2/\sigma_1^2 > 1$ **
>
> **Note change for "less than", reverse order.

10.111 a. $H_0: \sigma_A^2 = \sigma_B^2$ vs. $H_a: \sigma_A^2 \neq \sigma_B^2$

b. $H_0: \sigma_I/\sigma_{II} = 1$ vs. $H_a: \sigma_I/\sigma_{II} > 1$

c. $H_0: \sigma_A^2 / \sigma_B^2 = 1$ vs. $H_a: \sigma_A^2 / \sigma_B^2 \neq 1$

d. $H_0: \sigma_C^2 / \sigma_D^2 = 1$ vs. $H_a: \sigma_C^2 / \sigma_D^2 < 1$
or equivalently,
$H_0: \sigma_D^2 / \sigma_C^2 = 1$ vs. $H_a: \sigma_D^2 / \sigma_C^2 > 1$

10.113 Divide both sides of the original inequality by σ_p^2

F-DISTRIBUTION

Key facts about the F-distribution:
1. The total area under the F-distribution is 1.
2. It is zero or positively valued.
3. The shape is skewed right (much like χ^2).
4. A different curve exists for each pair of sample sizes.
5. Critical values are determined based on α and degrees of freedom in the numerator (df_n) and degrees of freedom in the denominator (df_d).
6. Degrees of freedom = df = n - 1.

Notation: $F(df_n, df_d, \alpha)$ =

F(df for numerator, df for denominator, area to the right)
↑ ↑ ↑ ↑
Table 9   column id #   row id #   Table 9a,b,c

ex: F(10,12,0.05) means df_n = 10(column), df_d = 12(row) and α = 0.05 (Table 9A)

Using Table 9A, F(10,12,0.05) = 2.75   (df = n-1 for each sample)

Explore the F- distribution using the Chapter 10 Skillbuilder Applets:
  'Properties of F-distribution' and 'F-distribution Probabilities'

10.115 a. F(9,11,0.025)   b. F(24,19,0.01)
c. F(8,15,0.01)   d. F(15,9,0.05)

10.117 a. 2.51 b. 2.20 c. 2.91 d. 4.10
e. 2.67 f. 3.77 g. 1.79 h. 2.99

Estimate the p-value using Tables 9A, B or C, (Appendix B, ES11-pp-722-727).
$P = P(F > F^* | df_n, df_d)$

Locate critical values for the given df_n and df_d on each of the Tables. Compare F^* to each and give an interval estimate of P using one or two of the following values: 0.05, 0.025, 0.01. The p-value can be calculated using the cumulative probability commands in ES11-p524.

Draw a diagram of a F-curve and label the regions with the given information.
For a <u>one-sided test</u>, be sure that the alternative hypothesis has a "greater than", otherwise adjust the order of the numerator and denominator (i.e. $\sigma_1 < \sigma_2 \Rightarrow \sigma_2 > \sigma_1$).
For a <u>two-sided test</u>, divide α by 2 and use the largest sample variance in the numerator for the calculated test statistic. Use Table 9A, B or C, based on $\alpha/2$.

10.119 $F(6,9,0.05) = \underline{3.37}$

10.121 $F^* = s_1^2 / s_2^2 = (3.2)^2/(2.6)^2 = 1.514 = \underline{1.51}$

10.123 $F^* = s_1^2 / s_2^2 = 14.44/29.16 = \underline{0.495}$
The smaller variance is in the numerator.

Hypothesis Test for the Ratio Between Two Standard Deviations
(or Variances), Independent Samples

Review the parts to a hypothesis test as outlined in: ES11-pp371&388, IRM-pp343&357, if needed.
Slight changes will occur in:
1. **the hypotheses**: (see box before exercise 10.111)
2. **the calculated test statistic**
$F = s_1^2 / s_2^2$, $df_1 = df_n = n_1 - 1$, $df_2 = df_d = n_2 - 1$
3. If H_o is rejected, a significant difference between the standard deviations (variances) is indicated.
If H_o is not rejected, no significant difference between the standard deviations (variances) is indicated.

Use subscripts on the sample (or population) variables that identify the source.
Estimate the p-value using Tables 9A, B or C, (Appendix B, ES11-pp-722-727).
$P = P(F > F^* | df_n, df_d)$

> Locate critical values for the given df_n and df_d on each of the Tables. Compare F* to each and give an interval estimate of P using one or two of the following values: 0.05, 0.025, 0.01.
>
> The p-value can be calculated using the cumulative probability commands in ES11-p524.

> Computer and/or calculator commands to perform a hypothesis test between two standard deviations or two variances can be found in ES11-pp528-529.

10.125 Step 1: a. The ratio of the standard deviations of 3-inch line measurements from English and Chemistry classes
  b. $H_0: \sigma_e = \sigma_c$
  $H_a: \sigma_e \neq \sigma_c$
Step 2: a. normality assumed, independence exists
  b. F   c. $\alpha = 0.05$
Step 3: a. $n_e = 25$, $s_e = 0.617$, $n_c = 12$, $s_c = 0.522$
  b. $F^* = s_e^2 / s_c^2 = (0.617)^2/(0.522)^2 = 1.40$
Step 4: -- using p-value approach --------------------
  a. $P = 2P(F > 1.40 | df = 24, 24)$;
      Using Tables 9, Appendix B, ES11-pp-722-727: **P** > 2(0.05); **P** > 0.10
      Using computer: 0.429
  b. **P** > α
  -- using classical approach -----------------
  a. F(24, 24, 0.025) = 2.27

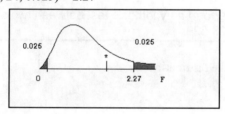

  b. F* falls in the noncritical region, see Step 4a

Step 5: a. Fail to reject H_0
  b. At the 0.05 level of significance, there is not sufficient evidence to conclude that a difference in standard deviations of 3-inch lines exist for the two classes.

10.127 Step 1: a. The ratio of the variances for two ovens
  b. $H_0: \sigma_k^2 = \sigma_m^2$
  $H_a: \sigma_k^2 \neq \sigma_m^2$
Step 2: a. normality indicated, independence exists
  b. F   c. $\alpha = 0.02$
Step 3: a. $n_m = 16$, $s_m^2 = 2.4$, $n_k = 12$, $s_k^2 = 3.2$
  b. $F^* = s_k^2/s_m^2 = 3.2/2.4 = 1.33$

Step 4: -- using p-value approach --------------------
a. $P = 2P(F > 1.33 | df = 11,15)$;
Using Tables 9, Appendix B, ES11-pp-722-727: $P > 2(0.05)$; $P > 0.10$
Using computer: 0.593
b. $P > \alpha$

-- using classical approach ------------------
a. $F(11, 15, 0.01) = 3.73$

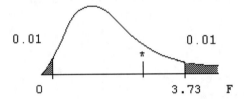

b. F* falls in the noncritical region, see Step 4a

Step 5: a. Fail to reject H_o
b. At the 0.02 level of significance, there is not sufficient evidence to conclude that a difference in variances exist for the two ovens.

The p-value can be calculated using the cumulative probability commands in ES11-p524.

Computer and/or calculator commands to perform a hypothesis test between two standard deviations or two variances can be found in ES11-pp528-529.

10.129 Multiply it by 2 to cover both sides of the distribution.

10.131 $F^* = s_a^2 / s_b^2 = (4.43)^2/(3.50)^2 = \underline{1.60}$

10.133 a.
Step 1: a. The ratio of the variances for endorsements from NBA and MLB players
b. $H_o : \sigma_{NBA}^2 = \sigma_{MLB}^2$
$H_a : \sigma_{NBA}^2 \neq \sigma_{MLB}^2$
Step 2: a. normality indicated, independence exists
b. F   c. $\alpha = 0.05$
Step 3: a. $n_{NBA} = 12$, $s_{NBA} = 7.048$, $n_{MLB} = 8$, $s_{MLB} = 2.948$
b. $F^* = s_{NBA}^2 / s_{MLB}^2 = (7.048)^2/(2.948)^2 = 5.72$
Step 4: -- using p-value approach --------------------
a. $P = 2P(F > 5.72 | df = 11,7)$;
Using Tables 9, Appendix B, ES11-pp-722-727: $P > 2(0.01 < P < 0.025)$;
$0.02 < P < 0.05$
Using a computer: $P = 0.029$

      b. **P** < α
        -- using classical approach ------------------
      a. F(11, 7, 0.025) = 4.76

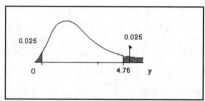

      b. F* falls in the critical region, see Step 4a

Step 5:   a. Reject H_o

        b. At the 0.05 level of significance, there is sufficient evidence to conclude that there is a difference in variability of endorsement amounts between NBA and MLB players.

b.

Step 1:   a. The mean endorsement amounts for NBA and MLB players, $\mu_{NBA} - \mu_{MLB}$

        b. H_o: $\mu_{NBA} - \mu_{MLB} = 0$
           H_a: $\mu_{NBA} - \mu_{MLB} > 0$

Step 2:   a. normality assumed
        b. t   c. α = 0.05

Step 3:   a. $n_{NBA} = 8$, $\bar{x}_{NBA} = 10.83$, $s_{NBA} = 7.048$
           $n_{MLB} = 10$, $\bar{x}_{MLB} = 2.70$, $s_{MLB} = 2.948$

        b. $t^* = (10.83 - 2.70)/\sqrt{((7.048^2/12) + (2.948^2/8))}$
           = 8.13/2.286 = 3.56

Step 4: -- using p-value approach --------------------
      a. **P** = P(t > 3.56|df = 7);
           Using Table 6, Appendix B, ES11-p719: **P** < 0.005
           Using Table 7, Appendix B, ES11-p720: 0.004 < **P** < 0.005
           Using computer: P = 0.001
      b. **P** < α
     -- using classical approach ------------------
      a. critical region: t ≥ 1.89
      b. t* falls in the critical region

Step 5:   a. Reject H_o

        b. At the 0.05 level of significance, there is sufficient evidence to show that the mean endorsement amount for NBA players is significantly higher than the mean endorsement amount for MLB players.

10.135   a. $n_1 = 8$, $\bar{x}_1 = 0.01525$, $s_1 = 0.00547$ $n_2 = 25$, $\bar{x}_2 = 0.02856$, $s_2 = 0.00680$

     b.   Step 1:   a. The ratio of the variances for the two lots
                  b. H_o: $\sigma_1^2 = \sigma_2^2$
                    H_a: $\sigma_1^2 \neq \sigma_2^2$

Step 2: a. normality indicated, independence exists
b. F c. $\alpha = 0.05$

Step 3: a. $n_1 = 8$, $s_1 = 0.00547$; $n_2 = 25$, $s_2 = 0.00680$
b. $F^* = s_2^2/s_1^2 = (0.00680)^2/(0.00547)^2 = 1.545 = 1.55$

Step 4: -- using p-value approach --------------------
a. $P = 2P(F > 1.55 | df = 24,7)$;
Using Tables 9, Appendix B, ES11-pp-722-727: $P > 2(0.05)$; $P > 0.10$
Using a computer: $P = 0.575$
b. $P > \alpha$
-- using classical approach ------------------
a. $F(24, 7, 0.025) = 4.42$

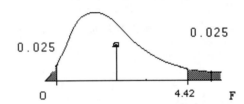

b. F* falls in the noncritical region, see Step 4a

Step 5: a. Fail to reject H_o
b. At the 0.05 level of significance, there is no sufficient evidence to conclude that a difference in variances exist for the two lots.

c.
Step 1: a. The mean levels of Critical Feature A in two lots, $\mu_2 - \mu_1$
b. $H_o: \mu_2 - \mu_1 = 0$
$H_a: \mu_2 - \mu_1 \neq 0$

Step 2: a. normality assumed
b. t c. $\alpha = 0.05$

Step 3: a. $n_1 = 8$, $\bar{x}_1 = 0.01525$, $s_1 = 0.00547$
$n_2 = 25$, $\bar{x}_2 = 0.02856$, $s_2 = 0.00680$
b. $t^* = (0.02856 - 0.01525)/\sqrt{((0.00680^2/25) + (0.00547^2/8))}$
$= 0.01331/0.00236 = 5.64$

Step 4: -- using p-value approach --------------------
a. $P = 2P(t > 5.64 | df = 7)$;
Using Table 6, Appendix B, ES11-p719: $P < 2(0.005)$; $P < 0.01$
Using Table 7, Appendix B, ES11-p720: $P < 2(0.003)$; $P < 0.006$
Using computer: $P = 0.0008$
b. $P < \alpha$
-- using classical approach ------------------
a. critical region: $t \leq -2.36$, $t \geq 2.36$
b. t* falls in the critical region

Step 5: a. Reject H_o
b. At the 0.05 level of significance, there is sufficient evidence to show that the meal levels of Critical Feature A is different between the two lots.

10.137 a. Men: $\bar{x}_m = \$68.14$, $s_m = \$47.95$ Women: $\bar{x}_w = \$85.90$, $s_w = \$63.50$

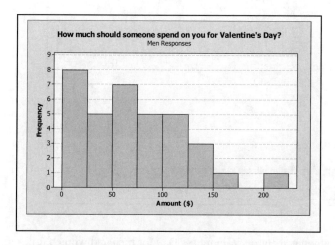

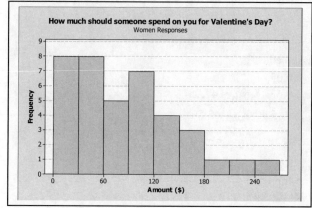

b. The two distributions appear to be very similar with regards to shape, center and spread.
c. Both samples appear to be skewed right distributions but the normality tests conducted below indicate that the samples might very well have come from a population with a normal distributed. The p-values are 0.216 and 0.052.

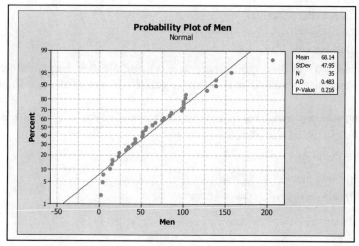

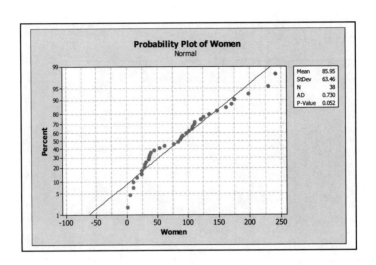

d. Step 1: a. The mean amount spent on Valentine's Day for men and women, $\mu_w - \mu_m$

b. $H_o: \mu_w - \mu_m = 0$

$H_a: \mu_w - \mu_m > 0$

Step 2: a. normality assumed

b. t   c. $\alpha = 0.05$

Step 3: a. $n_m = 35$, $\bar{x}_m = 68.14$, $s_m = 47.95$

$n_w = 38$, $\bar{x}_w = 85.90$, $s_w = 63.50$

b. $t^* = (85.90 - 68.14)/\sqrt{(63.50^2 / 38) + (47.95^2 / 35)}$

$= 17.76/13.107 = 1.36$

Step 4: -- using p-value approach --------------------

a. $P = P(t > 1.36 | df = 34)$;

Using Table 6, Appendix B, ES11-p719: $0.05 < P < 0.10$

Using Table 7, Appendix B, ES11-p720: $0.085 < P < 0.102$

Using computer: $P = 0.089$

b. $P > \alpha$

-- using classical approach ------------------

a. critical region: $t \geq 1.70$

b. t^* falls in the noncritical region

Step 5: a. Fail to reject H_o

b. At the 0.05 level of significance, there is insufficient evidence to show that the mean amount that should be spent on Valentine's Day is greater for women than men.

e. Step 1: a. The standard deviations for amounts that should be spent for women and men on Valentine's Day

b. $H_o: \sigma_w = \sigma_m$

$H_a: \sigma_w \neq \sigma_m$

Step 2: a. normality assumed, independence exists

b. F   c. $\alpha = 0.05$

Step 3: a. $n_m = 35$, $s_m = 47.95$; $n_w = 38$, $s_w = 63.50$

b. $F^* = s_w^2 / s_m^2 = (63.50)^2/(47.95)^2 = 1.75$

Step 4: -- using p-value approach --------------------
a. $P = 2P(F > 1.75|df = 37,34)$;
Using Tables 9, Appendix B, ES11-pp-722-727: $P \approx 2(0.05)$; $P \approx 0.10$
Using a computer: $P = 0.1026$
b. $P > \alpha$
-- using classical approach ------------------
a. critical region: $F(30, 30, 0.025) = 2.07$
b. F^* falls in the noncritical region

Step 5: a. Fail to reject H_o

b. At the 0.05 level of significance, there is insufficient evidence that the standard deviations of amounts that should be spent for Valentine's Day is different between men and women.

f. The results of the hypothesis tests in parts (d) & (e) show that even though there was a difference in the sample statistics, the difference was not a significant difference

10.139 Everybody will get different results, but they can all be expected to look very similar to the following.
a. Adjust the Minitab and Excel commands from Exercise 10.79a.
N(100,20)

N(120,20)

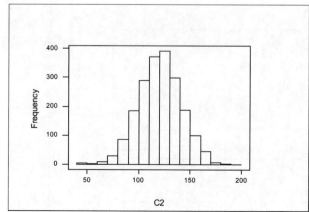

These two very large samples strongly suggest that we are sampling two normal populations.

b. Minitab commands:
Use commands from Exercise 10.79c substituting standard deviation for mean in the row statistics and (C20/C11)**2 for the expression in C21.

Excel commands:
Use commands from Exercise 10.79c substituting standard deviation for average in the paste function and (T1/K1)**2 for the expression in U1.

c. Use the Minitab and Excel commands from Exercise 10.79d using cutpoint classes from 0 to 5 in increments of 0.2 for the column of F's.

An empirical distribution of 100 F* values

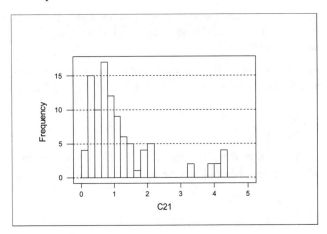

d. Minitab commands:
Use patterned data commands to list from 0 to 5 in increments of 0.2 into C22. Use cumulative probability commands on page 524 of the text with 7 degrees of freedom for both the numerator and denominator, C22 as the input column and C23 as the output column. The ScatterPlot command, with C23 and C22 and Connect, will draw the graph.

Excel commands:
Use patterned data commands to list from 0 to 5 in increments of 0.2 into column V. Use cumulative probability commands on page 524 of the text with 7 degrees of freedom for both the numerator and denominator, column V as the F values and column W as the ouput range. Use Edit formula (=) to calculate column X = 1 – col. W. Use Chart Wizard for the graph of the data in columns X and V.

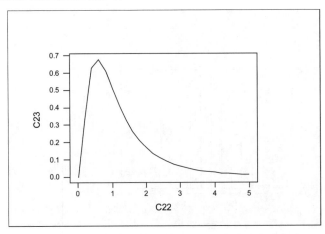

The observed distribution of F*s seems to be very similar to the theoretical distribution. Remember these samples were drawn from populations that were both normal and had the same standard deviations.

CHAPTER EXERCISES

10.141 Dependent samples; the same set of 18 diamonds are appraised by two appraisers.

10.143 Step 1: The mean difference in IQ scores for oldest and youngest members of a family (d = O - Y)
Step 2: a. normality indicated
b. t   c. $1-\alpha = 0.95$
Step 3: $n = 12$, $\bar{d} = 3.583$, $s_d = 19.58$
Step 4: a. $\alpha/2 = 0.05/2 = 0.025$; df = 11; t(11, 0.025) = 2.20
b. $E = t(df,\alpha/2) \cdot (s_d/\sqrt{n}) = (2.20)(19.58/\sqrt{12})$
$= (2.20)(5.65) = 12.435$
c. $\bar{d} \pm E = 3.583 \pm 12.435$
Step 5: <u>-8.85 to 16.02</u>, the 0.95 interval for μ_d

10.145 Step 1: The mean difference in readings for filter 1 and filter 2 (d = $F_1 - F_2$)
Step 2: a. normality indicated
b. t   c. $1-\alpha = 0.90$
Step 3: $n = 20$, $\bar{d} = 2.00$, $s_d = 2.714$
Step 4: a. $\alpha/2 = 0.10/2 = 0.05$; df = 19; t(19, 0.05) = 1.73
b. $E = t(df,\alpha/2) \cdot (s_d/\sqrt{n}) = (1.73)(2.714/\sqrt{20})$
$= (1.73)(0.607) = 1.05$
c. $\bar{d} \pm E = 2.00 \pm 1.05$
Step 5: <u>0.95 to 3.05</u>, the 0.90 interval for μ_d

10.147 Sample statistics: d = Before – After $n = 8$, $\bar{d} = 1.0$, $s_d = 2.39$
a. Point estimate = 1.0
b.
Step 1: a. The mean difference in the reduction of diastolic blood pressure
b. $H_o: \mu_d = 0$ (no difference)
$H_a: \mu_d > 0$ (salt-free diet reduced blood pressure)
Step 2: a. normality indicated
b. t   c. $\alpha = 0.01$
Step 3: a. $n = 8$, $\bar{d} = 1.0$, $s_d = 2.39$
b. $t^* = (\bar{d} - \mu_d)/(s_d/\sqrt{n})$
$= (1.0 - 0.0)/(2.39/\sqrt{8}) = 1.18$
Step 4: -- using p-value approach --------------------
a. $\mathbf{P} = P(t > 1.18|df=7)$;
Using Table 6, Appendix B, ES11-p719: $0.10 < \mathbf{P} < 0.25$
Using Table 7, Appendix B, ES11-p720: $0.135 < \mathbf{P} < 0.154$
Using computer: P = 0.138

     b. $P > \alpha$
     -- using classical approach ------------------
     a. critical region: $t \geq 3.00$
     b. t* falls in the noncritical region

  Step 5: a. Fail to reject H_o

     b. At the 0.01 level of significance, there is insufficient evidence that the salt-free diet produced a significant mean reduction in diastolic blood pressure.

10.149 Step 1: The difference between the mean anxiety scores for males and females, μ_f-μ_m

   Step 2: a. normality assumed, CLT with $n_f = 50$ and $n_m = 50$.
     b. t   c. $1-\alpha = 0.95$
   Step 3: $n_f = 50$, $\bar{x}_f = 75.7$, $s_f = 13.6$
     $n_m = 50$, $\bar{x}_m = 70.5$, $s_m = 13.2$
     $\bar{x}_f - \bar{x}_m = 75.7 - 70.5 = 5.2$
   Step 4: a. $\alpha/2 = 0.05/2 = 0.0025$; df = 49; t(49, 0.025) = 2.02
     b. $E = t(df,\alpha/2) \cdot \sqrt{(s_f^2 / n_f) + (s_m^2 / n_m)}$
      $= (2.02) \cdot \sqrt{(13.6^2/50) + (13.2^2/50)}$
      $= (2.02)(2.68) = 5.41$
     c. $(\bar{x}_f - \bar{x}_m) \pm E = 5.2 \pm 5.41$
   Step 5: <u>-0.21 to 10.61</u>, the 0.95 interval for $\mu_f - \mu_m$

10.151 $\bar{x}_1 = 278.4/36 = 7.73$, $s_1^2 = 0.3086$; $\bar{x}_2 = 310.8/42 = 7.4$, $s_2^2 = 0.7888$

   Step 1: The difference between the mean designs for two rocket nozzles, μ_1-μ_2
   Step 2: a. normality assumed, CLT with $n_1 = 36$ and $n_2 = 42$.
     b. t c. $1-\alpha = 0.99$
   Step 3: sample information given in exercise
     $\bar{x}_1 - \bar{x}_2 = 7.73 - 7.4 = 0.33$
   Step 4: a. $\alpha/2 = 0.01/2 = 0.005$; df = 35; t(35, 0.005) = 2.72
     b. $E = t(df,\alpha/2) \cdot \sqrt{(s_1^2 / n_1) + (s_2^2 / n_2)}$
      $= (2.72) \cdot \sqrt{(0.3086/36) + (0.7888/42)}$
      $= (2.72)(0.165) = 0.45$
     c. $(\bar{x}_1 - \bar{x}_2) \pm E = 0.33 \pm 0.45$
   Step 5: <u>-0.12 to 0.78</u>, the 0.99 interval for $\mu_1 - \mu_2$

10.153 Step 1: The difference between means of two methods used in ice fusion, μ_A-μ_B
   Step 2: a. normality assumed
     b. t c. $1-\alpha = 0.95$

Step 3: $n_A = 13$, $\bar{x}_A = 80.021$, $s_A^2 = 0.0005744$, $s_A = 0.02397$
$n_B = 8$, $\bar{x}_B = 79.979$, $s_B^2 = 0.0009839$, $s_B = 0.03137$
$\bar{x}_A - \bar{x}_B = 80.021 - 79.979 = 0.042$

Step 4: a. $\alpha/2 = 0.05/2 = 0.025$; df = 7; t(7, 0.025) = 2.36
b. $E = t(df,\alpha/2) \cdot \sqrt{(s_A^2 / n_A) + (s_B^2 / n_B)}$
$= (2.36) \cdot \sqrt{(0.0005744/13) + (0.0009839/8)}$
$= (2.36)(0.0129) = 0.030$
c. $(\bar{x}_A - \bar{x}_B) \pm E = 0.042 \pm 0.030$

Step 5: <u>0.012 to 0.072</u>, the 0.95 confidence interval for $\mu_A - \mu_B$

10.155 Step 1: a. The difference between the mean 40-yard sprint time recorded by football players on artificial turf and grass, $\mu_2 - \mu_1$
b. H_o: $\mu_2 - \mu_1 = 0$ (no difference)
H_a: $\mu_2 - \mu_1 > 0$ (artif. turf yields a lower time)

Step 2: a. normality assumed, CLT with $n_c = 22$ and $n_h = 22$.
b. t   c. $\alpha = 0.05$

Step 3: a. sample information given in exercise
b. $t = [(\bar{x}_2 - \bar{x}_1) - (\mu_2 - \mu_1)] / \sqrt{(s_2^2/n_2) + (s_1^2/n_1)}$
$t^* = [(4.96 - 4.85) - 0] / [\sqrt{(0.42^2/22) + (0.31^2/22)}]$
$= 0.988$

Step 4: -- using p-value approach --------------------
a. $P = P(t > 0.988 | df = 21)$;
Using Table 6, Appendix B, ES11-p719: $0.10 < P < 0.25$
Using Table 7, Appendix B, ES11-p720: $0.164 < P < 0.189$
Using computer: $P = 0.1672$
b. $P > \alpha$

-- using classical approach ------------------
a. critical region: $t \geq 1.72$
b. t* falls in the noncritical region

Step 5: a. Fail to reject H_o
b. At the 0.05 level of significance, there is insufficient evidence to show that the artificial turf had faster mean sprint times than that of grass.

10.157 Step 1: a. The difference between the mean number of days missed at work by people receiving CSM treatment and those undergoing physical therapy, CSM(1) and Therapy(2), $\mu_2 - \mu_1$
b. H_o: $\mu_2 - \mu_1 = 0$ (no difference)
H_a: $\mu_2 - \mu_1 > 0$ (physical therapy less effective)

Step 2: a. normality assumed

b. t   c. α = 0.01

Step 3:   a. $n_1 = 32$, $\bar{x}_1 = 10.6$, $s_1 = 4.8$; $n_2 = 28$, $\bar{x}_2 = 12.5$, $s_2 = 6.3$

b. $t^\star = [(\bar{x}_2 - \bar{x}_1) - (\mu_2 - \mu_1)]/\sqrt{(s_2^2/n_2) + (s_1^2/n_1)}$

$= [(12.5 - 10.6) - 0]/[\sqrt{(6.3^2 / 28) + (4.8^2 / 32)}]$

$= 1.30$

Step 4:   -- using p-value approach --------------------
a. $P = P(t > 1.30 | df = 27)$;
Using Table 6, Appendix B, ES11-p719: $0.10 < P < 0.25$
Using Table 7, Appendix B, ES11-p720: $0.102 < P < 0.103$
Using computer: $P = 0.1023$

b. $P > \alpha$

-- using classical approach ------------------
a. critical region: $t \geq 2.47$
b. $t^\star$ is in the noncritical region

Step 5:   a. Fail to reject H_0

b. At the 0.01 level of significance, there is insufficient evidence to show that people treated by chiropractors using CSM miss fewer days of work due to acute lower back pain than people undergoing physical therapy.

10.159 a. $n_A = 15$, $\bar{x}_A = \underline{15.53}$, $s_A^2 = \underline{1.98}$, $s_A = \underline{1.41}$

b. $n_B = 15$, $\bar{x}_B = \underline{12.53}$, $s_B^2 = \underline{1.98}$, $s_B = \underline{1.41}$

c.   Step 1:   a. The difference between mean torques required to remove screws from two different materials, $\mu_A - \mu_B$

b. H_0: $\mu_A - \mu_B = 0$
H_a: $\mu_A - \mu_B \neq 0$

Step 2:   a. normality indicated
b. t   c. α = 0.01

Step 3:   a. sample information given above

b. $t^\star = [(\bar{x}_A - \bar{x}_B) - (\mu_A - \mu_B)]/\sqrt{(s_A^2 / n_A) + (s_B^2 / n_B)}$

$= [(15.53 - 12.53) - 0]/[\sqrt{(1.98/15) + (1.98/15)}]$

$= 5.84$

Step 4:   a. $P = 2P(t^\star > 5.84 | df = 14)$;
Using Table 6, Appendix B, ES11-p719: $P < 0.01$
Using Table 7, Appendix B, ES11-p720: $2(\tfrac{1}{2}P < 0.001)$, $P < 0.002$
Using computer: $P = 0.00004$

b. $P < \alpha$

-- using classical approach ------------------
a. critical region: $t \leq -2.98$, $t \geq 2.98$
b. $t^\star$ falls in the critical region

Step 5:   a. Reject H_0

b. At the 0.01 level of significance, the mean torques are not the same for Material A and Material B.

10.161 a.   M: $\bar{x} = 74.69$, s = 10.19,   F: $\bar{x} = 79.83$, s = 8.80

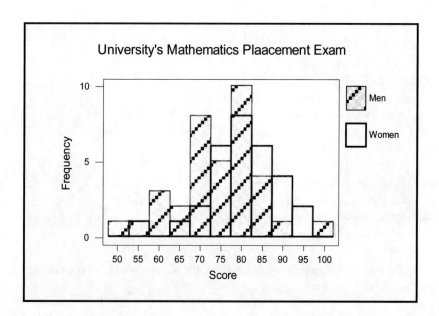

b.
Step 1: a. The mean mathematics placement score for all men
         b. H_o: $\mu = 77$
            H_a: $\mu \neq 77$
Step 2: a. normality indicated
         b. t   c. $\alpha = 0.05$
Step 3: a. n = 36, $\bar{x} = 74.69$, s = 10.19
         b. $t = (\bar{x} - \mu)/(s/\sqrt{n})$
            $t^* = (74.69 - 77.0)/(10.19/\sqrt{36}) = -1.36$
Step 4: -- using p-value approach --------------------
         a. **P** = 2P(t > 1.36|df = 35);
            Using Table 6, Appendix B, ES11-p719: 0.10 < **P** < 0.20
            Using Table 7, Appendix B, ES11-p720: 0.085 < ½**P** < 0.101;
                                                   0.170 < **P** < 0.202
            Using a computer: **P** = 0.183
         b. **P** > α
         -- using classical approach ------------------
         a. t ≤ -2.03, t ≥ 2.03
         b. t* falls in the noncritical region, see Step 4a

Step 5: a. Fail to reject H_o
         b. The sample does provide sufficient evidence that the mean score for men is 77, at the 0.05 level of significance.

Step 1: a. The mean mathematics placement score for all women
         b. H_o: $\mu = 77$
            H_a: $\mu \neq 77$
Step 2: a. normality indicated

    b. t   c. $\alpha = 0.05$

Step 3: a. n = 30, $\bar{x} = 79.83$, s = 8.80
    b. $t = (\bar{x} - \mu)/(s/\sqrt{n})$
     $t^* = 79.83 - 77.0/(8.80/\sqrt{30}) = 1.76$

Step 4: -- using p-value approach --------------------
    a. $\mathbf{P} = 2P(t > 1.76 | df = 29)$;
       Using Table 6, Appendix B, ES11-p719: $0.05 < \mathbf{P} < 0.10$
       Using Table 7, Appendix B, ES11-p720: $0.041 < \tfrac{1}{2}\mathbf{P} < 0.050$;
                       $0.082 < \mathbf{P} < 0.100$

       Using a computer: $\mathbf{P} = 0.089$
    b. $\mathbf{P} > \alpha$
    -- using classical approach ------------------
    a. $t \leq -2.05, t \geq 2.05$
    b. t^* falls in the noncritical region, see Step 4a

Step 5: a. Fail to reject H_o
    b. The sample does provide sufficient evidence that the mean score for women is 77, at the 0.05 level of significance.

c. They're both not significantly different than 77. It is easy to jump to the wrong conclusion.

d.

Step 1: a. The difference between mean scores for male and female students, $\mu_F - \mu_M$
    b. H_o: $\mu_F - \mu_M = 0$
     H_a: $\mu_F - \mu_M \neq 0$

Step 2: a. normality indicated
    b. t   c. $\alpha = 0.05$

Step 3: a. sample information given above
    b. $t^* = (79.83 - 74.7)/\sqrt{(8.80^2 / 30) + (10.2^2 / 36)}$
     = 2.19

Step 4: a. $\mathbf{P} = 2P(t^* > 2.19 | df = 29)$;
       Using Table 6, Appendix B, ES11-p719: $0.02 < \mathbf{P} < 0.05$
       Using Table 7, Appendix B, ES11-p720: $2(0.018 < \tfrac{1}{2}\mathbf{P} < 0.022)$,
                       $0.036 < \mathbf{P} < 0.044$

       Using computer: $\mathbf{P} = 0.0368$
    b. $\mathbf{P} < \alpha$
    -- using classical approach ------------------
    a. critical region: $t \leq -2.05, t \geq 2.05$
    b. t^* falls in the critical region

Step 5: a. Reject H_o
    b. At the 0.05 level of significance, the mean scores for men and women are not equal.

e. & f. No, a significant difference was found. The questions of (b) and (d) are asking different questions. In (b), individually two hypothesis tests are testing if the means are different than 77. In this case, the two sample means are on opposite sides of 77, but not significantly far from 77. Yet the two sample means are themselves far enough apart to be significantly different.

10.163 Step 1: The difference in proportions requiring service from two manufacturers, $p_1 - p_2$
Step 2: a. n's > 20, np's and nq's all > 5
b. z   c. $1 - \alpha = 0.95$
Step 3: sample information given in exercise
$p_1' - p_2' = 0.15 - 0.09 = 0.060$
Step 4: a. $\alpha/2 = 0.05/2 = 0.025$; $z(0.025) = 1.96$
b. $E = z(\alpha/2) \cdot \sqrt{(p_1' \cdot q_1' / n_1) + (p_2' \cdot q_2' / n_2)}$
$= 1.96 \cdot \sqrt{(0.15 \cdot 0.85/75) + (0.09 \cdot 0.91/75)}$
$= (1.96)(0.0528) = 0.104$
c. $(p_1' - p_2') \pm E = 0.060 \pm 0.104$
Step 5: <u>-0.044 to 0.164</u>, the 0.95 interval for $p_1 - p_2$

10.165 Find the p-value for each sample size for the situation:
H_o: $p_w - p_m = 0$
H_a: $p_w - p_m \neq 0$
a. n = 100
If samples are of same size, then
$p_p' = (p_w + p_m)/2 = (0.88 + 0.75)/2 = 0.815$
$q_p' = 1 - p_p' = 1.000 - 0.815 = 0.185$
$z = [(p_w' - p_m') - (p_w - p_m)] / \sqrt{(p_p')(q_p')[(1/n_w) + (1/n_m)]}$
$z^* = (0.88 - 0.75) / \sqrt{(0.815)(0.185)[(1/100) + (1/100)]}$
$= 0.13/0.0549 = 2.37$
P $= 2P(z > 2.37)$;
Using Table 3, Appendix B, ES11-pp716-717: **P** $= 2(1.0000 - 0.9911)$
$= 2(0.0089) = 0.0178$
Using Table 5, Appendix B, ES11-pp718: $2(0.0082 <$ **P** $< 0.0094)$;
$0.0164 <$ **P** < 0.0188
Using a computer: p-value = 0.0178
There is a significant difference for $\alpha \geq 0.02$.

b. n = 150
If samples are of same size, then
$p_p' = (p_w + p_m)/2 = (0.88 + 0.75)/2 = 0.815$
$q_p' = 1 - p_p' = 1.000 - 0.815 = 0.185$
$z = [(p_w' - p_m') - (p_w - p_m)] / \sqrt{(p_p')(q_p')[(1/n_w) + (1/n_m)]}$
$z^* = (0.88 - 0.75) / \sqrt{(0.815)(0.185)[(1/150) + (1/150)]}$
$= 0.13/0.0448 = 2.90$
P $= 2P(z > 2.90)$;
Using Table 3, Appendix B, ES11-pp716-717: **P** $= 2(1.0000 - 0.9981)$
$= 2(0.0019) = 0.0038$
Using Table 5, Appendix B, ES11-pp718: ½**P** $= 0.0019$; **P** $= 0.0038$

Using a computer: p-value = 0.0038
There is a significant difference for $\alpha \geq 0.01$.

c. n = 200
If samples are of same size, then
$p'_p = (p_w + p_m)/2 = (0.88+0.75)/2 = 0.815$
$q'_p = 1 - p'_p = 1.000 - 0.815 = 0.185$
$z = [(p'_w - p'_m)-(p_w - p_m)]/\sqrt{(p'_p)(q'_p)[(1/n_w) + (1/n_m)]}$
$z^* = (0.88 - 0.75)/\sqrt{(0.815)(0.185)[(1/200) + (1/200)]}$
$= 0.13/0.0388 = 3.35$

$P = 2P(z > 3.35)$;
Using Table 3, Appendix B, ES11-pp716-717: $P = 2(1.0000 - 0.9996)$
$= 2(0.0004) = 0.0008$
Using Table 5, Appendix B, ES11-pp718: $2(P = 0.0004)$; $P = 0.0008$
Using a computer: p-value = 0.0008
There is a significant difference for $\alpha \geq 0.001$.

d. As the sample size increases, the standard error for the difference of two proportions became smaller; this in turn meant that the 88% and 75% become more standard errors apart as reflected in the increase in z*. As n increased, the test became more sensitive meaning that the p-value kept decreasing.

10.167 Step 1: a. The difference in the proportion of accountants and lawyers who believe that the new burden-of-proof tax rules will cause an increase in taxpayer wins in court, $p_a - p_l$

b. H_o: $p_a - p_l = 0$
H_a: $p_a - p_l \neq 0$

Step 2: a. n's > 20, np's and nq's all > 5
b. z   c. $\alpha = 0.01$

Step 3: a. $n_a = 175$, $x_a = 101$, $p'_a = 101/175 = 0.5771$
$n_l = 165$, $x_l = 84$, $p'_l = 84/165 = 0.5091$
$p'_p = (x_a + x_l)/(n_a + n_l) = (101+84)/(175+165) = 0.5441$
$q'_p = 1 - p'_p = 1.000 - 0.5441 = 0.4559$

b. $z = [(p'_a - p'_l)-(p_a - p_l)]/\sqrt{(p'_p)(q'_p)[(1/n_a) + (1/n_l)]}$
$z^* = (0.5771 - 0.5091)/\sqrt{(0.5441)(0.4559)[(1/175) + (1/165)]}$
$= 0.068/0.054 = 1.26$

Step 4: -- using p-value approach --------------------
a. $P = 2P(z > 1.26)$;
Using Table 3, Appendix B, ES11-pp716-717: $P = 2(1.0000 - 0.8962)$
$= 2(0.1038) = 0.2076$
Using Table 5, Appendix B, ES11-pp718: $0.0968 < \frac{1}{2}P < 0.1056$;
$0.1936 < P < 0.2112$
Using a computer: p-value = 0.2077
b. $P > \alpha$

-- using classical approach ------------------
a. critical region: $z \leq -2.58$ and $z \geq 2.58$
b. z^* falls in the noncritical region

Step 5: a. Fail to reject H_o

b. There is not sufficient evidence to show that accountants and lawyers differ in their beliefs about the burden of proof, at the 0.01 level.

10.169 Step 1: a. The ratio of the variance of time needed by men to that needed by women to assemble a product

b. H_o: $\sigma_m^2 = \sigma_f^2$
H_a: $\sigma_m^2 > \sigma_f^2$

Step 2: a. normality indicated, independence exists
b. F   c. $\alpha = 0.05$

Step 3: a. $n_m = 15$, $s_m = 4.5$, $n_f = 15$, $s_f = 2.8$
b. $F^* = s_m^2 / s_f^2 = (4.5)^2/(2.8)^2 = 2.58$

Step 4: -- using p-value approach --------------------
a. $P = P(F > 2.58 | df = 14,14)$;
Using Tables 9, Appendix B, ES11-pp-722-727: $0.025 < P < 0.05$
Using computer: $p = 0.0435$
b. $P < \alpha$
-- using classical approach ------------------
a. critical region: $F \geq 2.53$
b. F^* falls in the critical region

Step 5: a. Reject H_o

b. At the 0.05 level of significance, there is sufficient evidence to conclude that male assembly times are more variable.

10.171 Step 1: a. The difference between the variances in the threads of lug nuts and studs

b. H_o: $\sigma_n^2 = \sigma_s^2$
H_a: $\sigma_n^2 \neq \sigma_s^2$

Step 2: a. normality indicated, independence exists
b. F   c. $\alpha = 0.05$

Step 3: a. $n_n = 60$, $s_n^2 = 0.00213$, $n_s = 40$, $s_s^2 = 0.00166$
b. $F^* = s_n^2 / s_s^2 = 0.00213/0.00166 = 1.28$

Step 4: -- using p-value approach --------------------
a. $P = P(F > 1.28 | df = 59,39)$;
Using Tables 9, Appendix B, ES11-pp-722-727: $P > 0.10$
Using computer: $P = 0.4160$
b. $P > \alpha$
-- using classical approach ------------------
a. critical region: $F \geq 2.01$
b. F^* falls in the noncritical region

Step 5: a. Fail to reject H_o

b. At the 0.05 level of significance, there is no difference in the variances of lug nuts and studs.

10.173 a.

	N	Mean	StDev
Cont	50	0.005459	0.000763
Test	50	0.003507	0.000683

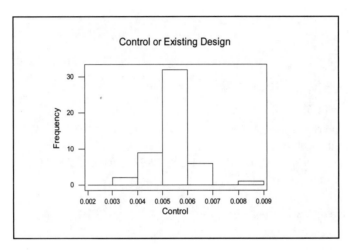

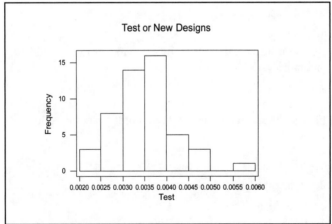

b. Both sets of data have an approximately normal distribution, therefore the assumptions are satisfied.

c. One-tailed – looking for a reduction

d.

Step 1: a. The difference between the variances in two different designs

b. $H_o: \sigma_c^2 = \sigma_t^2$

$H_a: \sigma_c^2 > \sigma_t^2$ (variance for control is greater)

Step 2: a. normality indicated, independence exists

b. F   c. $\alpha = 0.05$

Step 3: a. $n_c = 50$, $s_c = 0.000763$, $n_t = 50$, $s_t = 0.000683$

b. $F^* = s_c^2 / s_t^2 = (0.000763^2 / 0.000683^2) = 1.248$

Step 4: -- using p-value approach --------------------
a. $P = P(F > 1.25 | df = 49, 49)$;
   Using Tables 9, Appendix B, ES11-pp-722-727: $P > 0.05$
   Using a computer: $P = 0.44$
b. $P > \alpha$
-- using classical approach ------------------
a. critical region: $F \geq 1.69$
b. F^* falls in the noncritical region

Step 5: a. Fail to reject H_o

b. At the 0.05 level of significance, there is no difference in the variances for the two designs. The new design has not significantly reduced the variability.

e.
Step 1: a. The difference between mean amount of force for two designs, $\mu_c - \mu_t$

b. H_o: $\mu_c - \mu_t = 0$
   H_a: $\mu_c - \mu_t > 0$

Step 2: a. normality indicated
b. t   c. $\alpha = 0.05$

Step 3: a. sample data given above

b. $t^* = [(0.005459 - 0.003507) - 0] / \sqrt{(0.000763^2 / 50) + (0.000683^2 / 50)}$
   $= 13.48$

Step 4: a. $P = P(t > 13.48 | df = 49)$;
   Using Table 6, Appendix B, ES11-p719: $P = +0.000$
   Using Table 7, Appendix B, ES11-p720: $P = +0.000$
   Using a computer: $P = 0.000$
b. $P < \alpha$
-- using classical approach ------------------
a. critical region: $t \geq 1.68$
b. t^* falls in the critical region

Step 5: a. Reject H_o

b. At the 0.05 level of significance, there is sufficient evidence to show that the new design has reduced the mean amount of force

b. The mean force has been reduced, but not the variability.

CHAPTER 11 ∇ APPLICATIONS OF CHI-SQUARE

Chapter Preview

Chapter 11 demonstrates hypothesis tests, as did Chapters 8, 9, and 10. The difference lies in the type of data that is to be analyzed. Enumerative type data, that is, data which can be counted and placed into categories, will be discussed and investigated in three types of tests. Each test will compare actual (observed) results with expected (theoretical) results. One will use the comparisons to determine whether a "claimed" relationship exists, one will determine whether two factors or variables are independent, and the third will determine whether the proportions per variable are the same. Also, the chi square statistic, χ^2, will be reintroduced and utilized in performing these tests.

A graphic appearing in USA Today showing the ways that people cool their mouth after eating spicy hot food is used in the Chapter 11 opening section "Cooling a Great Hot Taste".

SECTION 11.1 EXERCISES

11.1 a. The name of their preferred way to "cool" their mouth after eating a delicious spicy favorite.

b. Population: US adults professing to love eating hot spicy food. Variable: method of cooling the heat.

c. Water: 73/200 = 0.365 = 36.5%;
Bread: 29/200 = 0.145 = 14.5%;
Milk: 35/200 = 0.175 = 17.5%;
Beer: 19/200 = 0.095 = 9.5%
Soda: 20/200 = 0.10 = 10%;
Nothing: 13/200 = 0.065 = 6.5%
Other: 11/200 = 0.055 = 5.5%;

11.3 a. 23.2 b. 23.3 c. 3.94 d. 8.64

11.5 a. $\chi^2(14, 0.01) = 29.1$ b. $\chi^2(25, 0.025) = 40.6$, $\chi^2(25, 0.975) = 13.1$

SECTION 11.2 EXERCISES

Characteristics of a Multinomial Experiment

1. There are **n** identical independent trials.
2. The outcome of each trial fits into exactly one of the **k** possible categories or cells.
3. The number of times a trial outcome falls into a particular cell is given by O_i
    (O - for observed, i for 1 → k).
4. $O_1 + O_2 + O_3 + ... + O_k = n$
5. There is a constant probability associated with each of the k cells in such a way that $p_1 + p_2 + ... + p_k = 1$.

NOTE: Variable - the characteristic about each item that is of interest.

Various levels of the variable - the k possible outcomes or responses.

11.7 a. One trial: asking one person b. Variable: birth day of the week

c. Levels: 7 for the 7 days of the week

Writing Hypotheses for Multinomial Experiments

The null hypothesis is written in a form to show that there is no difference between the experimental (observed) frequencies and the theoretical (expected) frequencies.
The alternative hypothesis is the "opposite" of the null hypothesis. It is written in a form to show that a difference does exist.

> ex.: The marital status distribution for New York state is 21%, 64%, 8%, and 7% for the possible categories of single, married, widowed, and divorced.
> H_o: P(S) = .21, P(M) = .64, P(W) = .08, P(D) = .07
> H_a: The percentages are different than specified in H_o.
>
> ex.: A gambler thinks that a die may be *loaded*.
> H_o: P(1)=P(2)=P(3)=P(4)=P(5)=P(6) = 1/6   (not loaded)
> H_a: The probabilities are different.   (is loaded)

11.9 a. H_o: P(1) = P(2) = P(3) = P(4) = P(5) = 0.2 H_a: The numbers are not equally likely.

b. H_o: P(1) = 2/8, P(2) = 3/8, P(3) = 2/8, P(4) = 1/8
   H_a: At least one probability is different from H_o.

c. H_o: P(E) = 0.16, P(G) = 0.38, P(F) = 0.41, P(P) = 0.05
   H_a: The percentages are different than specified in H_o.

> The p-value is <u>estimated</u> using Table 8 (Appendix B, ES11-p721):
> a) Locate df row.
> b) Locate χ^{2*} between two critical values in the df row; the p-value is in the interval between the two corresponding probabilities at the top of the columns labeled *area to the right*.
> OR
> The p-value can be calculated using computer and/or calculator commands found in ES11-pp455-456.

11.11 a. **P** = $P(\chi^2 > 12.25 | df=3)$;
  Using Table 8: $0.005 < \mathbf{P} < 0.01$
  Using computer/calculator: **P** = 0.0066

b. **P** = $P(\chi^2 > 5.98 | df=2)$;
  Using Table 8: $0.05 < \mathbf{P} < 0.10$
  Using computer/calculator: **P** = 0.0503

> ### Determining the Test Criteria
>
> 1. Draw a picture of the χ^2 distribution (skewed right, starting at 0).
> 2. Locate the critical region (based on α and H_a).
>   Since we are testing H_o: "no difference" versus H_a: "difference", all of the α is placed in the right tail to represent a significant or large *difference* between the observed and expected values.
> 3. Shade in the critical region (the area where you will reject H_o, the right-hand tail)
> 4. Find the appropriate critical value from Table 8 (Appendix B, ES11-p721), using $\chi^2(df,\alpha)$, where df = k - 1. k is equal to the number of cells or categories the data are classified into.

> Remember this critical or boundary value divides the area under the χ^2 distribution curve into critical and noncritical regions and is part of the critical region.

11.13 9 parts + 3 parts + 3 parts + 1 part = 16 parts total.

Steps for a Hypothesis/Significance Test for Multinomial Experiments

Follow the steps outlined for p-value and classical hypothesis tests in: ES11-pp371&388, and/or IRM-pp343&357.
The variations are noted below.

1. Distribute the sample information (observed frequencies) into the appropriate cells (data may be already categorized).

2. Calculate the expected frequencies using probabilities determined by H_O and the formula:
    $E_i = np_i$, where E_i is the expected frequency for cell i, n is the sample size and p_i is the probability for the ith cell

3. Use the observed and expected frequencies from each cell to calculate the test statistic, χ^2.
    Use the formula $$\chi^2 = \sum_{all\ cells} \frac{(O-E)^2}{E}$$

4.a) p-value approach:
    Since all of the α is placed in the right tail, the p-value = $P(\chi^2 > \chi^{2*})$.
    The p-value is <u>estimated</u> using Table 8 (Appendix B, ES11-p721):
    a) Locate df row.
    b) Locate χ^{2*} between two critical values in the df row; the p-value is in the interval between the two corresponding probabilities at the top of the columns labeled *area to the right*.
    (ctn)

OR
    The p-value can be calculated using computer and/or calculator commands found in ES11-pp553-554.

b) Classical approach:
    Follow the steps in determining the test criteria found in: ES11-p548, IRM-p579.
    Then locate χ^{2*} on the χ^2 curve with respect to the critical value.

5. Make a decision and interpret it.
    a) If P < α or the calculated test statistic falls into the critical region, then reject H_O. There is sufficient evidence to indicate that there is a *difference* between the observed and expected frequencies.
    b) If P > α or the calculated test statistic falls into the noncritical region, fail to reject H_O. There is <u>not</u> sufficient evidence to indicate a *difference* between the observed and expected frequencies.

11.15 a. H_o: $P(A) = P(B) = P(C) = P(D) = P(E) = 0.2$ b. χ^2

c. (1) & (2)
Step 1: a. Preference of floor polish, the probability that a particular type is preferred
b. H_o: Equal preference
H_a: preferences not all equal

Step 2: a. Assume that the 100 consumers represent a random sample.
b. χ^2 with df = 4   c. $\alpha = 0.10$

Step 3: a. sample information given in exercise
b. $\chi^2 = \Sigma[(O-E)^2/E]$ (as found on accompanying table)
$E = n \cdot p = 100(1/5) = 20$, for all cells

Polish	A	B	C	D	E	Total
Observed	27	17	15	22	19	100
Expected	20	20	20	20	20	100
$(O-E)^2/E$	49/20	9/20	25/20	4/20	1/20	88/20

$\chi^{2*} = 4.40$

Step 4: -- using p-value approach --------------------
a. $P = P(\chi^2 > 4.40 | df=4)$;
Using Table 8: $0.25 < P < 0.50$
Using computer/calculator: $P = 0.355$
b. $P > \alpha$
-- using classical approach -----------------
a. $\chi^2(4, 0.10) = 7.78$
b. χ^{2*} falls in the noncritical region, see Step 4a

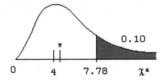

Step 5: a. Fail to reject H_o
b. At the 0.10 level of significance, there is not sufficient evidence to show the preferences are not all equal.

11.17 Step 1: a. The proportions: P(Yes), P(No), P(Not sure)
b. H_o: The opinions of students with respect to the civility issue are in the proportions of 0.51, 0.37, 0.12
H_a: The proportions are different than listed

Step 2: a. Assume that the 300 opinions represent a random sample.
b. χ^2 with df = 2   c. $\alpha = 0.01$

Step 3: a. sample information given in exercise
b. $\chi^2 = \Sigma[(O-E)^2/E]$ (as found on accompanying table)
$E(\text{Yes}) = n \cdot p = 300(0.51) = 153$; $E(\text{No}) = 300(0.37) = 111$;
$E(\text{Not sure}) = 300(0.12) = 36$

Quality	Yes	No	Not sure	Total
Observed	126	118	56	300
Expected	153	111	36	300
$(O-E)^2/E$	4.765	0.441	11.111	16.317

$$\chi^{2*} = 16.317$$

Step 4: -- using p-value approach --------------------
a. $P = P(\chi^2 > 16.317 | df=2)$;
Using Table 8: $P < 0.005$
Using computer/calculator: $P = 0.000$
b. $P < \alpha$

-- using classical approach ------------------
a. critical region: $\chi^2 \geq 9.21$
b. χ^{2*} falls in the critical region

Step 5: a. Reject H_o
b. At the 0.01 level of significance, there is sufficient evidence to show the proportions for student opinions are not the same as the published poll.

11.19 a. E(magenta) = n·p = 100[6/(6+3+1)] = <u>60</u>
b. <u>2</u>
c. (1) & (2)
Step 1: a. The proportions: P(magenta), P(chartreuse), P(ochre)
b. H_o: 6:3:1 ratio vs. H_a: ratio other than 6:3:1
Step 2: a. Assume that the 100 seeds represent a random sample.
b. χ^2 with df = 2   c. $\alpha = 0.10$
Step 3: a. sample information given in exercise
b. $\chi^2 = \Sigma[(O-E)^2/E]$ (as found on accompanying table)
E(magenta) = n·P(m) = 100(0.6) = 60
E(chartreuse) = n·P(c) = 100(0.3) = 30
E(orche) = n·P(o) = 100(0.1) = 10

Color	magenta	chartreuse	orche	Total
Observed	52	36	12	100
Expected	60	30	10	100
$(O-E)^2/E$	64/60	36/30	4/10	160/60

$$\chi^{2*} = 160/60 = 2.67$$

Step 4: -- using p-value approach --------------------
a. $P = P(\chi^2 > 2.67 | df=2)$;
Using Table 8: $0.25 < P < 0.50$
Using computer/calculator: $P = 0.263$
b. $P > \alpha$

-- using classical approach ------------------
a. critical region: $\chi^2 \geq 4.61$
b. χ^{2*} falls in the noncritical region

Step 5: a. Fail to reject H_o
b. At the 0.10 level of significance, there is not sufficient evidence to show the ratio is other than 6:3:1.

11.21 Step 1: a. The proportions: P(Quality I), P(Quality II), P(Quality III), P(Quality IV).
b. H_o: The qualities of meats purchased are in the proportions of 0.10, 0.30, 0.35, 0.25
H_a: The proportions are different than listed
Step 2: a. Assume that the 500 purchases represent a random sample.
b. χ^2 with df = 3   c. $\alpha = 0.05$
Step 3: a. sample information given in exercise
b. $\chi^2 = \Sigma[(O-E)^2/E]$ (as found on accompanying table)
E(I) = n·p = 500(0.10) = 50; E(II) = 500(0.30) = 150;
E(III) = 500(0.35) = 175; E(IV) = 500(0.25) = 125

Quality	I	II	III	IV	Total
Observed	46	162	191	101	500
Expected	50	150	175	125	500
$(O-E)^2/E$	16/50	144/150	256/175	576/125	7.35

$\chi^{2*} = 7.35$

Step 4: -- using p-value approach --------------------
a. $P = P(\chi^2 > 7.35 | df=3)$;
Using Table 8: $0.05 < P < 0.10$
Using computer/calculator: $P = 0.062$
b. $P > \alpha$

-- using classical approach -----------------
a. critical region: $\chi^2 \geq 7.81$
b. χ^{2*} falls in the noncritical region

Step 5: a. Fail to reject H_o
b. At the 0.05 level of significance, there is not sufficient evidence to show the proportions of meat qualities bought are different than listed.

11.23 Step 1: a. The proportions of staffing situations: P(1), P(2), P(3), P(4), P(5)
b. H_o: Opinions are distributed 0.12, 0.32, 0.38, 0.12, 0.06
H_a: Opinions are distributed differently.
Step 2: a. Assume that the 500 nurses surveyed represent a random sample.
b. χ^2 with df = 4   c. $\alpha = 0.05$
Step 3: a. sample information given in exercise
b. $\chi^2 = \Sigma[(O-E)^2/E]$ (as found on accompanying table)
E(1) = n·p = 500(0.12) = 60; E(2) = 500(0.32) = 160;
E(3) = 500(0.38) = 190; E(4) = 500(0.12) = 60;
E(5) = 500(0.06) = 30

```
         Opinion    1     2     3     4     5     Total
         ---------------------------------------------------
         Observed   165   140   125   50    20    500
         Expected   60    160   190   60    30    500
         ---------------------------------------------------
         (O-E)²/E   183.75 2.50  22.24 1.67  3.33  213.49
                    χ²* = 213.49
```

Step 4: -- using p-value approach ------------------
      a. $P = P(\chi^2 > 213.49 | df=4)$;
                              Using Table 8:   P < 0.005
                              Using computer/calculator: P = 0+
      b. $P < \alpha$
      -- using classical approach ------------------
      a. critical region: $\chi^2 \geq 9.49$
      b. χ²* falls in the critical region

Step 5: a. Reject H_o

b. The opinions of the 500 nurses are significantly different than the opinions of the original 1800 nurses, at the 0.05 level of significance.

> One of the properties of random numbers is that they occur with equal probabilities. Since there are 10 values (0-9), each integer will have the same probability 1/10 = 0.1.

11.25   Step 1: a. The proportions of most valuable steps for climbing out of debt:
                  P(1), P(2), P(3), P(4), P(5), P(6), P(7)
            b. H_o: Opinions are distributed 0.30, 0.21, 0.17, 0.12, 0.07, 0.05, 0.08
              H_a: Opinions are distributed differently.

Step 2: a. Assume that the 60 financial planners surveyed represent a random sample.
        b. χ^2 with df = 6   c. $\alpha = 0.05$

Step 3: a. sample information given in exercise
        b. $\chi^2 = \Sigma[(O-E)^2/E]$ (as found on accompanying table)
        $E(1) = n \cdot p = 60(0.30) = 18$; $E(2) = 60(0.21) = 12.6$;
        $E(3) = 60(0.17) = 10.2$; $E(4) = 60(0.12) = 7.2$;
        $E(5) = 60(0.07) = 4.2$; $E(6) = 60(0.05) = 3$;
        $E(7) = 60(0.08) = 4.8$

```
         Opinion    1     2     3     4     5     6     7     Total
         -----------------------------------------------------------
         Observed   10    13    13    8     9     3     4     60
         Expected   18    12.6  10.2  7.2   4.2   3     4.8   60
         -----------------------------------------------------------
         (O-E)²/E   3.56  0.013 0.768 0.089 5.49  0     0.133 10.05
                    χ²* = 10.05
```

Step 4: -- using p-value approach ------------------
      a. $P = P(\chi^2 > 10.05 | df=6)$;
                              Using Table 8: 0.10 < P < 0.25
                              Using computer/calculator: P = 0.123

b. $P > \alpha$
-- using classical approach -----------------
a. critical region: $\chi^2 \geq 12.6$
b. χ^{2*} falls in the noncritical region

Step 5: a. Fail to reject H_o
b. At the 0.05 level of significance, there is not sufficient evidence to show the opinions are distributed differently than listed.

11.27 Step 1: a. The proportions of preferred method of cooling a hot taste:
P(water), P(bread), P(milk), P(beer), P(soda), P(don't), P(other)
b. H_o: sample proportions are not significantly different from "Some like it hot" graph
H_a: proportions different.

Step 2: a. Assume that the 200 adults surveyed represent a random sample.
b. χ^2 with df = 6   c. $\alpha = 0.05$

Step 3: a. sample information given in exercise
b. $\chi^2 = \Sigma[(O-E)^2/E]$ (as found on accompanying table)
E(w) = n·p = 200(0.43) = 86; E(b) = 200(0.19) = 38;
E(m) = 200(0.15) = 30; E(be) = 200(0.07) = 14;
E(s) = 200(0.07) = 14; E(d) = 200(0.06) = 12;
E(o) = 200(0.03) = 6

Opinion	w	b	m	be	s	d	o	Total
Observed	73	29	35	19	20	13	11	200
Expected	86	38	30	14	14	12	6	200

(O-E)²/E 1.965 2.132 0.8333 1.786 2.571 0.083 4.167 13.537
$\chi^{2*} = 13.537$

Step 4: -- using p-value approach ------------------
a. $P = P(\chi^2 > 13.537 | df=6)$;
Using Table 8: $0.025 < P < 0.05$
Using computer/calculator: $P = 0.0352$
b. $P < \alpha$
-- using classical approach -----------------
a. critical region: $\chi^2 \geq 12.6$
b. χ^{2*} falls in the critical region

Step 5: a. Reject H_o
b. At the 0.05 level of significance, there is sufficient evidence to show that the 200 adult distribution is different from that presented in the graphic.

11.29 a.
Step 1: a. The proportions of types and number of guns in a household:
P(rifle, shotgun, pistol), P(2 of the 3 types), P(1 of the 3 types), P(decline).

b. H_0: The proportions for all gun owners: 41%, 27%, 29%, 3%

H_a: The proportions are different than listed

Step 2: a. Assume that the 2000 individuals surveyed represent a random sample.
b. χ^2 with df = 3   c. $\alpha = 0.05$

Step 3: a. sample information given in exercise

b. $\chi^2 = \Sigma[(O-E)^2/E]$ (as found on accompanying table)
$$E(3 \text{ types}) = n \cdot p = 2000(0.41) = 820,$$
$$E(2 \text{ types}) = 2000(0.27) = 540,$$
$$E(1 \text{ type}) = 2000(0.29) = 580, \ E(\text{decline}) = 2000(0.03) = 60$$

Handgun	3 types	2 types	1 type	decline	Total
Observed	780	550	560	110	2000
Expected	820	540	580	60	2000
$(O-E)^2/E$	1.9512	0.1852	0.6897	41.6667	44.4928

$\chi^{2*} = 44.4928$

Step 4: -- using p-value approach --------------------

a. $P = P(\chi^2 > 44.4928 | df=3)$;

Using Table 8: $P < 0.005$

Using computer/calculator: $P = 0+$

b. $P < \alpha$

-- using classical approach -----------------

a. critical region: $\chi^2 \geq 7.81$

b. χ^{2*} falls in the critical region

Step 5: a. Reject H_0

b. The distribution of the number of types of guns owned in Memphis is different from those nationally at the 0.05 level of significance.

b. The 4th cell caused the calculated value of chi-square to be very large.
The "decline" category would need to be broken down into more specific answer categories.

SECTION 11.3 EXERCISES

Contingency Tables

A contingency table is a table consisting of rows and columns used to summarize and cross-classify data according to two variables. Each row represents the categories for one of the variables, and each column represents the categories for the other variable. The intersections of these rows and columns produce cells. The data will be in a form where two varieties of hypothesis tests are possible. These are:

1. Tests of independence
    - to determine if one variable is independent of the other variable, and
2. Tests of homogeneity
    - to determine if the proportion distribution for one of the variables is the same for each the categories of the second variable.

> Writing Hypotheses for Tests of Independence and/or Homogeneity
>
> The null hypothesis is written in a form to show that there is no difference between the experimental (observed) frequencies and the theoretical (expected) frequencies. The alternative hypothesis is the opposite of the null hypothesis. It is written in a form to show that a difference does exist.
>
> Therefore, in tests of independence:
> H_o: One variable is independent of the other variable
>
> and in tests of homogeneity:
> H_o: The proportions for one variable are distributed the same for all categories of the second variable.

11.31 a. H_o: Voters preference and voters party affiliation are independent.
   H_a: Voters preference and party affiliation are not independent.

b. H_o: The distribution is the same for all three
   H_a: The distribution is not the same for all three

c. H_o: The proportion of yeses is the same in all categories sampled.
   H_a: The proportion of yeses is not the same in all categories.

11.33 E = (40)(50)/200 = 2000/200 = <u>10</u>

11.35 a. sample information given in exercise
   $\chi^2 = \Sigma[(O-E)^2/E]$ (as found on accompanying table)
   Expected values:

	Boat-Rel	Non-boat
Lee	18.84	29.16
Collier	12.16	18.84

   $\chi^{2*} = 0.921 + 0.595 + 1.426 + 0.921 = 3.862$

b. **P** = $P(\chi^2 > 3.862 | df=1)$;
   Using Table 8: $0.025 < P < 0.05$
   Using computer/calculator: **P** = 0.049

c. **P** < α; Reject H_o
   The proportion of boat-related deaths is not independent of the county at the 0.05 level of significance.

> The computer or calculator can perform a hypothesis test for independence or homogeneity. Since these tests are completed in the same fashion, the same command may be used.
> See ES11-pp565-566 for more information.

11.37 Step 1: a. The proportions of direction: P(continued direction per vehicle),
P(reverse direction per vehicle).
b. H_o: Direction for white-tailed deer is independent of type of vehicle.
H_a: Direction for white-tailed deer is not independent of type of vehicle.
Step 2: a. Assume deer a random sample.
b. χ^2 with df = 1
c. $\alpha = 0.01$
Step 3: a. sample information given in exercise
b. $\chi^2 = \Sigma[(O-E)^2/E]$ (as found on accompanying table)
Expected counts are printed below observed counts
Chi-Square contributions are printed below expected counts

	continued	reversed	Total
1	315	73	388
	272.08	115.92	
	6.771	15.893	
2	84	97	181
	126.92	54.08	
	14.516	34.069	
Total	399	170	569

Chi-Sq = 71.249, DF = 1, P-Value = 0.000

Step 4: -- using p-value approach --------------------
a. $P = P(\chi^2 > 71.249|df=1)$;
Using Table 8:   **P < 0.005**
Using computer/calculator: **P = 0.000**
b. **P < α**
-- using classical approach ------------------
a. critical region: $\chi^2 \geq 6.63$
b. χ^{2*} falls in the critical region

Step 5: a. Reject H_o
b. At the 0.01 level of significance, there is sufficient evidence to show that direction is not independent of the vehicle type for white-tailed deer.

11.39 Step 1: a. The proportions of having Tourette's: P(Tourette's for Hispanic), P(Tourette's for Non-Hispanic White), P(Tourette's for Non-Hispanic Black).
b. H_o: Having Tourette's is independent of ethnicity and race.
H_a: Having Tourette's is not independent of ethnicity and race.
Step 2: a. Given a random sample.
b. χ^2 with df = 2
c. $\alpha = 0.05$

Step 3: a. sample information given in exercise
b. $\chi^2 = \Sigma[(O-E)^2/E]$ (as found on accompanying table)

```
Expected counts are printed below observed counts
Chi-Square contributions are printed below expected counts
```

```
                Non-Hisp    Non-Hisp
     Hispanic    White       Black     Total
  1     26         164         18       208
       26.55      158.16      23.29
       0.011      0.216       1.202

  2    7321       43602       6427     57350
      7320.45    43607.84    6421.71
       0.000      0.001       0.004

Total  7347       43766       6445     57558

Chi-Sq = 1.434, DF = 2, P-Value = 0.488
```

Step 4: -- using p-value approach -------------------
    a. $P = P(\chi^2 > 1.434 | df=2)$;
        Using Table 8: $0.25 < P < 0.50$
        Using computer/calculator: $P = 0.488$
    b. $P > \alpha$
-- using classical approach -----------------
    a. critical region: $\chi^2 \geq 5.99$
    b. χ^{2*} falls in the noncritical region

Step 5: a. Fail to reject H_o
       b. At the 0.05 level of significance, there is sufficient evidence to indicate that having Tourette's is independent of ethnicity and race.

11.41 Step 1: a. The proportions supervisor expectation responses: P(true for length of employment category), P(not true for length of employment category).
       b. H_o: Response is independent of years of service.
         H_a: Response is not independent of years of service.

Step 2: a. Assume a random sample.
       b. χ^2 with df = 3
       c. $\alpha = 0.10$

Step 3: a. sample information given in exercise
       b. $\chi^2 = \Sigma[(O-E)^2/E]$ (as found on accompanying table)
Expected values:

	True	Not true
Less than 1 yr	21.94	9.06
1 to 3 years	19.82	8.18
3 to 10 years	26.18	10.82
10 or more years	24.06	9.94

$\chi^{2*} = 0.707 + 1.712 + 0.002 + 0.004 + 0.126 + 0.305 + 0.156 + 0.378 = 3.390$

Step 4: -- using p-value approach -------------------
    a. $P = P(\chi^2 > 3.390 | df=3)$;
        Using Table 8: $0.25 < P < 0.50$
        Using computer/calculator: $P = 0.335$
    b. $P > \alpha$

-- using classical approach -----------------
    a. critical region: $\chi^2 \geq 6.25$

    b. χ^{2*} is not in the critical region

Step 5: a. Fail to reject H_o
    b. At the 0.10 level of significance, there is not sufficient evidence to conclude a lack of independence between the length of employment and response.

11.43 Step 1: a. The proportions of community size that married men were reared in: P(under 10,000 for present size categories), P(10,000 to 49,999 for present size categories), P(50,000 or over for present size categories).
    b. H_o: Size of community of present residence is independent of the size of community reared in.
    H_a: Size of community of present residence is not independent of the size of community reared in.
Step 2: a. Given a random sample.
    b. χ^2 with df = 4
    c. $\alpha = 0.01$
Step 3: a. sample information given in exercise
    b. $\chi^2 = \Sigma[(O-E)^2/E]$ (as found on accompanying table)
    Expected values:

	less than 10,000	10,000 - 49,999	50,000 or over
less than 10,000	14.36	37.16	62.47
10,000 – 49,999	19.15	49.55	83.30
50,000 or over	29.48	76.28	128.23

$\chi^{2*} = 6.464 + 1.652 + 4.887 + 0.069 + 4.213 = 2.122 + 2.441 + 6.510 + 7.382 = 35.741$

Step 4: -- using p-value approach ------------------
    a. $P = P(\chi^2 > 35.741 | df=4)$;
        Using Table 8: $P < 0.005$
        Using computer/calculator: $P = 0.0+$
    b. $P < \alpha$
    -- using classical approach ------------------

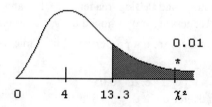

    a. $\chi^2(4, 0.01) = 13.3$
    b. χ^{2*} falls in the critical region

Step 5: a. Reject H_o
    b. The sample evidence does present significant evidence to show the lack of independence, at the 0.01 level of significance.

11.45 Step 1: a. The proportions: P(nondefective for 5 days of the week), P(defective for 5 days of the week)
b. H_o: The number of defective items is independent of the day of the week

H_a: The number of defectives is not independent of day

Step 2: a. Assume a random sample.
b. χ^2 with df = 4   c. $\alpha = 0.05$

Step 3: a. sample information given in exercise
b. $\chi^2 = \Sigma[(O-E)^2/E]$ (as found on accompanying table)
Expected values:

	Mon	Tue	Wed	Thu	Fri	Total
Nondefective	91	91	91	91	91	455
Defective	9	9	9	9	9	45
						500

$\chi^{2*} = \Sigma[(O-E)^2/E] = 0.396 + 0.011 + 0.176 + 0.176 + 0.011 + 4.000 + 0.111 + 1.778 + 1.778 + 0.111$

$\chi^{2*} = 8.548$

Step 4: -- using p-value approach --------------------
a. $P = P(\chi^2 > 8.548 | df=4)$;
Using Table 8: $0.05 < P < 0.10$
Using computer: $P = 0.074$

b. $P > \alpha$
-- using classical approach -----------------
a. critical region: $\chi^2 \geq 9.49$
b. χ^{2*} is in the noncritical region

Step 5: a. Fail to reject H_o
b. At the 0.05 level of significance, there is not sufficient evidence to show a lack of independence between the number of defective articles and the day of the week.

11.47 Step 1: a. The proportions: P(Blog creators) and P(Blog readers) are the same for all months.
b. H_o: Blog creators and Blog readers are proportioned the same for each month.

H_a: Blog creators and Blog readers are not proportioned the same for each month.

Step 2: a. Assume a random sample.
b. χ^2 with df = 2   c. $\alpha = 0.05$

Step 3: a. sample information given in exercise
b. $\chi^2 = \Sigma[(O-E)^2/E]$
Expected counts are printed below observed counts
Chi-Square contributions are printed below expected counts

	Blog creators	Blog readers	Total
1	74	205	279
	62.78	216.23	

```
                2.007    0.583

        2        93      316     409
                92.03   316.98
                0.010    0.003

        3       130      502     632
               142.20   489.80
                1.047    0.304

   Total       297     1023    1320
```

Chi-Sq = 3.954, DF = 2, P-Value = 0.138
$\chi^{2*} = 3.954$

Step 4: -- using p-value approach ------------------
a. $P = P(\chi^2 > 3.954 | df=2)$;
Using Table 8: $0.10 < P < 0.25$
Using computer: $P = 0.138$
b. $P > \alpha$
-- using classical approach ---------------
a. critical region: $\chi^2 \geq 5.99$
b. χ^{2*} is in the noncritical region

Step 5: a. Fail to reject H_o
b. At the 0.05 level of significance, there is not sufficient evidence to show that the distribution of blog creators and blog readers are not proportioned the same for each month throughout the months listed.

11.49 Step 1: a. The proportions of injuries per sports: P(type of injury with playing basketball), P(type of injury with playing volleyball).
b. H_o: There is no difference between the distributions
H_a: distributions different
Step 2: a. Assume a random sample.
b. χ^2 with df = 3
c. $\alpha = 0.05$
Step 3: a. sample information given in exercise
b. $\chi^2 = \Sigma[(O-E)^2/E]$ (as found on accompanying table)

Expected counts are printed below observed counts
Chi-Square contributions are printed below expected counts

```
        Basketball  Volleyball   Total
   1        28          19         47
          25.34       21.66
          0.279       0.326

   2        11           7         18
           9.71        8.29
          0.173       0.202

   3         6           8         14
           7.55        6.45
```

```
                    0.318      0.372
          4          10         13       23
                    12.40      10.60
                    0.465      0.544
     Total           55         47      102

     Chi-Sq = 2.678, DF = 3, P-Value = 0.444
```

Step 4: -- using p-value approach ------------------
a. $P = P(\chi^2 > 2.678 | df=3)$;
Using Table 8: $0.25 < P < 0.50$
Using computer/calculator: $P = 0.444$

b. $P > \alpha$
-- using classical approach ----------------
a. critical region: $\chi^2 \geq 7.81$
b. χ^{2*} is not in the critical region

Step 5: a. Fail to reject H_o
b. At the 0.05 level of significance, there is not sufficient evidence to conclude that there is a significant difference between the two sports.

11.51 Step 1: a. The proportions: P(who fear darkness) and P(who do not fear darkness) are the same for all age groups.
b. H_o: Fear and Do Not Fear darkness are proportioned the same for each age group.
H_a: Fear and Do Not Fear darkness are not proportioned the same for each age group.

Step 2: a. Assume a random sample.
b. χ^2 with df = 4   c. $\alpha = 0.01$

Step 3: a. sample information given in exercise
b. $\chi^2 = \Sigma[(O-E)^2/E]$

Expected values:

	Elem	J. H.	S. H.	Coll.	Adult
Fear	70.8	70.8	70.8	70.8	70.8
Do not	129.2	129.2	129.2	129.2	129.2

$\chi^{2*} = 2.102 + 0.020 + 6.712 + 17.105 + 26.359$
$\quad + 1.152 + 0.011 + 3.678 + 9.373 + 14.445 \approx 80.959$

Step 4: -- using p-value approach ------------------
a. $P = P(\chi^2 > 80.959 | df=4)$;
Using Table 8: $P < 0.005$
Using computer: $P = 0+$

b. $P < \alpha$
-- using classical approach ----------------
a. critical region: $\chi^2 \geq 13.3$
b. χ^{2*} is in the critical region

Step 5: a. Reject H_o

b. There is sufficient evidence to show that the age groups have different proportions which fear darkness, at the 0.01 level of significance.

11.53 Step 1: a. The proportions: P(females) and P(males) are the same for all drug dosage groups.
b. H_o: Females and males are proportioned the same for each drug dosage.
H_a: Females and males are not proportioned the same for each drug dosage.

Step 2: a. Assume a random sample.
b. χ^2 with df = 2   c. $\alpha = 0.01$

Step 3: a. sample information given in exercise
b. $\chi^2 = \Sigma[(O-E)^2/E]$
Expected counts are printed below observed counts
Chi-Square contributions are printed below expected counts

	10 mg drug	20 mg drug	placebo	Total
1	54	56	60	170
	57.33	55.33	57.33	
	0.194	0.008	0.124	
2	32	27	26	85
	28.67	27.67	28.67	
	0.388	0.016	0.248	
Total	86	83	86	255

$\chi^{2*} = 0.978$

Step 4: -- using p-value approach ------------------
a. $P = P(\chi^2 > 0.978 | df=2)$;
Using Table 8: $0.50 < P < 0.75$
Using computer: $P = 0.613$
b. $P > \alpha$
-- using classical approach ----------------
a. critical region: $\chi^2 \geq 9.21$
b. χ^{2*} is in the noncritical region

Step 5: a. Fail to reject H_o
b. There is not sufficient evidence to show the distribution is not proportioned the same for each drug, at the 0.01 level of significance.

Step 1: a. The proportions: P(ages 40-49), P(ages 50-59) and P(ages 60-69) are the same for all drug dosage groups.
b. H_o: Age groups are proportioned the same for each drug dosage.
H_a: Age groups are not proportioned the same for each drug dosage.

Step 2: a. Assume a random sample.
b. χ^2 with df = 4   c. $\alpha = 0.01$

Step 3: a. sample information given in exercise
b. $\chi^2 = \Sigma[(O-E)^2/E]$
Expected counts are printed below observed counts

Chi-Square contributions are printed below expected counts

	10 mg drug B	20 mg drug B	placebo B	Total
1	18	20	19	57
	19.22	18.55	19.22	
	0.078	0.113	0.003	
2	48	41	57	146
	49.24	47.52	49.24	
	0.031	0.895	1.223	
3	20	22	10	52
	17.54	16.93	17.54	
	0.346	1.521	3.239	
Total	86	83	86	255

$\chi^{2*} = 7.449$

Step 4: -- using p-value approach ------------------
a. $P = P(\chi^2 > 7.449 | df=4)$;
   Using Table 8: $0.10 < P < 0.25$
   Using computer: $P = 0.114$
b. $P > \alpha$
-- using classical approach ----------------
a. critical region: $\chi^2 \geq 13.3$
b. χ^{2*} is in the noncritical region

Step 5: a. Fail to reject H_0
b. There is not sufficient evidence to show the distribution is not proportioned the same for each drug, at the 0.01 level of significance.

11.55 a.

	West	N.Cent.	South	N.East	Total
1	55	46	47	41	190
	47.25	47.25	47.25	47.25	
2	45	54	53	59	210
	52.75	52.75	52.75	52.75	
Total	100	100	100	100	400

$\chi^{2*} = 1.271 + 0.033 + 0.001 + 0.827 + 1.139 + 0.030 + 0.001 + 0.741 = \underline{4.043}$

$P = P(\chi^{2*} > 4.043 | df = 3) = \underline{0.257}$

b.

	West	N.Cent.	South	N.East	Total
1	110	92	94	82	380
	94.5	94.5	94.5	94.5	

2	90	108	106	118	420
	105.5	105.5	105.5	105.5	
Total	200	200	200	200	800

$\chi^{2*} = 2.542 + 0.066 + 0.003 + 1.653 + 2.277 + 0.059 + 0.002 + 1.481 = \underline{8.083}$

$P = P(\chi^{2*} > 8.083 | df = 3) = \underline{0.044}$

	West	N.Cent.	South	N.East	Total
1	165	138	141	123	570
	141.75	141.75	141.75	141.75	
2	135	162	159	177	630
	158.25	158.25	158.25	158.25	
Total	300	300	300	300	1200

$\chi^{2*} = 3.813 + 0.099 + 0.004 + 2.480 + 3.416 + 0.089 + 0.004 + 2.222 = \underline{12.127}$

$P = P(\chi^{2*} > 12.127 | df = 3) = \underline{0.007}$

   c.   Yes, as the sample sizes increase, the relative differences are the same but the actual counts are much larger and further apart.

CHAPTER EXERCISES

11.57 Step 1: a. The proportions: P(1st type), P(2nd type), P(3rd type)
            b. H_o: 1:3:4 proportions
               H_a: proportions are other than 1:3:4

Step 2: a. Assume that the 80 hybrids represent a random sample.
          b. χ^2 with df = 2   c. $\alpha = 0.05$

Step 3: a. sample information given in exercise
          b. $\chi^2 = \Sigma[(O-E)^2/E]$ (as found on accompanying table)
          E(1st) = n·p = 800[1/(1+3+4)] = 100
          E(2nd) = 800[3/(1+3+4)] = 300
          E(3rd) = 800[4/(1+3+4)] = 400

	1st	2nd	3rd	Total
Observed	80	340	380	800
Expected	100	300	400	800
$(O-E)^2/E$	4.00	5.33	1.00	10.33

$\chi^{2*} = 10.33$

Step 4: -- using p-value approach --------------------
          a. $P = P(\chi^2 > 10.33 | df=2)$;

Using Table 8: $0.005 < P < 0.01$
Using computer: $P = 0.006$

b. $P < \alpha$

-- using classical approach ----------------

a. $\chi^2(2, 0.05) = 5.99$ $\chi^2 \geq 5.99$

b. χ^{2*} is in the critical region.

Step 5: a. Reject H_o

b. We have sufficient evidence to show that the ratio is not the hypothesized 1:3:4 ratio, at the 0.05 level of significance.

11.59 Step 1: a. The proportions: P(parent Home), P(campus), P(off-campus), P(own home), P(other)

a. H_o: P(parent home) = 0.46, P(campus) = 0.26, P(off-campus) = 0.18,
P(own home) = 0.09, P(other) = 0.02

H_a: The percentages are different than listed

Step 2: a. Assume that the 1000 individuals represent a random sample.

b. χ^2 with df = 4   c. $\alpha = 0.05$

Step 3: a. sample information given in exercise

b. $\chi^2 = \Sigma[(O-E)^2/E]$ (as found on accompanying table)

E(parent) = n·p = 1000(0.46) = 460,
E(campus) = 1000(0.26) = 260,
E(off-campus) = 1000(0.18) = 180,
E(own home) = 1000(0.09) = 90,
E(other) = 1000(0.02) = 20

	Par.	Cam.	Off.	Own.	Oth.	Total
Observed	484	230	168	96	22	1000
Expected	458	258	178	88	18	1000
$(O-E)^2/E$	1.476	3.039	0.562	0.727	0.889	6.693

$\chi^{2*} = 6.693$

Step 4: -- using p-value approach -------------------

a. $P = P(\chi^2 > 6.693 | df=4)$;

Using Table 8: $0.10 < P < 0.25$
Using computer: $P = 0.153$

b. $P > \alpha$

-- using classical approach ----------------

a. $\chi^2(4, 0.05) = 9.49$ $\chi^2 \geq 9.49$

b. χ^{2*} is in the noncritical region.

Step 5: a. Fail to reject H_o

b. We do not have sufficient evidence to show that the sample distribution is different than the newspaper distribution, at the 0.05 level of significance.

11.61 Step 1: a. The largest number of holes played by golfers in one day.

b. H_o: P(18) = 0.05, P(19-27) = 0.12, P(28-36) = 0.28, P(37-45) = 0.20,
P(46-54) = 0.18, P(55 or more) = 0.17
H_a: The percentages are different than listed

Step 2: a. Assume that the 200 individuals represent a random sample.
b. χ^2 with df = 5   c. $\alpha = 0.01$

Step 3: a. sample information given in exercise
b. $\chi^2 = \Sigma[(O-E)^2/E]$

$$\chi^{2*} = \frac{(12-10)^2}{10} + \frac{(35-24)^2}{24} + \frac{(60-56)^2}{56} + \frac{(44-40)^2}{40} + \frac{(35-36)^2}{36} + \frac{(14-34)^2}{34}$$

$= 0.40 + 5.04 + 0.29 + 0.40 + 0.03 + 11.76 = 17.92$
$\chi^{2*} = 17.92$

Step 4: -- using p-value approach --------------------
a. $P = P(\chi^2 > 17.92 | df=5)$;
Using Table 8: $P < 0.005$
Using computer: $P = 0.003$
b. $P < \alpha$
-- using classical approach ----------------
a. $\chi^2(5, 0.01) = 15.1$ $\chi^2 \geq 15.1$
b. χ^{2*} is in the critical region.

Step 5: a. Reject H_o
b. We do have sufficient evidence to show that the sample distribution is different than the Golf magazine distribution, at the 0.01 level of significance.

11.63 The probability associated with each weight interval can be found using the standard normal distribution on Table 3 (Appendix B, ES11-pp716-717). (slight discrepancies based on type of table)

$P(x < 130) = P[z < (130-160)/15] = P(z < -2.00) = \underline{0.0228}$

$P(130 \leq x < 145) = P[-2.00 < z < (145-160)/15] = P(-2.00 < z < -1.00) = 0.1587 - 0.0228 = \underline{0.1359}$

$P(145 \leq x < 160) = P[-1.00 < z < (160-160)/15] = P(-1.00 < z < 0.00) = 0.5000 - 0.1587 = \underline{0.3413}$

$P(160 \leq x < 175) = P[0.00 < z < (175-160)/15] = P(0.00 < z < 1.00) = 0.8413 - 0.5000 = \underline{0.3413}$

$P(175 \leq x < 190) = P[1.00 < z < (190-160)/15] = P(1.00 < z < 2.00) = 0.9773 - 0.8413$
$= \underline{0.1360} \approx 0.1359$

$P(x \geq 190) = P(z > 2.00) = 1.0000 - 0.9773 = \underline{0.0227} \approx 0.0228$

Step 1: a. The proportions of weight: P(< 130), P(130-144), P(145-159), P(160-174), P(175-189), (190 and over).
b. H_o: The weights are normally distributed about a mean of 160 with a standard deviation of 15 pounds
H_a: The weights are not N(160,15)

Step 2: a. Assume that the 300 adult males represent a random sample.
b. χ^2 with df = 5   c. $\alpha = 0.05$

Step 3: a. sample information given in exercise
b. $\chi^2 = \Sigma[(O-E)^2/E]$ (as found on accompanying table)

Expected values = n·p = 300·p

	<130	130-145	145-160	160-175	175-190	>190
Observed	7	38	100	102	40	13
Expected	6.84	40.77	102.39	102.39	40.77	6.84
$(O-E)^2/E$	0.004	0.188	0.056	0.001	0.015	5.548

$\chi^{2*} = 5.812$

Step 4: -- using p-value approach --------------------
a. $P = P(\chi^2 > 5.812 | df=5)$;
Using Table 8: $0.25 < P < 0.50$
Using computer: $P = 0.325$
b. $P > \alpha$
-- using classical approach -----------------
a. $\chi^2(5, 0.05) = 11.1$ $\chi^2 \geq 11.1$
b. χ^{2*} is in the noncritical region.

Step 5: a. Fail to reject H_o
b. We do not have sufficient evidence to show that this data is not normally distributed with $\mu = 160$ and $\sigma = 15$, at the 0.05 level of significance.

11.65 a.
Step 1: a. The distribution of colors in a bag of M&M's.
b. H_o: P(brown) = 0.30, P(red & yellow) = 0.20, P(blue, green & orange) = 0.10
H_a: Distributions are different
Step 2: a. Assume a random sample.
b. χ^2 with df = 5   c. $\alpha = 0.05$
Step 3: a. sample information given in exercise
b. $\chi^2 = \Sigma[(O-E)^2/E]$ (as found on accompanying table)

Expected	red	green	blue	orange	yellow	brown
Probability	0.20	0.10	0.10	0.10	0.20	0.30
Expected Frequency	11.6	5.8	5.8	5.8	11.6	17.4
Observed Frequency	15	9	3	3	9	19

$\chi^{2*} = 0.99655 + 1.76552 + 1.35172 + 1.35172 + 0.58276 + 0.14713 = \underline{6.1954}$

Step 4: -- using p-value approach --------------------
a. $P = P(\chi^2 > 6.1954 | df=5)$;
Using Table 8: $P > 0.25$
Using computer/calculator: $P = 0.2877$
b. $P > \alpha$
-- using classical approach -----------------
a. critical region: $\chi^2 \geq 11.1$

b. χ^{2*} falls in the noncritical region

Step 5: a. Fail to reject H_o
b. There is not significant evidence to show the distribution is different than the target distribution, at the 0.05 level of significance.

b.
Step 1: a. The distribution of colors in a bag of M&M's.
b. H_o: P(brown) = 0.30, P(red & yellow) = 0.20, P(blue, green & orange) = 0.10
H_a: Distributions are different
Step 2: a. Assume a random sample.
b. χ^2 with df = 5   c. $\alpha = 0.05$
Step 3: a. sample information given in exercise
b. $\chi^2 = \Sigma[(O-E)^2/E]$ (as found on accompanying table)

Expected	red	green	blue	orange	yellow	brown
Probability	0.20	0.10	0.10	0.10	0.20	0.30
Expected Frequency	23.4	11.7	11.7	11.7	23.4	35.1
Observed Frequency	24	26	22	6	12	27

$\chi^{2*} = 0.0154 + 17.4778 + 9.0675 + 2.7769 + 5.5538 + 1.8692 = \underline{36.761}$

Step 4: -- using p-value approach --------------------
a. $P = P(\chi^2 > 36.761 | df=5)$;
Using Table 8: $P < 0.005$
Using computer/calculator: $P = 0+$
b. $P < \alpha$
-- using classical approach -----------------
a. critical region: $\chi^2 \geq 11.1$
b. χ^{2*} falls in the critical region

Step 5: a. Reject H_o
b. There is significant evidence to show the distribution is different than the target distribution, at the 0.05 level of significance.

c.
Step 1: a. The distribution of colors in a bag of M&M's.
b. H_o: P(brown) = 0.30, P(red & yellow) = 0.20, P(blue, green & orange) = 0.10
H_a: Distributions are different
Step 2: a. Assume a random sample.
b. χ^2 with df = 5   c. $\alpha = 0.05$

Step 3: a. sample information given in exercise
b. $\chi^2 = \Sigma[(O-E)^2/E]$ (as found on accompanying table)

Expected	red	green	blue	orange	yellow	brown
Probability	0.20	0.10	0.10	0.10	0.20	0.30
Expected Frequency	342.6	171.3	171.3	171.3	342.6	513.9
Observed Frequency	288	222	217	199	413	374

$\chi^{2*} = 08.7016 + 15.0058 + 12.1920 + 4.4792 + 14.4663 + 38.0853 = \underline{92.93}$

Step 4: -- using p-value approach ------------------
a. $P = P(\chi^2 > 92.93 | df=5)$;
Using Table 8: $P < 0.005$
Using computer/calculator: $P = 0+$

b. $P < \alpha$

-- using classical approach ----------------
a. critical region: $\chi^2 \geq 11.1$
b. χ^{2*} falls in the critical region

Step 5: a. Reject H_o
b. There is significant evidence to show the distribution is different than the target distribution, at the 0.05 level of significance.

d. Chi-square is not very sensitive to the variability within small samples, and becomes more sensitive to those variations as the sample size gets larger.

Exercise 11.67 is a good situation for a discussion about potential effect of round-off errors

FYI: In exercise 11.67, do not use rounded values.   Note – use of a rounded p-value will lead to the opposite decision of what the classical method produces.

11.67 a. yes, equal proportions mean no effect due to shift (independent of shift)
b. shift had an effect, defects depend on shifts
c.
Step 1: a. The proportions of type of defects and shift times
b. H_o: proportions are not different from shift to shift (independence)
H_a: proportions are different from shift to shift
Step 2: a. Assume a random sample.
b. χ^2 with df = 6   c. $\alpha = 0.01$
Step 3: a. sample information given in exercise
b. $\chi^2 = \Sigma[(O-E)^2/E]$
Expected counts are printed below observed counts
Chi-Square contributions are printed below expected counts

```
              A      B      C      D    Total
    1        17     23     43     17     100
           22.94  24.16  39.45  13.46
           1.536  0.056  0.320  0.934

    2        27     37     33      9     106
           24.31  25.61  41.82  14.26
           0.297  5.067  1.859  1.942

    3        31     19     53     18     121
           27.75  29.23  47.73  16.28
           0.380  3.582  0.581  0.181

    Total    75     79    129     44     327

    Chi-Sq = 16.734, DF = 6, P-Value = 0.010
```

Step 4: -- using p-value approach -------------------
a. $P = P(\chi^2 > 16.734 | df=6)$;

Using Table 8: $0.01 < P < 0.025$
Using computer: $P = 0.0103$

b. $P > \alpha$

-- using classical approach -----------------

a. $\chi^2(6, 0.01) = 16.8 \qquad \chi^2 \geq 16.8$

b. χ^{2*} is in the noncritical region.

Step 5: a. Fail to reject H_o

b. The data show sufficient evidence to conclude that the type of defect is independent of the shift in which it occurred, at the 0.01 level of significance.

11.69 Step 1: a. The proportion of each political preference for each age group and the proportion of each age group who answer with each political preference.

b. H_o: Political preference is independent of age.
H_a: Political preference is not independent of age.

Step 2: a. Assume a random sample.

b. χ^2 with df = 4 \qquad c. $\alpha = 0.01$

Step 3: a. sample information given in exercise

b. $\chi^2 = \Sigma[(O-E)^2/E]$ (as found on accompanying table)

Expected values:

	20-35	36-50	Over 50
Conservative	28.00	30.00	22.00
Moderate	73.50	78.75	57.75
Liberal	38.50	41.25	30.25

$\chi^{2*} = 2.286 + 3.333 + 0.182 + 0.575 + 0.496 + 2.815 + 0.058 + 6.402 + 7.192 = 23.339$

Step 4: -- using p-value approach ------------------

a. $P = P(\chi^2 > 23.339 | df=4)$;

Using Table 8: $P < 0.005$
Using computer/calculator: $P = 0.000+$

b. $P < \alpha$

-- using classical approach -----------------

a. $\chi^2(4, 0.01) = 13.3 \qquad \chi^2 \geq 13.3$

b. χ^{2*} falls in the critical region

Step 5: a. Reject H_o

b. There is sufficient evidence to reject the null hypothesis and conclude that political preference is not independent of age, at the 0.01 level of significance.

11.71 Step 1: a. The distributions of housing starts: P(northeast for years category), P(south for years category), P(midwest for years category), P(west for years category).

b. H_o: The distribution of housing starts across the regions is the same for all years.

H_a: The distributions are different.

Step 2: a. Assume a random sample.

b. χ^2 with df = 6 \qquad c. $\alpha = 0.05$

Step 3: a. sample information given in exercise

b. $\chi^2 = \Sigma[(O-E)^2/E]$ (as found on accompanying table)
Expected counts are printed below observed counts

	1996-2000	2001-2005	2006-2010	Total
1	145	161	170	476
	160.20	158.05	157.75	
2	710	687	688	2085
	701.71	692.31	690.97	
3	331	314	313	958
	322.42	318.10	317.48	
4	382	385	373	1140
	383.67	378.53	377.80	
Total	1568	1547	1544	4659

Chi-Sq = 1.442 + 0.055 + 0.952 + 0.098 + 0.041 + 0.013 + 0.228 + 0.053 + 0.063 + 0.007 + 0.111 + 0.061 = <u>3.123</u>

$\chi^{2*} = 3.123$

Step 4: -- using p-value approach ------------------
a. $P = P(\chi^2 > 3.123 | df=6)$;
Using Table 8: $0.75 < P < 0.90$
Using computer: $P = 0.793$

b. $P > \alpha$

-- using classical approach ----------------
a. $\chi^2(6, 0.05) = 12.6$
b. χ^{2*} is in the noncritical region.

Step 5: a. Fail to reject H_0
b. There is not sufficient evidence to show the distribution of housing starts across the regions are different, at the 0.05 level of significance.

11.73 Step 1: a. The proportions of popcorn that popped and did not pop:
P(Brand A), P(Brand B), P(Brand C), P(Brand D)
b. H_0: Proportion of popcorn that popped is the same for all brands.
H_a: The proportions are not the same for all brands.

Step 2: a. Assume a random sample.
b. χ^2 with df = 3   c. $\alpha = 0.05$

Step 3: a. sample information given in exercise
b. $\chi^2 = \Sigma[(O-E)^2/E]$
Expected values:

	A	B	C	D	Total
Popped	88	88	88	88	352
Not popped	12	12	12	12	48
Totals	100	100	100	100	400

$\chi^{2*} = \Sigma[(O-E)^2/E] = 0.333 + 1.333 + 0.083 + 0.750 + 0.045 + 0.182 + 0.011 + 0.102$
$\chi^{2*} = 2.839$

Step 4: -- using p-value approach ------------------
a. $P = P(\chi^2 > 2.839 | df=3)$;
Using Table 8: $0.25 < P < 0.50$
Using computer: $P = 0.417$
b. $P > \alpha$
-- using classical approach -----------------
a. $\chi^2(3, 0.05) = 7.81$ $\chi^2 \geq 7.81$
b. χ^{2*} falls in the noncritical region

Step 5: a. Fail to reject H_o
b. There is not sufficient evidence to show that the proportions of popped corn are not the same for all brands, at the 0.05 level of significance.

11.75 a. 2003: $18650/(18650+6812) = 0.732 = 73.2\%$ deceased donors
2004: $20018/(20018+6966) = 0.742 = 74.2\%$ deceased donors
These percentages do not seem significantly different.

b.
Step 1: a. The proportions for organ donors from a deceased person and living person, over 2003 and 2004.
b. H_o: The ratio of deceased donor to living donor is the same.
H_a: The ratio of organ donors is not the same.

Step 2: a. Given a random sample.
b. χ^2 with df = 1   c. $\alpha = 0.05$

Step 3: a. sample information given in exercise
b. $\chi^2 = \Sigma[(O-E)^2/E]$ (as found on accompanying table)
Expected counts are printed below observed counts
Chi-Square contributions are printed below expected counts

	From a deceased donor	From a living donor	Total
1	18650	6812	25462
	18772.92	6689.08	
	0.805	2.259	
2	20018	6966	26984
	19895.08	7088.92	
	0.759	2.131	
Total	38668	13778	52446

$\chi^{2*} = 5.955$

Step 4: -- using p-value approach ------------------
a. $P = P(\chi^2 > 5.955 | df=1)$;
Using Table 8: $0.01 < P < 0.025$
Using computer/calculator: $P = 0.015$
b. $P < \alpha$

-- using classical approach ----------------
a. $\chi^2(1,0.05) = 3.84$ $\chi^2 \geq 3.84$
b. χ^{2*} falls in the critical region

Step 5: a. Reject H_o
b. The ratio of deceased versus living donors did change significantly between 2003 and 2004, at the 0.05 level of significance.

c. The decision reached in the hypothesis is the opposite of what was predicted in part a. With very large sample sizes, differences in proportions must be very small to be considered non-existent.

11.77 Conditions for rolling a balanced die 600 times:
The critical value is $\chi^2(5,0.05) = 11.1$ (6 possible outcomes)
With 600 rolls, the expected value for each cell is 100.
Many combinations of observed frequencies are possible to cause us to reject the equally likely hypothesis. The combinations will have to have a calculated χ^2 value greater than 11.1 or a p-value less than 0.05. Two possibilities are presented.

1. If each observed frequency is different from the expected by the same amount, then $11.1/6 = 1.85$ is the amount of chi-square that would come from each cell.

$(O-E)^2/E = (O - 100)^2/100 = 1.85$

$(O - 100)^2 = 185$
$O - 100 = \pm 13.6$
$O = 86 \text{ or } 114$

That is, if three of the observed frequency values are 86 and the other three are 114, the faces of the die will be declared not to be equally likely.

Row	P	OBS	EXP	CHI-SQ
1	0.166667	86	100.000	1.96005
2	0.166667	86	100.000	1.96005
3	0.166667	86	100.000	1.96005
4	0.166667	114	100.000	1.95994
5	0.166667	114	100.000	1.95994
6	0.166667	114	100.000	1.95994

Row	SUM(P)	SUN(OBS)	SUM(EXP)	CHI-SQ*
1	1.00000	600	600.001	11.7600

DF = 5   p-value 0.038
Note χ^2 value and p-value

2. Now suppose just one is different and the other five all occur with the same frequency:

Remember, the total observed most be 600. Therefore, for every five one outcome is different from the expected, the other five each must be different by one to balance. If the five are each different from the expected by x, then the one that is very different is off by 5x. The sum of 5 - x's squared and 5x squared is $30x^2$. Thus,

$30x^2 = 11.1$

$x^2 = 11.1/30 = 0.37$

$(O-E)^2/E = (O-100)^2/100 = 0.37$

$(O-100)^2 = 37$

O - 100 = ±6.08 (round-up)

O = either 93 or 107 for the five cells, and

O for the other cell must be off by 5(7) or 35; it is either 65 or 135.

Row	P	OBS	EXP	CHI-SQ
1	0.166667	93	100.000	0.4900
2	0.166667	93	100.000	0.4900
3	0.166667	93	100.000	0.4900
4	0.166667	93	100.000	0.4900
5	0.166667	93	100.000	0.4900
6	0.166667	135	100.000	12.2498

Row	SUM(P)	SUN(OBS)	SUM(EXP)	CHI-SQ*
1	1.00000	600	600.001	14.7000

DF = 5   p-value 0.012

Note χ^2 and p-value.

11.79   Answers will vary.

11.81   Answers will vary.

11.83   Answers will vary.

CHAPTER 12 ∇ ANALYSIS OF VARIANCE

Chapter Preview

In Chapters 8 and 9, hypothesis tests were demonstrated for testing a single mean. Hypothesis tests between two means were subsequently demonstrated in Chapter 10. To continue in this fashion, Chapter 12 introduces the concept of the analysis of variance technique (ANOVA) so that a hypothesis test for the equality of several means can be completed. The F-distribution will be utilized in this test, since we will be comparing the measures of variation (variance) among the different sets of data and the measure of variation (variance) within the sets the data.

Information on average one-way commute times in major U.S. cities is featured in the chapter's opening section "The Morning Rush".

SECTION 12.1 EXERCISES

12.1 With respect to average: 3 lowest – Dallas, Seattle, St. Louis; 3 highest – Atlanta, Boston, Philadelphia. Largest difference between St. Louis and Philadelphia, both in average and limits.

12.3 Use a two-sample t-test to compare the lowest and highest commute times.

12.5
 a. Atlanta: 16.52 to 32.81; Boston: 24.16 to 41.84
 b. Means appear different but intervals overlap by half.
 c. Dallas: 23.85 to 37.87
 d. Dallas very close to Boston and also overlaps Atlanta by half.
 e. Mean commute times are the same.
 f. Computed confidence intervals are all wider and computed averages are all higher than those given in "The Morning Rush"

12.7 Units produced per hour at each temperature level

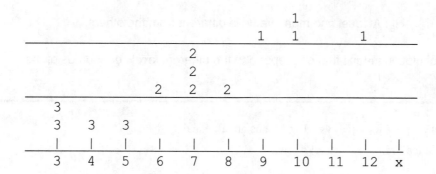

Yes, there appears to be a difference between the three sets.

12.9
 a. df(error) = 20 – 3 = 17; SS(Factor) = 164.2 – 40.4 = 123.8
    MS(Factor) = 123.8/3 = 41.2667
    MS(Error) = 40.4/17 = 2.3765

 b. F* = 41.2667/2.3765 = 17.36

SECTION 12.2 EXERCISES

12.11 The boxplot shows greater variability between the four levels. The means for each are all quite different.

12.13
 a. Yes, cities listed are quite different than national average.
 b. The cities are independent and each parking establishment should be independent within the city. Random samples of parking establishments were selected for each city and averages calculated.

Each would be part of the large set of all parking establishment costs that was used to calculate the national average of $15.

SECTION 12.3 EXERCISES

12.15   a. 0   b. 2   c. 4   d. 31   e. 393

WRITING HYPOTHESES FOR THE DIFFERENCE AMONG SEVERAL MEANS

null hypothesis: $H_o: \mu_1 = \mu_2 = \mu_3 = \ldots = \mu_n$
or
H_o: The mean values for all n levels of the experiment are the same.
[factor has no effect]

alternative hypothesis: H_a: The means are not all equal.
[factor has an effect]
or
H_a: At least one mean value is different from the others.

Use subscripts on the population means that correspond to the different levels or sources of the experiment.

12.17   a. $H_o: \mu_1 = \mu_2 = \mu_3 = \mu_4 = \mu_5$ vs. H_a: Means not all equal
[mean scores are all same]

b. $H_o: \mu_1 = \mu_2 = \mu_3 = \mu_4$ vs. H_a: Means not all equal
[mean scores are all same]

c. $H_o: \mu_1 = \mu_2 = \mu_3 = \mu_4$ vs. H_a: Means not all equal
[factor has no effect]   [has an effect]

d. $H_o: \mu_1 = \mu_2 = \mu_3$ vs. H_a: Means not all equal
[no effect]   [has an effect]

Review the rules for calculating the p-value in: ES11-p527, IRM-p537, if necessary. Remember to use the F-distribution, therefore either Tables 9A, B or C will be used to find probabilities. Review of the use of these tables can be found in: ES11-pp523-524, IRM-pp537-539.

Use Tables 9A, B or C (Appendix B, ES11-pp722-727) depending on α. Locate the critical value using the degrees of freedom for the numerator (factor) and the degrees of freedom for the denominator(error). Reviewing how to determine the test criteria in: ES11-pp523-524, IRM-pp537-539, as it is applied to the F-distribution may be helpful.

12.19 a. b.
c.

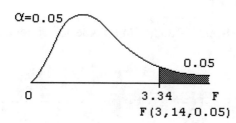

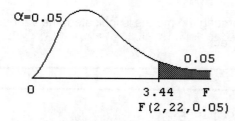

12.21 a. 0.04 of the probability distribution associated with F and a true null hypothesis is more extreme than F*. That is, area under the curve and to the right of F*.

b. Reject the null hypothesis; since the p-value is less than the previously set value for alpha.

c. Fail to reject the null hypothesis; since the p-value is greater than the previously set value for alpha.

12.23 a. The test factor has no effect on the mean at the tested levels or the mean value of the variable is the same at all levels of the test factor.

b. The test factor does have an effect on the mean at the tested levels

c. For the p-value approach, $\mathbf{P} = P(F > F^*)$ must be $\leq \alpha$.
For the classical approach, the calculated value of F must fall in the critical region; that is, the variance between levels of the factor must be significantly larger than variance within the levels.

d. The tested factor has a significant effect on the variable.

e. For the p-value approach, $\mathbf{P} = P(F > F^*)$ must be $> \alpha$.
For the classical approach, the calculated value of F must fall in the non-critical region; that is, the variance between levels of the factor must not be significantly larger than variance within the levels.

f. The tested factor does not have a significant effect on the variable.

Hypothesis Test for the Difference Among Several Means, Independent Samples

Review the parts to a hypothesis test as outlined in: ES11-pp371&388, IRM-pp343&357, if needed. Slight changes will occur in:

1. **the hypotheses**: (see box before exercise 12.17)
2. **the calculated test statistic**

$$F = \frac{MS(factor)}{MS(error)}$$

3. If H_o is rejected, a significant difference among the means is indicated, the various levels of the factor do have an effect.

   If H_o is not rejected, no significant difference among the means is indicated, the various levels of the factors do not have an effect.

FORMULAS FOR ANOVA TABLE

$$SS(total) = \sum(x^2) - \frac{(\sum x)^2}{n}$$

$$SS(factor) = \left(\frac{C_1^2}{k_1} + \frac{C_2^2}{k_2} + \frac{C_3^2}{k_3} + \ldots\right) - \frac{(\sum x)^2}{n}$$, where C_i = column total

k_i = number of data values in the ith column

(check: $n = \sum k_i$)

$$SS(error) = \sum x^2 - \left(\frac{C_1^2}{k_1} + \frac{C_2^2}{k_2} + \frac{C_3^2}{k_3} + \ldots\right) \text{ or } = SS(total) - SS(factor)$$

$df_{total} = n-1$, $df_{factor} = c-1$, $df_{error} = n - c$

$df_{total} = df_{factor} + df_{error}$

ANOVA Table

Source	df	SS	MS
Factor			
Error			
Total			

MS(factor) = SS(factor)/df(factor)

MS(error) = SS(error)/df(error)

12.25 Verify -- answers given in exercise.

12.27 a. $n = 17$
  b. df(Group) $= 2 = c - 1$; $c = 3$ groups
  c. MS(Group) $= 11.0/2 = 5.50$
     MS(Error) $= 36.53/14 = 2.609 = 2.61$
  d. $F = 5.50/2.61 = 2.107 = 2.11$
  e. $P = P(F > 2.11 | df_n = 2, df_d = 14)$
     using Tables 9: $P > 0.05$
     using a computer/calculator: $P = 0.1581 \approx 0.159$
  f. Most likely with a p-value of 0.159, the decision would be to Fail to Reject Ho. The conclusion would be that there is not sufficient evidence to show the group means are not all the same.

Computer and/or calculator commands to ANOVA hypothesis tests can be found in ES11-p594. Explanation of the output is found in ES11-p595.

12.29 Step 1: a. The mean level of work for a new worker, the mean level of work for worker A, the mean level of work for worker B.
  b. H_o: The mean values for workers are all equal.
  H_a: The mean values for workers are not all equal.

Step 2: a. Assume the data were randomly collected and are independent, and the effects due to chance and untested factors are normally distributed.
  b. F   c. $\alpha = 0.05$

Step 3: a. $n = 15$, $C_1 = 46$, $C_2 = 58$, $C_3 = 57$, $T = 161$, $\Sigma x^2 = 1771$

Source	df	SS	MS	F*
Work	2	17.73	8.87	4.22
Error	12	25.20	2.10	
Total	14	42.93		

$F^* = 8.87/2.10 = 4.22$

Step 4: -- using p-value approach ---------------
  a. $P = P(F > 4.22 | df_n = 2, df_d = 12)$;
     Using Table 9: $0.025 < P < 0.05$
     Using computer: $P = 0.041$
  b. $P < \alpha$
  -- using classical approach -------------
  a. critical region: $F \geq 3.89$
  b. F* falls in the critical region

Step 5: a. Reject H_o.
  b. There is significant difference between the workers with regards to the mean amount of work produced, at the 0.05 level of significance.

12.31 Step 1: a. The mean level of mpg for a 4 Cyl engine, the mean level of mpg for a 5 Cyl engine, the mean level of mpg for a 6 Cyl engine, the mean level of mpg for a 8 Cyl engine.
b. H_o: The mean mpg values for engines are all equal.
H_a: The mean mpg values for engines are not all equal.
Step 2: a. Assume the data were randomly collected and are independent, and the effects due to chance and untested factors are normally distributed.
b. F   c. $\alpha = 0.01$
Step 3: a. n = 38, $C_1 = 234$, $C_2 = 207$, $C_3 = 193$, $C_4 = 148$, T = 782, $\Sigma x^2 = 16292$

Source	df	SS	MS	F*
Factor	3	130.66	43.55	21.59
Error	34	68.60	2.02	
Total	37	199.26		

$F^* = 43.55/2.02 = 21.59$

Step 4: -- using p-value approach ---------------
a. $P = P(F > 21.59 | df_n = 3, df_d = 34)$;
Using Table 9: $P < 0.01$
Using computer: $P = 0.000$
b. $P < \alpha$
-- using classical approach -------------
a. critical region: $F \geq 4.51$
b. F* falls in the critical region

Step 5: a. Reject H_o.
b. There is significant evidence to show that the mpg for pickup trucks is not the same for all four engine sizes, at the 0.01 level of significance.

12.33 Step 1: a. The mean ratings obtained by the restaurants in each of the three categories.
b. H_o: $\mu_F = \mu_D = \mu_S$ (no difference in ratings)
H_a: The means of the ratings obtained by the restaurants in the three categories are not all equal.
Step 2: a. Assume the data were randomly collected and are independent, and the effects due to chance and untested factors are normally distributed.
b. F   c. $\alpha = 0.05$
Step 3: a. n = 15, $C_1 = 95$, $C_2 = 82$, $C_3 = 84$, T = 261, $\Sigma x^2 = 4593$
b.

Source	DF	SS	MS	F	P
Factor	2	19.60	9.80	3.68	0.057
Error	12	32.00	2.67		
Total	14	51.60			

$F^* = 9.80/2.67 = 3.68$

Step 4: -- using p-value approach ---------------
a. $P = P(F > 3.68 | df_n = 2, df_d = 12)$;

Using Table 9: **P** > 0.05
Using computer: **P** = 0.057
b. **P** > α
-- using classical approach ------------
a. critical region: F ≥ 3.89
b. F* is in the noncritical region

Step 5: a. Fail to reject H_o.

b. The data shows no significant evidence that the means of the three categories of ratings given to the restaurants are different, at the 0.05 level of significance.

12.35 a. Step 1: a. The mean age of three test groups: the mean age for the TTS group, the mean age for the Antivert group, the mean age for the placebo group.
b. H_o: The mean age for groups are all equal.
H_a: The mean age for groups are not all equal.

Step 2: a. Assume the data were randomly collected and are independent, and the effects due to chance and untested factors are normally distributed.
b. F   c. α = 0.05

Step 3: a. n = 58, k_1 = 18, C_1 = 846, k_2 = 21, C_2 = 894, k_3 = 19, C_3 = 805, T = 2545,
Σx^2 = 120,549

Source	df	SS	MS	F*
Group	2	255	127	0.81
Error	55	8622	157	
Total	57	8876		

F* = 127/157 = 0.81

Step 4: -- using p-value approach ---------------
a. **P** = P(F > 0.81|df_n = 2, df_d = 55);
Using Table 9: **P** > 0.05
Using computer: **P** = 0.449
b. **P** > α
-- using classical approach -------------
a. F(2,55,0.05) = 3.23   F ≥ 3.23
b. F* is not in the critical region

Step 5: a. Fail to reject H_o.

b. The data does not show a significant difference between the mean ages of the groups, at the 0.05 level of significance.

12.37 a.

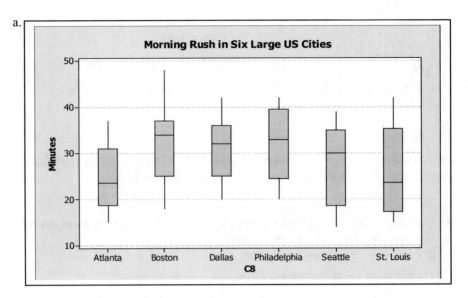

b. No, each boxplot looks about the same as the others

c.

Step 1: a. The mean commute times for Atlanta, Boston, Dallas, Philadelphia, Seattle, St. Louis

b. H_o: The average morning commute time in 6 U.S. cities are all equal.

H_a: The average morning commute time in 6 U.S. cities is different.

Step 2: a. Assume the data were randomly collected and are independent, and the effects due to chance and untested factors are normally distributed.

b. F   c. $\alpha = 0.05$

Step 3: a. n = 36

b.

Source	DF	SS	MS	F
Factor	5	372.5	74.5	0.96
Error	30	2329.0	77.6	
Total	35	2701.6		

$F^* = 74.5/77.6 = 0.96$

Step 4: -- using p-value approach ---------------

a. $P = P(F > 0.96 | df_n = 5, df_d = 30)$;

Using Table 9: $P > 0.05$

Using computer: $P = 0.458$

b. $P > \alpha$

-- using classical approach -------------

a. critical region: $F \geq 2.53$

b. F^* falls in the noncritical region

Step 5: a. Fail to reject H_o.

b. There is no significant difference between the average commute times, at the 0.05 level of significance.

d. Yes, they agree, the centers and spreads for each city appear the same graphically and statistically.

e. No, no, no, just asked in a different format.

12.39 a.

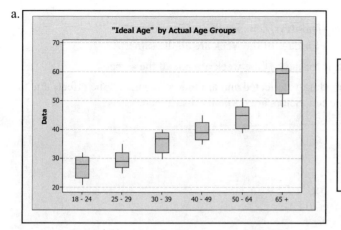

The graph suggests that each age group has its own "ideal age" and it increases with age, but each age group seems to have a range of ideal ages within a 10 year range.

b.
Step 1: a. The mean "ideal age" for different age groups.
b. H_o: The mean "ideal age" is the same for all age groups.
H_a: The mean "ideal age" is not the same for all age groups.

Step 2: a. Assume the data were randomly collected and are independent, and the effects due to chance and untested factors are normally distributed.
b. F   c. $\alpha = 0.05$

Step 3: a. n = 36, C_{18} = 163, C_{25} = 207, C_{30} = 250, C_{40} = 197, C_{50} = 222, C_{65} = 346, T = 1385, Σx^2 = 57583

b.
Source	DF	SS	MS	F	P
Factor	5	3765.3	753.1	42.33	0.000
Error	30	533.7	17.8		
Total	35	4299.0			

$F^* = 753.1/17.8 = 42.33$

Step 4: -- using p-value approach ---------------
a. $P = P(F > 42.33 | df_n = 5, df_d = 30)$;
Using Table 9: $P < 0.01$
Using computer: $P = 0.000$
b. $P < \alpha$
-- using classical approach -------------
a. critical region: $F \geq 2.53$
b. F^* is in the critical region

Step 5: a. Reject H_o.
b. The data shows significant evidence that the mean "ideal age" is not the same for each of the age groups, at the 0.05 level of significance.

c. Answers will vary but it appears that a person's "ideal age" depends on their current age.

d. The boxplot showed that the mean "ideal ages" were different since the means do not line up horizontally. The idea that each age group has its own ideal age and it increases with current age is suggested by the fact that the means do line up on a nearly straight diagonal line.

12.41 a.

Step 1: a. The mean percent tips for Tuesdays, Thursdays, and Saturdays
b. H_o: The mean percent tips for days of the week are all the same.
H_a: The mean percent tips for days of the week are not all the same.

Step 2: a. Assume the data were randomly collected and are independent, and the effects due to chance and untested factors are normally distributed.
b. F   c. $\alpha = 0.05$

Step 3: a. n = 36
b.

Source	df	SS	MS	F*
Factor	2	132.1	66.0	2.46
Error	33	886.8	26.9	
Total	35	1018.9		

Step 4: -- using p-value approach ---------------
a. $P = P(F > 2.46 | df_n = 2, df_d = 33)$;
Using Table 9: $P > 0.05$
Using computer: $P = 0.101$
b. $P > \alpha$
-- using classical approach -------------
a. critical region: $F \geq 3.32$
b. F* falls in the noncritical region

Step 5: a. Fail to reject H_o.
b. There is no significant evidence that day of the week has an effect on the percent tip received, at the 0.05 level of significance.

b.

Step 1: a. The mean percent tips for $0-$29, $30-$59, $60-$89 amounts
b. H_o: The mean percent tips for bill amounts are all the same.
H_a: The mean percent tip for bill amounts are not all the same.

Step 2: a. Assume the data were randomly collected and are independent, and the effects due to chance and untested factors are normally distributed.
b. F   c. $\alpha = 0.05$

Step 3: a. n = 36
b.

Source	df	SS	MS	F*
Factor	2	254.9	127.4	5.50
Error	33	764.0	23.2	
Total	35	1018.9		

Step 4: -- using p-value approach ---------------
a. $P = P(F > 5.50 | df_n = 2, df_d = 33)$;
Using Table 9: $P < 0.01$
Using computer: $P = 0.009$
b. $P < \alpha$
-- using classical approach -------------

a. critical region: $F \geq 3.32$
b. F^* falls in the critical region

Step 5: a. Reject H_o.

b. There is significant evidence that the bill amount has an effect on the percent tip received, at the 0.05 level of significance.

12.43 a.

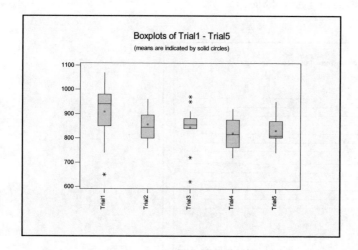

Trial 1 appears to have more variation and a higher mean.

b.
Step 1: a. The mean measurement of the velocity of light in air for five trials.
b. H_o: The mean trial result is the same for each of the five trials.
H_a: The mean trial result is not the same for each of the five trials.
Step 2: a. Assume the data were randomly collected and are independent, and the effects due to chance and untested factors are normally distributed.
b. F   c. $\alpha = 0.05$
Step 3: a. n = 100
b.

Source	df	SS	MS	F*
Factor	4	94514	23629	4.29
Error	95	523510	5511	
Total	99	618024		

Step 4: -- using p-value approach ---------------
a. $P = P(F > 4.29 | df_n = 4, df_d = 95)$;
Using Table 9: $P < 0.01$
Using computer: $P = 0.003$
b. $P < \alpha$
-- using classical approach -------------
a. $F(4, 95, 0.05) \approx 2.49$
b. F^* is in the critical region

Step 5: a. Reject H_o.

b. There is sufficient evidence to show that at least one mean is significantly different from the others at the 0.05 level of significance.

c. The first trial had the highest mean and largest standard deviation. These results may have been due to start up procedures and methods.

12.45 a. Answers will vary but it would make sense that the number of items purchased would increase with an increase in number of customers. This being the case, the months of November and December should have the highest volume.

b.
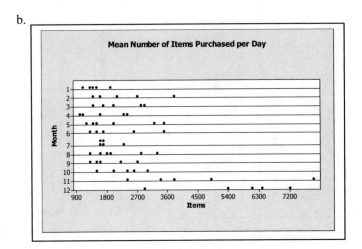

c. Answers may vary but yes, the graph supports the conjecture that the months with the larger number of customers would be the same months for the large number of items purchased per day.

d.
Step 1: a. The mean number of items purchased per day by month.
   b. H_o: The mean number of items purchased per day is the same for each of the twelve months.
   H_a: The mean number of items purchased per day is not the same for each of the twelve months.
Step 2: a. Assume the data were randomly collected and are independent, and the effects due to chance and untested factors are normally distributed.
   b. F   c. $\alpha = 0.05$

Step 3: a. n = 62

Source	DF	SS	MS	F	P
Month	11	84869019	7715365	7.56	0.000
Error	50	51003447	1020069		
Total	61	135872465			

Step 4: -- using p-value approach ---------------
   a. $P = P(F > 7.56 | df_n = 11, df_d = 50)$;
      Using Table 9: $P < 0.01$
      Using computer: $P = 0.000$
   b. $P < \alpha$

-- using classical approach -------------
a. $F(11, 50, 0.05) = 2.08$
b. F^* is in the critical region

Step 5: a. Reject H_o.

b. There is sufficient evidence to show that at least one monthly mean is significantly different from the others, that is that the month does effect the mean number of items purchased per day, at the 0.05 level of significance.

e. Answers may vary but based on answers given above, the reasoning in part a is supported by the F-test results.

CHAPTER EXERCISES

12.47 a. H_o: The mean amount of salt is the same in all tested brands of peanut butter.

H_a: The mean amount of salt is not the same in all tested brands of peanut butter.

b. Assumptions: samples were randomly selected and are independent, and the effects due to chance and untested factors are normally distributed.

$\alpha = 0.05$;   test statistic: F

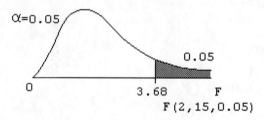

c. Fail to reject H_o. There is no evidence to show a significant difference in the mean amounts of salt in the tested brands.

d. Since the p-value is quite large (much larger than 0.05), it tells us the sample data is quite likely to have occurred under the assumed conditions and a true null hypothesis. Therefore, we 'fail to reject H_o.'

12.49 a. H_o: The mean amount spent is the same for all four supermarkets.

H_a: The mean amount spent is not the same for all four supermarkets.

b. Fail to reject H_o. There is not sufficient evidence to reject the null hypothesis – that is, there is not sufficient evidence to show the mean amount spent is not the same for all four supermarkets. In other words, with regards to the mean amount spent, it does not matter financially where you grocery shop.

c. No, there does not seem to be a difference among the mean grocery cost for these four supermarkets - the means range form $2.47 to $2.72, that is only 25 cent difference.

d. No, the 4 standard deviations listed are all between 2.17 and 2.93 which seem very close in value, therefore showing little difference.

e. The means are nearly identical and the standard deviations are also very close in value. Almost all of the variation in this data is within the stores (factor levels) thus totally supporting the results of the hypothesis test.

12.51 Step 1: a. The mean amounts of soft drink dispensed: the mean amount for machine A, the mean amount for machine B, the mean amount for machine C, the mean amount for machine D, the mean amount for machine E.

b. H_o: The mean amounts dispensed by the machines are all equal.

H_a: The mean amounts dispensed by the machines are not all equal.

Step 2: a. Assume the data were randomly collected and are independent, and the effects due to chance and untested factors are normally distributed.

b. F c. $\alpha = 0.01$

Step 3: a. n = 18, C_A = 16.5, C_B = 20.6, C_C = 16.9, C_D = 19.1, C_E = 21.8, T = 94.9, Σx^2 = 523.49

Source	df	SS	MS	F*
Machine	4	20.998	5.2495	31.6
Error	13	2.158	0.166	
Total	17	23.156		

Step 4: -- using p-value approach ---------------
a. $P = P(F > 31.6 | df_n = 4, df_d = 13)$;

Using Table 9: $P < 0.01$

Using computer: $P = 0.000$

b. $P < \alpha$

-- using classical approach -------------

a. critical value: $F(4,13,0.01) = 5.21$ critical region: $F \geq 5.21$

b. F* is in the critical region

Step 5: a. Reject H_o.

b. There is a significant difference between the machines with regards to mean amount of soft drink dispensed, at the 0.01 level of significance.

12.53 Step 1: a. The mean stopping distance on wet pavement for each brand of tire.

b. H_o: The mean stopping distance is not affected by the brand of tire.

H_a: The mean stopping distance is affected by the brand of tire.

Step 2: a. Assume the data were randomly collected and are independent, and the effects due to chance and untested factors are normally distributed.

b. F c. $\alpha = 0.05$

Step 3: a. n = 23, C_A = 217, C_B = 194, C_C = 216, C_D = 245, T = 872, Σx^2 = 33,282

Source	df	SS	MS	F*

Brand	3	95.36	31.79	4.78
Error	19	126.47	6.66	
Total	22	221.83		

Step 4: -- using p-value approach --------------
  a. $P = P(F > 4.78 | df_n = 3, df_d = 19)$;
  Using Table 9: $0.01 < P < 0.025$
  Using computer: $P = 0.012$
  b. $P < \alpha$
  -- using classical approach ------------
  a. critical value: $F(3,19,0.05) = 3.13$ critical region: $F \geq 3.13$
  b. F^* is in the critical region

Step 5: a. Reject H_o.
  b. There is a significant difference between the mean stopping distances for the different tire brands, at the 0.05 level of significance.

12.55 Step 1: a. The mean durability for different brands of golf balls: the mean durability for Brand A, the mean durability for Brand B, the mean durability for Brand C, the mean durability for Brand D, the mean durability for Brand E, the mean durability for Brand F.
  b. H_o: The six different brands of golf balls withstood the durability test equally well, as measured by the mean number of hits before failure.
  H_a: The six different brands of golf balls do not withstand the durability test equally well.

Step 2: a. Assume the data were randomly collected and are independent, and the effects due to chance and untested factors are normally distributed.
  b. F   c. $\alpha = 0.05$

Step 3: a.

Source	df	SS	MS	F*
Brand	5	75047	15009.4	5.30
Error	36	101899	2830.5	
Total	41	176946		

Step 4: -- using p-value approach --------------
  a. $P = P(F > 5.30 | df_n = 5, df_d = 36)$;
  Using Table 9: $P < 0.01$
  Using computer: $P = 0.001$
  b. $P < \alpha$
  -- using classical approach ------------
  a. critical value: $F(5,36,0.05) \approx 2.48$ critical region: $F \geq 2.48$
  b. F^* is in the critical region

Step 5: a. Reject H_o.
  b. There is a significant difference between the mean number of hits before failure for the six brands of golf balls tested, at the 0.05 level of significance.

12.57 a.
  Step 1: a. The mean points scored by teams representing each division.
  b. H_o: $\mu_E = \mu_N = \mu_S = \mu_W$

H_a: The mean points scored by teams representing each division are not all equal.

Step 2: a. Assume the data were randomly collected and are independent, and the effects due to chance and untested factors are normally distributed.

b. F   c. $\alpha = 0.05$

Step 3: a. n = 32,

Source	df	SS	MS	F*
Division	3	16885	5628	0.85
Error	28	184551	6591	
Total	31	201436		

Step 4: -- using p-value approach ---------------
a. **P** = P(F > 0.85|df_n = 3, df_d = 28);

Using Table 9: **P** > 0.05
Using computer: **P** = 0.476

b. **P** > α

-- using classical approach -------------
a. F(3, 28, 0.05) ≈ 2.95
b. F* is not in the critical region

Step 5: a. Fail to reject H_0.

b. At the 0.05 level of significance, the data does not show sufficient evidence to conclude that the points scored by the teams in each division are not all equal.

b.
Step 1: a. The mean points scored by the opponents of the teams representing each division.

b. H_0: $\mu_E = \mu_N = \mu_S = \mu_W$

H_a: The mean points scored by the opponents of the teams representing each division are not all equal.

Step 2: a. Assume the data were randomly collected and are independent, and the effects due to chance and untested factors are normally distributed.

b. F   c. $\alpha = 0.05$

Step 3: a. n = 32,

Source	df	SS	MS	F*
Division	3	5718	1906	0.53
Error	28	99912	3568	
Total	31	105630		

Step 4: -- using p-value approach ---------------
a. **P** = P(F > 0.53|df_n = 3, df_d = 28);

Using Table 9: **P** > 0.05
Using computer: **P** = 0.663

b. **P** > α

-- using classical approach -------------
a. F(3, 28, 0.05) ≈ 2.95
b. F* is not in the critical region

Step 5: a. Fail to reject H_0.

b. The data does not show sufficient evidence to conclude the mean points scored by the opponents of each division are not all equal, at the 0.05 level of significance.

12.59 Step 1: a. The mean car sale profits for each season of the year.

b. H_o: $\mu_{SP} = \mu_{SU} = \mu_{FA} = \mu_{WI}$

H_a: The mean car sale profits for each season are not all equal.

Step 2: a. Assume the data were randomly collected and are independent, and the effects due to chance and untested factors are normally distributed.

b. F   c. $\alpha = 0.05$

Step 3: a. n = 72,

Source	DF	SS	MS	F
Season	3	145652222	48550741	3.83
Error	68	862272222	12680474	
Total	71	1007924444		

Step 4: -- using p-value approach ---------------

a. $P = P(F > 0.85 | df_n = 3, df_d = 68)$;

Using Table 9: $0.01 < P < 0.025$

Using computer: $P = 0.014$

b. $P < \alpha$

-- using classical approach -------------

a. $F(3, 68, 0.05) = 2.76$

b. F* is in the critical region

Step 5: a. Reject H_o.

b. At the 0.05 level of significance, the data does show sufficient evidence to conclude that seasons do have an effect on car sales profits.

12.61 a.

Step 1: a. The mean nominal comparison for five competitors: the mean nominal comparison for A, the mean nominal comparison for B, the mean nominal comparison for C, the mean nominal comparison for D, the mean nominal comparison for E.

b. H_o: The mean nominal comparison is the same for all five competitors.

H_a: The mean nominal comparison is not the same for all five competitors.

Step 2: a. Assume the data were randomly collected and are independent, and the effects due to chance and untested factors are normally distributed.

b. F   c. $\alpha = 0.01$

Step 3: a.

Source	DF	SS	MS	F	P
Factor	4	0.001830	0.000458	1.05	0.385
Error	105	0.045732	0.000436		
Total	109	0.047563			

Step 4: -- using p-value approach ---------------

a. $P = P(F > 1.05 | df_n = 4, df_d = 105)$;

Using Table 9: $P > 0.05$

Using computer: $P = 0.385$

b. $P > \alpha$

-- using classical approach -------------
a. critical value: $F(4, 105, 0.01) = 3.65$ critical region: $F \geq 3.65$
b. F* is not in the critical region

Step 5: a. Fail to reject H_o.

b. There is no significant difference between the mean nominal comparisons for the five competitors, at the 0.01 level of significance.

b.

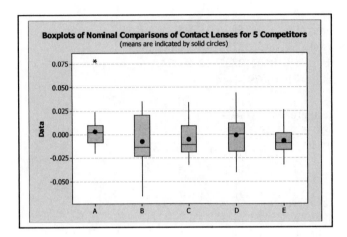

c. The graph portrays the results found in the F-test, the fact the all of the means are the same. The solid circles shown on the graph are the means for each of the competitors. It shows that the means are all at about the same level of nominal comparison (they all lay close to a horizontal straight line across).

12.63 a.

Step 1: a. The mean petal width for three species of irises.

b. H_o: The mean petal width is the same for each specie of iris

H_a: The mean petal width is not the same for each specie of iris.

Step 2: a. Assume the data were randomly collected and are independent, and the effects due to chance and untested factors are normally distributed.

b. F   c. $\alpha = 0.05$

Step 3: a.

Source	df	SS	MS	F
Specie	2	1671.56	835.78	118.06
Error	27	191.14	7.08	
Total	29	1862.70		

Step 4: -- using p-value approach ---------------
a. $P = P(F > 118.06 | df_n = 2, df_d = 27)$;
   Using Table 9: $P < 0.01$
   Using computer: $P = 0.000$
b. $P < \alpha$
-- using classical approach -------------
a. critical value: $F(2,27,0.05) \approx 3.37$
b. F^* is in the critical region

Step 5: a. Reject H_o.
b. There is sufficient evidence to show that at least one specie's petal width is significantly different from the others, at the 0.05 level of significance.

b.
Step 1: a. The mean sepal width for three species of irises.
b. H_o: The mean sepal width is the same for each specie of iris
   H_a: The mean sepal width is not the same for each specie of iris.

Step 2: a. Assume the data were randomly collected and are independent, and the effects due to chance and untested factors are normally distributed.
b. F   c. $\alpha = 0.05$

Step 3: a.

Source	df	SS	MS	F
Specie	2	197.1	98.6	7.78
Error	27	342.2	12.7	
Total	29	539.4		

Step 4: -- using p-value approach ---------------
a. $P = P(F > 7.78 | df_n = 2, df_d = 27)$;
   Using Table 9: $P < 0.01$
   Using computer: $P = 0.002$
b. $P < \alpha$
-- using classical approach -------------
a. critical value: $F(2,27,0.05) \approx 3.37$
b. F^* is in the critical region

Step 5: a. Reject H_o.
b. There is sufficient evidence to show that at least one specie's sepal width is significantly different from the others, at the 0.05 level of significance.

c. Type 0 has the shortest PW and the longest SW. Type 1 has the longest PW and the middle SW. Type 2 has the middle PW and the shortest SW.

12.65 a. The graphical evidence that not all days of the week are the same is illustrated by the various ranges of data values per day of the week. Mondays appears to have the smallest range and number of customers, whereas Tuesday and Wednesdays have some of the highest values and largest spreads of data values.

b. The 5 points located to the right and separate from the rest of the data all occurred during the months of November and December. The Tuesday and Wednesday both had November months, the days right before Thanksgiving.

c. Based on the F statistic shown on the ANOVA table, not all of the days of the week are the same when it comes to the number of customers if a level of significance of 0.10 is used.

d. No, one can not tell which days of the week are different based on the ANOVA table output.

e. Verify – answers given in exercise.

12.67 a. Answers will vary but it would make sense that the total cost of items purchased would increase with an increase in number of customers and number of items purchased. This being the case, the months of November and December should have the highest amounts. Whether the day of the week will have a significant effect needs to be seen and will probably be borderline based on the other results.

b.

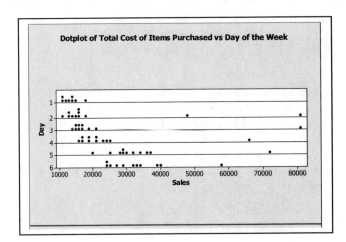

c. Answers may vary but yes, the graph supports the conjecture that the days of the week with the larger total cost of items would be the same days that were large for number of customers and number of items purchased per day. Again each of these are in the holiday months of November and December.

d.
Step 1: a. The mean total cost of items purchased per day by day of the week.
b. H_0: The mean total cost of items purchased per day is the same for each day of the week.
H_a: The mean total cost of items purchased per day is not the same for each day of the week. That is; day of the week does have effect on the mean number of items purchased.

Step 2: a. Assume the data were randomly collected and are independent, and the effects due to chance and untested factors are normally distributed.
b. F   c. $\alpha = 0.05$

Step 3: a.

Source	DF	SS	MS	F	P
Day	5	2657284622	531456924	2.24	0.063
Error	56	13311874185	237712039		
Total	61	15969158806			

Step 4: -- using p-value approach --------------
a. $P = P(F > 2.24 | df_n = 5, df_d = 56)$;

Using Table 9: $P > 0.05$
Using computer: $P = 0.063$

b. $P > \alpha$

-- using classical approach -------------
a. $F(5, 56, 0.05) \approx 2.45$
b. F^* is in the noncritical region

Step 5: a. Fail to reject H_0.

b. There is not sufficient evidence to show that the day of the week has an effect on the mean total cost of items purchased per day, at the 0.05 level of significance.

e. Answers may vary but based on answers given above, the reasoning in part a is not supported by the F-test results. The p-value is very close to the level of significance. With a higher level of significance, the F-test results would correspond closer to the answers given based on the dotplot.

f. The answers in exercises 12.65, 12.66 and 12.67 all support each other and make sense. The more customers, the larger the number of items purchased and the larger the total cost of items purchased per day. All of these variables are related to each other, but the day of the week does not have a significant effect unless a level of significance higher than 0.05 is used.

12.69 Sample information:
$k_1 = 3$, $k_2 = 3$, $k_3 = 3$, $n = 9$, $C_1 = 24$, $C_2 = 39$, $C_3 = 27$, $T = 90$, $\Sigma x^2 = 960$

Using formula (12.3):
SS(factor) = $[(24^2/3)+(39^2/3)+(27^2/3)]-(90^2/9) = 942 - 900 = \underline{42}$

Using formula given in exercise:
$x_1 = 24/3 = 8$, $x_2 = 39/3 = 13$, $x_3 = 27/3 = 9$,
$\bar{x} = 90/9 = 10$

SS(factor) = $3(8-10)^2 + 3(13-10)^2 + 3(9-10)^2 = 12 + 27 + 3 = \underline{42}$

CHAPTER 13 ∇ LINEAR CORRELATION AND REGRESSION ANALYSIS

Chapter Preview

Chapter 3 introduced the concepts of correlation and regression analysis for bivariate data. In Chapter 13, we will look at these concepts in a more detailed manner using confidence intervals and hypothesis tests. These inference tests will be utilized on the correlation coefficient, the slope of the regression line and the regression line.

Height compatibility between married couples is the focus of this chapter's opening section "Height Compatibility".

SECTION 13.1 EXERCISES

13.1 a. Wife's height; along x-axis
  b. Husband's height; along y-axis
  c. Yes; the elongated oval pattern of points suggest a linear relationship

13.3 a. $\bar{x}$ and $\bar{y}$ are the mean values of the two variables. In Chapter Two we learned that the summation of the deviations about the mean was zero.

Algebraically: $\Sigma(x - \bar{x}) = \Sigma x - \Sigma \bar{x} = \Sigma x - n \cdot \bar{x} = 0$
and $\Sigma(y - \bar{y}) = \Sigma y - \Sigma \bar{y} = \Sigma y - n \cdot \bar{y} = 0$

  b. divides data into 4 quadrants

  c. 1.) The set of data will be predominantly ordered pairs which have coordinates such that both the x and y values are larger than $\bar{x}$ and $\bar{y}$, and both smaller than $\bar{x}$ and $\bar{y}$; this will result in the product $(x-\bar{x})(y-\bar{y})$ being positive. Graphically, the points will be mostly located in the upper right and the lower left of the four quarters of the graph formed by the vertical line
    $x = \bar{x}$ and the horizontal line $y = \bar{y}$.

  2.) The set of data will be predominantly ordered pairs which have coordinates such that either the x value is larger than $\bar{x}$ and y is smaller than $\bar{y}$, or x is smaller than $\bar{x}$ and y is larger than $\bar{y}$; this will result in the product $(x-\bar{x})(y-\bar{y})$ being negative. Graphically, the points will be mostly located in the upper left and the lower right of the four quarters of the graph formed by the vertical line $x = \bar{x}$ and the horizontal line $y = \bar{y}$.

  3.) The set of data will be ordered pairs which have coordinates such that the product $(x-\bar{x})(y-\bar{y})$ being distributed between positive, negative and zero so that the sum is near zero. Graphically, the points will be approximately evenly distributed between the four quarters of the graph formed by the vertical line $x = \bar{x}$ and the horizontal line $y = \bar{y}$.

13.5 a.

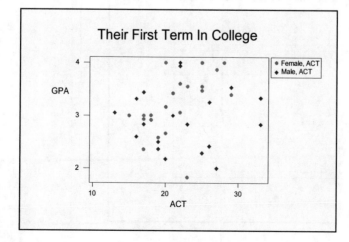

b. Patterns are somewhat similar in that the points pretty much cover the full area of the diagram. The females, with one exception, are all located in the top part of the diagram.

c.
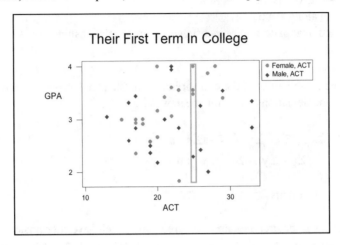

Prediction: any value from 1.8 to 4.0, not much of a prediction.

d. No, knowing the ACT score does not help at all.

Formulas for the covariance and correlation coefficient of bivariate data

Covariance of x and y - measure of linear dependency between x and y

$$\text{covar}(x,y) = \frac{\sum (x - \bar{x})(y - \bar{y})}{n - 1}$$

$s_x = \sqrt{SS(x)/(n-1)}$, where $SS(x) = \Sigma x^2 - [(\Sigma x)^2/n]$

$s_y = \sqrt{SS(y)/(n-1)}$, where $SS(y) = \Sigma y^2 - [(\Sigma y)^2/n]$

$$r = \frac{\text{covar}(x,y)}{s_x \cdot s_y} \quad \text{or} \quad r = \frac{SS(xy)}{\sqrt{SS(x) \cdot SS(y)}}, \text{ where } SS(xy) = \Sigma xy - [(\Sigma x)(\Sigma y)/n]$$

13.7 a.
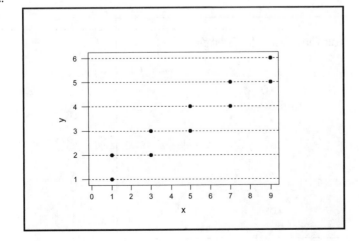

b.

x	y	x-$\bar{x}$	y-$\bar{y}$	(x-$\bar{x}$)(y-$\bar{y}$)
1	1	-4	-2.5	+10
1	2	-4	-1.5	+6
3	2	-2	-1.5	+3
3	3	-2	-0.5	+1
5	3	0	-0.5	0
5	4	0	+0.5	0
7	4	+2	+0.5	+1
7	5	+2	+1.5	+3
9	5	+4	+1.5	+6
9	6	+4	+2.5	+10
50	35	0	0.0	40

$\bar{x}$ = 50/10 = 5.0 and $\bar{y}$ = 35/10 = 3.5

covar(x,y) = [Σ(x-$\bar{x}$)(y-$\bar{y}$)]/(n-1) = 40/9 = <u>4.44</u>

Summary of data: n = 10, Σx = 50, Σy = 35, Σx^2 = 330, Σxy = 215, Σy^2 = 145

c. $s_x = \sqrt{[330 - (50^2/10)]/9} = \sqrt{8.889} = \underline{2.981}$

   $s_y = \sqrt{[145 - (35^2/10)]/9} = \sqrt{2.50} = \underline{1.581}$

d. r = 4.444/[(2.981)(1.581)] = <u>0.943</u>

e. SS(x) = 330 - (50²/10) = 80
    SS(y) = 145 - (35²/10) = 22.5
    SS(xy) = 215 - [(50)(35)/10] = 40
    r = 40/$\sqrt{(80)(22.5)}$ = <u>0.943</u>

13.9 Verify -- answers given in exercise.

Computer and/or calculator commands to calculate r, the correlation coefficient can be found in ES11-p139.

13.11 n = 32, Σx = 10991, Σy = 10991, Σx^2 = 3976501, Σxy = 3707964, Σy^2 = 3880695

SS(x) = 201435.9688, SS(y) = 105629.9688, SS(xy) = -67101.03125

a. r = SS(xy)/$\sqrt{SS(x) \cdot SS(y)}$ = -67101.03125/$\sqrt{(201435.9688)(105629.9688)}$ = <u>-0.460</u>

b. slight negative linear relationship

c.

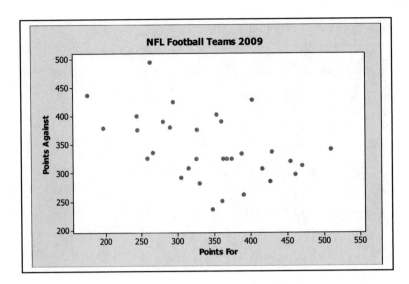

slight downward trend, as Points For increased, Points Against decreased

13.13 a.

x	y	x-$\bar{x}$	y-$\bar{y}$	(x-$\bar{x}$)(y-$\bar{y}$)
20	10	-50	-20	1000
30	50	-40	20	-800
60	30	-10	0	0
80	20	10	-10	-100
110	60	40	30	1200
120	10	50	-20	-1000
420	180	0	0	300

$\bar{x} = 420/6 = 70$, $\bar{y} = 180/6 = 30$
covar(x,y) = 300/5 = <u>60</u>

b. $s_x = 40.99$, $s_y = 20.98$

c. $r = 60/(40.99)(20.98) = 0.0698 = \underline{0.07}$

d. The value for r is the same.

SECTION 13.2 EXERCISES

13.15

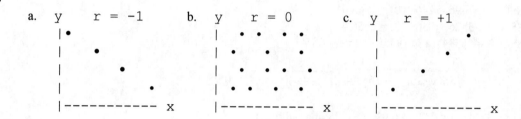

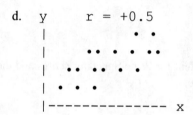 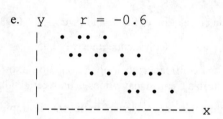

Estimating ρ - the population correlation coefficient

1. point estimate: r

2. confidence interval: use Table 10 in Appendix B, ES11-p728, to determine a 95% confidence interval.

   Locate the r value on the horizontal axis. Follow a vertical line* up until the corresponding sample size band intersects. This value is the lower limit. To find the upper limit, locate the r value on the vertical axis. Follow a horizontal line until it intersects the corresponding sample size.

* Use a thin rubber band stretched vertically over the table and the back cover of the textbook – located at the r value of concern. It will provide a straight line that is easy to read off of and also keep your place on the table.

13.17 a. 0.17 to 0.52
   b. The interval becomes more narrow.

13.19 a. 0.40 to 0.74   b. -0.78 to +0.15
   c. 0.05 to 0.93   d. -0.65 to -0.45

> Reminder: $r = SS(xy)/\sqrt{SS(x) \cdot SS(y)}$ where: $SS(x) = \Sigma x^2 - [(\Sigma x)^2/n]$
> $SS(y) = \Sigma y^2 - [(\Sigma y)^2/n]$
> $SS(xy) = \Sigma xy - [(\Sigma x)(\Sigma y)/n]$

13.21 Summary of data: $n = 10$, $\Sigma x = 746$, $\Sigma y = 736$, $\Sigma x^2 = 57{,}496$, $\Sigma xy = 56{,}574$, $\Sigma y^2 = 55{,}826$

$SS(x) = 57496 - (746^2/10) = 1844.4$
$SS(y) = 55826 - (736^2/10) = 1656.4$
$SS(xy) = 56574 - [(746)(736)/10] = 1668.4$

$r = SS(xy)/\sqrt{SS(x) \cdot SS(y)} = 1668.4/\sqrt{(1844.4)(1656.4)} = \underline{0.955}$

From Table 10: $\underline{0.78 \text{ to } 0.98}$, the 0.95 interval for ρ

13.23 a. Pearson correlation of Score and Price \$: $r = 0.937$

b. Using Table 10: Confidence interval: $0.55 < \rho < 1.00$

c. With 95% confidence, the population correlation coefficient for Wine Spectator scores and the prices of California Chardonnay wines is between 0.55 and 1.00.

d. The interval is very wide due to the small sample size.

> Hypotheses for the correlation coefficient are written with the same rules as before. Now in place of μ or σ, the population correlation coefficient, ρ, will be used. The standard form is using 0 as the test value, unless some other information is given in the exercise. 0 indicates that there is no linear relationship.
> (ex. $H_0: \rho = 0$ vs. $H_a: \rho \neq 0$)

13.25 a. $H_0: \rho = 0$ vs. $H_a: \rho > 0$

b. $H_0: \rho = 0$ vs. $H_a: \rho \neq 0$

c. $H_0: \rho = 0$ vs. $H_a: \rho < 0$

d. $H_0: \rho = 0$ vs. $H_a: \rho > 0$

13.27 a. $0.05 < P < 0.10$ b. $0.025 < P < 0.05$

> **Test criteria**
>
> 1. Draw a bell-shaped distribution locating 0 at the center, -1 at the far left and +1 at the far right.
> 2. Shade in the critical region(s) based on the alternative hypothesis (H_a).
> 3. Find the critical value(s) from Table 11, Appendix B, ES11-p729:
>   a. degrees of freedom (n - 2) is the row id #
>   b. α, is the column id #;
>   1) use the given α for a two-tailed test
>   2) use 2α for a one-tailed test
>   c. all values given in the table are positive. Negate the value if the critical region or part of the critical region is to the left of 0.

13.29 a. ±0.444
b. -0.378, if left tail critical region; 0.378, if right tail

13.31 a. The linear correlation coefficient, r = 0.58 is significant for all levels of $\alpha > 0.008$.
b. n = 25, df = 23: P < 0.01 for a two-tailed test. 0.008 is less than 0.01, therefore significant.
c. ± 0.537, using next smaller table value; ± 0.507, using interpolation
d. r is significant at the $\alpha = 0.01$ level.

> Hypothesis tests will be completed using the same format as before. You may want to review: ES11-pp371&388, IRM-pp343&357. The only differences are:
> 1. **writing hypotheses**: (see box before ex. 13.25)
> 2. **using Table 11**: (see box before ex. 13.29 for the classical approach and ES11-pp622-623 for the p-value approach)
> 3. **the calculated test statistic**: r*, the sample correlation coefficient

13.33 Step 1: a. The linear correlation coefficient for the population, ρ.
b. H_o: $\rho = 0.0$
H_a: $\rho \neq 0.0$
Step 2: a. random sample, assume normality for y at each x
b. r, df = n - 2 = 20 - 2 = 18
c. $\alpha = 0.10$
Step 3: a. n = 20, r = 0.43
b. r* = 0.43

Step 4: -- using p-value approach --------------------
   a. **P** = P(r < 0.43) + P(r > 0.43)
      = 2P(r > 0.43|df = 18);
         Using Table 11, ES11-p729: $0.05 < P < 0.10$
   b. **P** < α
-- using classical approach ------------------
   a. ±r(18, 0.10) = ±0.378

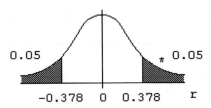

   b. r* falls in the critical region, see Step 4a.

Step 5: a. Reject H_o.
   b. There is sufficient reason to reject the null hypothesis, at the 0.10 level of significance.

13.35 Step 1: a. The linear correlation coefficient for the population, ρ.
   b. H_o: ρ = 0.0
      H_a: ρ > 0.0
Step 2: a. random sample, assume normality for y at each x
   b. r, df = n - 2 = 62 - 2 = 60
   c. α = 0.05
Step 3: a. n = 62, r = 0.24
   b. r* = 0.24
Step 4: -- using p-value approach --------------------
   a. **P** = P(r > 0.24)
      Using Table 11, ES11-p729: $0.025 < P < 0.05$
   b. **P** < α
-- using classical approach ------------------
   a. critical region: r ≥ 0.211
   b. r* falls in the critical region

Step 5: a. Reject H_o.
   b. There is sufficient reason to conclude that the correlation coefficient is positive, at the 0.05 level of significance.

13.37 Summary of data: n = 10, Σx = 26.2, Σy = 82.5, Σx² = 174.88, Σxy = 256.41, Σy² = 704.61
   SS(x) = 174.88 - (26.2²/10) = 106.236
   SS(y) = 704.61 - (82.5²/10) = 23.985
   SS(xy) = 256.41 - [(26.2)(82.5)/10] = 40.26

$$r = SS(xy)/\sqrt{SS(x) \cdot SS(y)} = 40.26/\sqrt{(106.236)(23.985)} = \underline{0.798}$$

Step 1: a. The linear correlation coefficient for size of a metropolitan area and its crime rate, ρ.
        b. $H_o: \rho = 0.0$
        $H_a: \rho \neq 0.0$

Step 2: a. random sample, assume normality for y at each x
        b. r, df = n - 2 = 10 - 2 = 8
        c. $\alpha = 0.05$

Step 3: a. n = 10, r = 0.798
        b. r* = 0.798

Step 4: -- using p-value approach --------------------
        a. **P** = P(r < 0.798) + P(r > 0.798)
           = 2P(r > 0.798);
              Using Table 11, ES11-p729: **P** < 0.01
              Using computer: **P** = 0.006
        b. **P** < α
        -- using classical approach ------------------
        a. critical region: r $\leq$ -0.632 and r $\geq$ 0.632
        b. r* falls in the critical region

Step 5: a. Reject H_o
        b. There is sufficient reason to conclude that the correlation coefficient is different than zero, at the 0.05 level of significance.

13.39 a. Summary of data: n = 7, r = $\underline{-0.861}$

b. Step 1: a. The linear correlation coefficient for a state's median household income and percent in poverty, ρ.
        b. $H_o: \rho = 0.0$
        $H_a: \rho \neq 0.0$

Step 2: a. random sample, assume normality for y at each x
        b. r, df = n - 2 = 7 - 2 = 5
        c. $\alpha = 0.05$

Step 3: a. n = 7, r = -0.861
        b. r* = -0.861

Step 4: -- using p-value approach --------------------
        a. **P** = P(r < -0.861) + P(r > 0.861)
           = 2P(r > 0.861);
              Using Table 11, ES11-p729: 0.01 < **P** < 0.02
              Using a computer: **P** = 0.013
        b. **P** < α
        -- using classical approach ------------------
        a. critical region: r $\leq$ -0.754 and r $\geq$ 0.754
        b. r* falls in the critical region

Step 5: a Reject H_o

b. There is sufficient reason to conclude that there is significant correlation between a state's median household income and its percent of population in poverty, at the 0.05 level of significance.

13.41 a. Many will not know, but with most agriculture products, higher yield usually does not improve the desired quality.

b.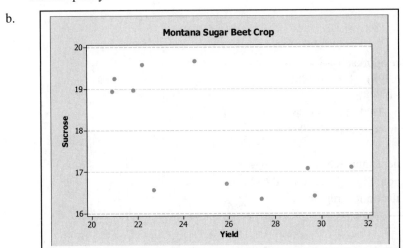

The relationship on the scatter diagram suggests that higher yields reduces the sucrose content of the sugar beets.

c. r = -0.686

d. Step 1: a. The linear correlation coefficient between yield per acre and sucrose percentage for sugar beets, ρ.
b. H_o: $\rho = 0.0$
H_a: $\rho \neq 0.0$
Step 2: a. random sample, assume normality for y at each x
b. r, df = n - 2 = 11 - 2 = 9
c. $\alpha = 0.05$
Step 3: a. n = 11, r = -0.686
b. r* = -0.686
Step 4: -- using p-value approach --------------------
a. **P** = P(r < -0.686) + P(r > 0.686)
= 2P(r > 0.686);
Using Table 11, ES11-p729: 0.01 < **P** < 0.02
Using a computer: **P** = 0.020
b. **P** < α
-- using classical approach -----------------
a. critical region: r ≤ -0.602 and r ≥ 0.602
b. r* falls in the critical region

Step 5: a. Reject H_o

b. At the 0.05 level of significance, the linear correlation coefficient is significantly different from zero.

e. The pattern on the scatter diagram will become two distinct groupings, one in the upper left and one in the lower right. The correlation coefficient will be stronger, and the p-value will decrease.

f.

Montana Sugar Beet Crop
without Carbon

(Scatter plot: Sucrose vs Yield)

Summary of data: n = 11, r = −0.813

Step 1: a. The linear correlation coefficient between yield per acre and sucrose percentage for sugar beets, ρ.

b. H_o: $\rho = 0.0$

H_a: $\rho \neq 0.0$

Step 2: a. random sample, assume normality for y at each x

b. r, df = n − 2 = 10 − 2 = 8

c. $\alpha = 0.05$

Step 3: a. n = 10, r = −0.813

b. r* = −0.813

Step 4: -- using p-value approach --------------------

a. **P** = P(r < −0.813) + P(r > 0.813)

= 2P(r > 0.813);

Using Table 11, ES11-p729: **P** < 0.01

Using a computer: **P** = 0.004

b. **P** < α

-- using classical approach -----------------

a. critical region: r ≤ −0.632 and r ≥ 0.632

b. r* falls in the critical region

Step 5: a. Reject H_o

b. At the 0.05 level of significance, the linear correlation coefficient is significantly different from zero.

Same answer as far as significant correlation with and without Carbon, but the relationship is stronger without Carbon. Due to the two separate clusters of points on the scatter diagram, further study of this situation is probably merited.

SECTION 13.3 EXERCISES

Sample Regression Line: $\hat{y} = b_0 + b_1 x$, where

$$b_1 = \frac{SS(xy)}{SS(x)} = \frac{\Sigma(xy) - [(\Sigma x)(\Sigma y)/n]}{\Sigma x^2 - (\Sigma x)^2/n} \quad \text{and} \quad b_0 = \frac{1}{n}(\Sigma y - b_1 \cdot \Sigma x)$$

Population Regression Line: $y = \beta_0 + \beta_1 x + \varepsilon$

Sample estimate of $\varepsilon = e = (y - \hat{y})$

Variance of y about the regression line = variance of the error e

$$s_e^2 = \frac{(\Sigma y^2) - [(b_0)(\Sigma y)] - [(b_1)(\Sigma xy)]}{n-2} = \frac{SSE}{n-2}$$

Standard Deviation of the error $= s_e = \sqrt{s_e^2}$

Explore the relationship between residuals and the line of best fit with the Chapter 13 Skillbuilder Applet "Residuals & Line of Best Fit".

Slight variations in sums of squares and further calculations can result from round-off errors.

13.43 Verify -- answers given in exercise.

$$r = \frac{SS(xy)}{\sqrt{SS(x) \cdot SS(y)}}$$

13.45

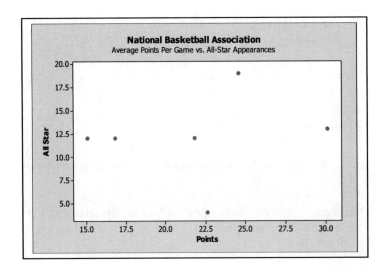

Summary of data: n = 6, $\Sigma x = 131$, $\Sigma y = 72$, $\Sigma x^2 = 3007.42$, $\Sigma xy = 1593.5$, $\Sigma y^2 = 978$

SS(x) = 3007.42 - (131²/6) = 147.253
SS(y) = 978 - (72²/6) = 114
SS(xy) = 1593.5 - [(131)72)/6] = 21.5

$r = SS(xy)/\sqrt{SS(x) \cdot SS(y)}$
 = 21.5/$\sqrt{(147.253)(114)}$ = <u>0.166</u>

Using formula 3.6: b_1 = 21.5/147.253 = 0.146

Using formula 3.7: b_0 = [72 - (0.146)(131)]/6 = 8.81

Best fit line: $\hat{y}$ = 8.81 + 0.146x

There is not a strong linear relationship shown by this data. The scatter diagram shows no definite trend or pattern to indicate a relationship. The correlation coefficient is close to zero and with a p-value of 0.753 would be considered insignificant with just about all levels of significance.

Computer and/or calculator commands to construct a scatter diagram can be found in ES11-pp129-130. Commands to calculate the equation of the line of best fit can be found in ES11-pp152-153 and pp638-640.

13.47   a.

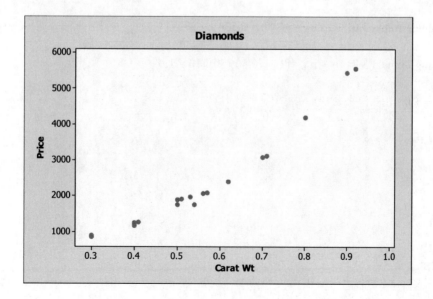

b.   There is a linear pattern to the data.

c. Only have data in this weight range, can not predict with confidence outside this range. Smaller values and larger value decrease and increase, respectively exponentially.

d. Summary of data: $n = 20$, $\Sigma x = 11.11$, $\Sigma y = 45428$
$\Sigma x^2 = 6.7657$, $\Sigma xy = 29696.1$, $\Sigma y^2 = 138722648$

$SS(x) = 6.7657 - (11.11^2/20) = 0.594095$
$SS(y) = 138722648 - (45428^2/20) = 35537488.8$
$SS(xy) = 29696.1 - [(11.11)(45428)/20] = 4460.846$

Using formula 3.6: $b_1 = 4460.846/0.594095 = 7508.64 = 7509$

Using formula 3.7: $b_0 = [45428 - (7508.64)(11.11)]/20 = -1899.64952 = -1900$

Best fit line: $\hat{y} = -1900 + 7509x$

e. $\hat{y} = -1900 + 7509x = -1900 + 7509(0.75) = \$3731.75$

f. $75.09, $x = 0.30$ carats to 0.92

g. $s = 336.914$, $s^2 = (336.914)^2 = 113511.0434$ (by computer)

The scatter diagram shows a sizeable amount of vertical distance between the top and bottom points along the line of best fit.

13.49 a. Summary of data: $n = 10$, $\Sigma x = 50$, $\Sigma y = 35$, $\Sigma x^2 = 330$, $\Sigma xy = 215$, $\Sigma y^2 = 145$

$SS(x) = 330 - (50^2/10) = 80$
$SS(xy) = 215 - [(50)(35)/10] = 40$

Using formula 3.6: $b_1 = 40/80 = 0.50$

Using formula 3.7: $b_0 = [35 - (0.50)(50)]/10 = 1.0$

$\hat{y} = 1.0 + 0.5x$

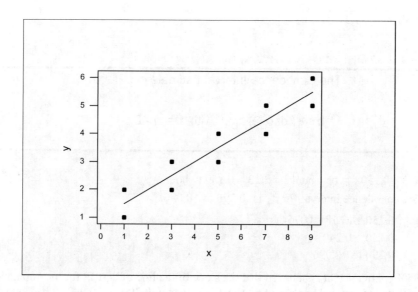

b. If x = 1, then $\hat{y}$ = 1.0 + 0.5(1) = <u>1.5</u>
    If x = 3, then $\hat{y}$ = 1.0 + 0.5(3) = <u>2.5</u>
    If x = 5, then $\hat{y}$ = 1.0 + 0.5(5) = <u>3.5</u>
    If x = 7, then $\hat{y}$ = 1.0 + 0.5(7) = <u>4.5</u>
    If x = 9, then $\hat{y}$ = 1.0 + 0.5(9) = <u>5.5</u>

c.
point	A	B	C	D	E	F	G	H	I	J
x	1	1	3	3	5	5	7	7	9	9
y	1	2	2	3	3	4	4	5	5	6
$\hat{y}$	1.5	1.5	2.5	2.5	3.5	3.5	4.5	4.5	5.5	5.5
e=y-$\hat{y}$	-.5	.5	-.5	.5	-.5	.5	-.5	.5	-.5	.5

d. s_e^2 = 2.50/8 = <u>0.3125</u>

e. s_e^2 = [145 - (1.0)(35) - (0.5)(215)]/8 = 2.50/8 = <u>0.3125</u>

SECTION 13.4 EXERCISES

13.51 From the scatter diagram, n = 72

a. S_1 = -3953.85 + 3.13(12,600) = 35484.15 ≈ 35,500

b. From the scatter diagram, n = 72
    $b_1 \pm t(df, \alpha/2) \cdot s_{b_1}$
    3.130 ± (1.99)(0.065) = 3.130 ± 0.129
    <u>3.001 to 3.259</u> (difference due to round-off)

13.53 $s_{b1} = \sqrt{s_e^2/SS(x)} = \sqrt{1.213/49.6} = \underline{0.1564}$

The Confidence Interval Estimate of β_1

$b_1 \pm t(df, \alpha/2) \cdot s_{b1}$ with df = n - 2

13.55 a. $\hat{y} = 20.3 - 0.150x$ or   books = 20.3 - 0.150 tv hours

  b. The 95% confidence interval for β_1 is -0.390 to +0.090.

  $-0.150 \pm (3.18)(0.0754)$
  -0.150 ± 0.240
  -0.390 to $+0.090$

  c. With 95% confidence, it is believed that the slope of the line of best fit for the population of 7th grade girls television and book reading times is between -0.390 and 0.090. The slope could be slightly negative or slightly positive.

13.57 a. $\hat{y} = -348 + 2.04x$

  b. The 95% confidence interval for β_1 is 1.60 to 2.48.

  $2.04 \pm (2.31)(0.1894)$
  2.04 ± 0.44
  1.60 to 2.48

  c. With 95% confidence, it is believed that the slope of the line of best fit for the population is between 1.60 and 2.48.

Hypotheses for the slope of the regression line are written with the same rules as before. Now in place of μ or σ, the population slope, β_1, will be used. The standard form is using 0 as the test value, unless some other information is given in the exercise. 0 indicates that the line has no value in predicting y for given x. H_a represents what the experimenter wants to show, i.e. that the slope is meaningful and valuable in predicting y for a given x.

(ex. H_o: $\beta_1 = 0$ vs. H_a: $\beta_1 \neq 0$)

To determine the p-value for the test of the slope of the regression line, the t-distribution is used. Review its use in: ES11-pp421-422, IRM-pp397-398, if necessary. The only difference required is **df = n - 2**, since the data is bivariate. The test statistic is $t^* = (b_1 - \beta_1)/s_{b1}$.

13.59 a. $P = P(t > 2.40 | df=16)$; $0.01 < P < 0.025$ or $P = \underline{0.0145}$

  b. $P = 2 \cdot P(t > 2.00 | df=13)$; $0.05 < P < 0.10$ or $P = 2(0.0334) = \underline{0.0668}$

  c. $P = P(t < -1.57 | df=22)$; $0.05 < P < 0.10$ or $P = \underline{0.0653}$

> Draw a picture of a t-distribution curve. Shade in the critical regions based on the alternative hypothesis (H_a).
>
> Using α and df = n - 2, find the critical value(s) using Table 6 (Appendix B, ES11-p719).

13.61 a. $\hat{y} = \underline{5936.79} + \underline{30.732}x$

b. $t = (30.732 - 0)/17.158 = 1.79$

c. $P = P(t > 1.79 | df = 8)$;
  Using Table 6, ES11-p719:
  $0.10 < P < 0.20$
  Using Table 7, ES11-p720:
  $0.11 < P < 0.128$
  Using computer: $P = 0.111$
  $P >$ most levels of significance
  Based on the p-value, horsepower does not appear to be an effective predictor of base price

d. $b_1 \pm t(df, \alpha/2) \cdot s_{b_1}$
  $30.732 \pm (2.31)(17.158) = 30.732 \pm 39.635$
  $\underline{-8.903}$ to $\underline{70.367}$, the 0.95 interval for β_1
  (discrepancies due to rounding)

> Hypothesis tests will be completed using the same format as before. You may want to review: ES11-pp371&388, IRM-pp343&357. The only differences are:
>
> 1. **writing hypotheses**: (see box before ex. 13.58)
> 2. **using Table 6 or 7**: df = n - 2
> 3. **the calculated test statistic**: $t^* = \dfrac{b_1 - \beta_1}{s_{b1}}$
>
> Computer and/or calculator commands to determine the line of best fit and also perform a hypothesis test concerning the slope can be found in ES11-pp638-640. Excel also constructs the confidence interval for the slope.

Slight variations in sums of squares and further calculations can result from round-off errors.

13.63 a.

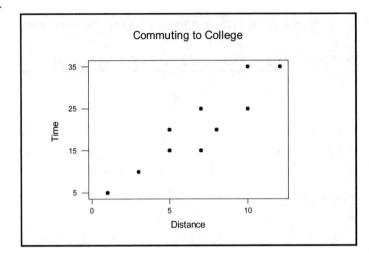

b. Summary of data: $n = 10$, $\Sigma x = 68$, $\Sigma y = 205$, $\Sigma x^2 = 566$, $\Sigma xy = 1670$, $\Sigma y^2 = 5075$

$SS(x) = 566 - (68^2/10) = 103.6$
$SS(xy) = 1670 - [(68)(205)/10] = 276.0$

$b_1 = 276.0/103.6 = 2.664$
$b_0 = [205 - (2.664)(68)]/10 = 2.38$

$\hat{y} = \underline{2.38 + 2.664x}$

c.
Step 1: a. The slope β_1 of the line of best fit for the population of distances and their corresponding times required for students to commute to college.
   b. $H_0: \beta_1 = 0$ (no value)
   $H_a: \beta_1 > 0$
Step 2: a. random sample, assume normality for y at each x
   b. t, df = 8   c. $\alpha = 0.05$
Step 3: a. $n = 10$, $b_1 = 2.664$,
   $s_e^2 = [5075 - (2.38)(205) - (2.664)(1670)]/8$
   $= 137.192/8 = 17.149$
   $s_{b1} = \sqrt{s_e^2/SS(x)} = \sqrt{17.149/103.6} = 0.407$
   b. $t = (b_1 - \beta_1)/s_{b1}$
   $t^* = (2.664 - 0)/0.407 = 6.55$
Step 4: -- using p-value approach --------------------
   a. $P = P(t > 6.55 | df = 8)$;
      Using Table 6, ES11-p719: $P < 0.005$
      Using Table 7, ES11-p720: $P < 0.002$
      Using computer: $P = 0.000$
   b. $P < \alpha$

-- using classical approach ------------------
a. critical region: $t \geq 1.86$
b. t* falls in the critical region

Step 5: a. Reject H_o.

b. The slope is significantly greater than zero, at the 0.05 level of significance.

d. $b_1 \pm t(df, \alpha/2) \cdot s_{b_1}$
$2.66 \pm (2.90)(0.407) = 2.66 \pm 1.18$
<u>1.48 to 3.84</u>, the 0.98 interval for β_1

13.65 a. $\hat{y} = 17.213 + 0.79741x$ or husband height = 17.2 + 0.797 wife height
Or $\hat{y} = 17.2 + 0.797x$

b.

Step 1: a. The slope β_1 of the line of best fit for the population of married couples heights.

b. $H_o: \beta_1 = 0$ (no value)
$H_a: \beta_1 > 0$

Step 2: a. random sample, assume normality for y at each x
b. t, df = 85   c. $\alpha = 0.05$

Step 3: a. n = 87, $b_1 = 0.7974$, $s_{b_1} = 0.0766$

b. $t = (b_1 - \beta_1)/s_{b_1}$
$t^* = (0.7974 - 0)/0.0766 = 10.41$

Step 4: -- using p-value approach --------------------
a. $\mathbf{P} = P(t > 10.41 | df = 85)$;
Using Table 6, ES11-p719: $\mathbf{P} < 0.005$
Using Table 7, ES11-p720: $\mathbf{P} = 0+$
Using computer: $\mathbf{P} = 0.000$

b. $\mathbf{P} < \alpha$

-- using classical approach ------------------
a. critical region: $t \geq 1.67$
b. t* falls in the critical region

Step 5: a. Reject H_o.

b. The slope is significantly greater than zero, at the 0.05 level of significance.

d. $b_1 \pm t(df, \alpha/2) \cdot s_{b_1}$

$0.797 \pm (1.99)(0.0766) = 0.797 \pm 0.152$

$\underline{0.645 \text{ to } 0.949}$, the 0.95 interval for β_1

13.67 a.

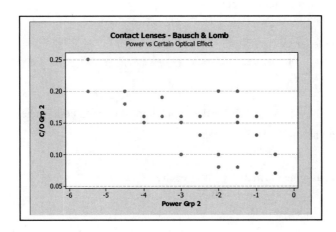

b. Pearson correlation of Power Grp 2 and C/O Grp 2 = -0.674
P-Value = 0.000

c. Step 1: a. The linear correlation coefficient between lens power and a certain optical effect, ρ.

b. $H_o: \rho = 0.0$

$H_a: \rho \neq 0.0$

Step 2: a. random sample, assume normality for y at each x
b. r, df = n - 2 = 30 - 2 = 28
c. $\alpha = 0.05$

Step 3: a. n = 30, r = -0.674
b. r* = -0.674
Step 4: -- using p-value approach --------------------
a. $P = P(r < -0.674) + P(r > 0.674) = 2P(r > 0.674)$;
Using Table 11, ES11-p729: $P < 0.01$
Using a computer: $P = 0.000$
b. $P < \alpha$
-- using classical approach ------------------
a. critical region: $r \leq -0.381$ and $r \geq 0.381$
b. r* falls in the critical region

Step 5: a. Reject H_o
b. At the 0.05 level of significance, the linear correlation coefficient is significantly different from zero.

d. The regression equation is: $\hat{y} = 0.0881 - 0.0221x$

e. Step 1: a. The slope β_1 of the line of best fit for lens power and a certain optical effect.
b. $H_o: \beta_1 = 0$ (no value)
$H_a: \beta_1 < 0$
Step 2: a. random sample, assume normality for y at each x
b. t, df = 28   c. $\alpha = 0.05$
Step 3: a. n = 30, $b_1 = -0.022120$, $s_{b1} = 0.004587$
b. $t = (b_1 - \beta_1)/s_{b1}$
$t^* = (-0.022120 - 0)/0.004587 = -4.82$
Step 4: -- using p-value approach --------------------
a. $P = P(t < -4.82 | df = 28) = P(t > 4.82 | df = 28)$;
Using Table 6, ES11-p719: $P < 0.005$
Using Table 7, ES11-p720: $P = 0+$
Using a computer/calculator: $P = 0.000$
b. $P < \alpha$
-- using classical approach ------------------
a. critical region: $t \leq -1.70$
b. t* falls in the critical region

Step 5: a. Reject H_o.
b. The slope is significantly less than zero, at the 0.05 level of significance.

SECTION 13.5 EXERCISES

Estimating $\mu_{y|x_o}$ and y_{x_o}

$\mu_{y|x_o}$ - the mean of the population y-values at a given x

y_{x_o} - the individual y-value selected at random for a given x

1. point estimate for $\mu_{y|x_o}$ and y_{x_o}: $\hat{y}$ ($\hat{y} = b_0 + b_1 x$)
2. confidence interval for $\mu_{y|x_o}$:

$$\hat{y} \pm t(n-2, \alpha/2) \cdot s_e \sqrt{(1/n) + [(x_o - \overline{x})^2 / SS(x)]}$$

3. prediction interval for y_{x_o}:

$$\hat{y} \pm t(n-2, \alpha/2) \cdot s_e \sqrt{1 + (1/n) + [(x_o - \overline{x})^2 / SS(x)]}$$

Review the Five-Step Procedure in: ES11-pp348-349, IRM-p327, if necessary.

Slight variations in sums of squares and further calculations can result from round-off errors.

13.69 Step 1: $\mu_{y|x=70}$, the mean crutch length for individuals who say they are 70 in. tall.

Step 2: a. random sample, normality assumed for y at each x
  b. t   c. $1 - \alpha = 0.95$

Step 3: $n = 107$, $x_o = 70$, $\overline{x} = 68.84$, $s_e = \sqrt{0.50} = 0.707$

$SS(x) = (n-1)s^2 = 106(7.35^2) = 5726.385$

$\hat{y} = 4.8 + 0.68(70) = 52.4$

Step 4: a. $\alpha/2 = 0.05/2 = 0.025$; df = 105; t(105, 0.025) = 1.96

b. $E = t(n-2, \alpha/2) \cdot s_e \sqrt{(1/n) + [(x_o - \overline{x})^2 / SS(x)]}$

$E = (1.96)(0.707) \cdot \sqrt{(1/107) + [(70 - 68.84)^2 / 5726.385]}$

$E = (1.96)(0.707) \sqrt{0.0095808} = 0.14$

c. $\hat{y} \pm E = 52.4 \pm 0.14$

Step 5: $\underline{52.3 \text{ to } 52.5}$, the 0.95 interval for $\mu_{y|x=70}$

Computer and/or calculator commands for calculating the regressions line can be found in ES11-pp152&153 and pp638-640.
MINITAB also provides confidence interval belts and prediction interval belts. See commands in ES11-p649.

13.71 From exercise 13.63:

$n = 10$, $\Sigma x = 68$, $SS(x) = 103.6$

$\hat{y} = 2.38 + 2.664x$, $s_e^2 = 17.149$

a. When $x = 4$, then $\hat{y} = 2.38 + 2.664(4) = 13.04$
Point estimate for $\mu_{y|x=4} = \underline{13.04}$

b. Step 1: $\mu_{y|x=4}$, the mean travel time required to commute four miles.

Step 2: a. random sample, normality assumed for y at each x
b. t	c. $1 - \alpha = 0.90$
Step 3: $n = 10$, $x_0 = 4$, $\bar{x} = \Sigma x/n = 68/10 = 6.8$,
$s_e^2 = 17.149$, $s_e = \sqrt{17.149} = 4.141$, $\hat{y} = 13.04$
Step 4: a. $\alpha/2 = 0.10/2 = 0.05$; df = 8; t(8,0.05) = 1.86
b. $E = t(n-2,\alpha/2) \cdot s_e \sqrt{(1/n)+[(x_0-\bar{x})^2/SS(x)]}$
$E = (1.86)(4.141)(\sqrt{(1/10)+[(4-6.8)^2/103.6]})$
$E = (1.86)(4.141)\sqrt{0.175676} = 3.23$
c. $\hat{y} \pm E = 13.04 \pm 3.23$
Step 5: <u>9.81 to 16.27</u>, the 0.90 interval for $\mu_{y|x=4}$

c. Step 1: $y_{x=4}$, the travel time required for one person to commute four miles.
Step 2: a. random sample, normality assumed for y at each x
b. t	c. $1 - \alpha = 0.90$
Step 3: $n = 10$, $x_0 = 4$, $\bar{x} = \Sigma x/n = 68/10 = 6.8$,
$s_e^2 = 17.149$, $s_e = \sqrt{17.149} = 4.141$
$\hat{y} = 13.04$
Step 4: a. $\alpha/2 = 0.10/2 = 0.05$; df = 8; t(8,0.05) = 1.86
b. $E = t(n-2,\alpha/2) \cdot s_e \sqrt{1+(1/n)+[(x_0-\bar{x})^2/SS(x)]}$
$E = (1.86)(4.141)\sqrt{1 + 0.175676} = 8.35$
c. $\hat{y} \pm E = 13.04 \pm 8.35$
Step 5: <u>4.69 to 21.39</u>, the 0.90 interval for $y_{x=4}$

d. When x = 9, then $\hat{y} = 2.38 + 2.664(9) = 26.36$
Point estimate for $\mu_{y|x=9}$ = <u>26.36</u>

$26.36 \pm (1.86)(\sqrt{17.149})(\sqrt{(1/10)+[(9-6.8)^2/103.6]})$
$26.36 \pm (1.86)(4.141)\sqrt{0.146718} = 26.36 \pm 2.95$

<u>23.41 to 29.31</u>, the 0.90 interval for $\mu_{y|x=9}$

$26.36 \pm (1.86)(4.141)\sqrt{1 + 0.146718} = 26.36 \pm 8.25$

<u>18.11 to 34.61</u>, the 0.90 interval for $y_{x=9}$

13.73 Summary of data: $n = 10$, $\Sigma x = 16.25$, $\Sigma y = 152$, $\Sigma x^2 = 31.5625$, $\Sigma xy = 275$, $\Sigma y^2 = 2504$

$SS(x) = 31.5625 - (16.25^2/10) = 5.15625$
$SS(xy) = 275 - [(16.25)(152)/10] = 28.0$

$b_1 = 28.0/5.15625 = 5.4303$
$b_0 = [152 - (5.4303)(16.25)]/10 = 6.3758$

$$\hat{y} = 6.3758 + 5.4303x$$

$$s_e^2 = [2504 - (6.3758)(152) - (5.4303)(275)]/8 = 5.19324$$

$$s_e = \sqrt{5.19324} = 2.279$$

When x = 2.0, then $\hat{y}$ = 6.3758 + 5.4303(2.0) = 17.24
t(8,0.025) = 2.31 and $\bar{x}$ = 16.25/10 = 1.625

a. Step 1: $\mu_{y|x=2.00}$, the mean heart-rate reduction for a dose of 2.00 mg.
Step 2: a. random sample, normality assumed for y at each x
   b. t   c. 1 - α = 0.95
Step 3: n = 10, x_o = 2.00, $\bar{x}$ = 1.625, s_e = 2.279
   $\hat{y}$ = 17.24
Step 4: a. α/2 = 0.05/2 = 0.025; df = 8; t(8,0.025) = 2.31
   b. $E = t(n-2,\alpha/2) \cdot s_e \cdot \sqrt{(1/n)+[(x_o-\bar{x})^2/SS(x)]}$
   $E = (2.31)(2.279) \cdot \sqrt{(1/10)+[(2.0-1.625)^2/5.15625]}$
   $E = (2.31)(2.279)\sqrt{0.127273} = 1.88$
   c. $\hat{y} \pm E = 17.24 \pm 1.88$
Step 5: <u>15.4 to 19.1</u>, the 0.95 interval for $\mu_{y|x=2}$

b. Step 1: $y_{x=2.00}$, the heart-rate reduction expected for an individual receiving a dose of 2.00 mg.
Step 2: a. random sample, normality assumed for y at each x
   b. t   c. 1 - α = 0.95
Step 3: n = 10, x_o = 2.00, $\bar{x}$ = 1.625, s_e = 2.279
   $\hat{y}$ = 17.24
Step 4: a. α/2 = 0.05/2 = 0.025; df = 8; t(8,0.025) = 2.31
   b. $E = t(n-2,\alpha/2) \cdot s_e \cdot \sqrt{1+(1/n)+[(x_o-\bar{x})^2/SS(x)]}$
   $E = (2.31)(2.279)\sqrt{1 + 0.127273} = 5.59$
   c. $\hat{y} \pm E = 17.24 \pm 5.59$
Step 5: <u>11.6 to 22.8</u>, the 0.95 interval for $y_{x=2}$

13.75 The standard error for $\bar{x}$'s is much smaller than the standard deviation for individual x's (CLT). Thus the confidence interval will be narrower in accordance.

13.77 a. The overall pattern is elongated in that as the number of customers increases, so does the number of items. The linearity stops with the last upper right 6 points.

b. In the regression analysis printout, the t-test for a significant slope is verified with a t = 27.71 and a corresponding p-value of 0.000. This would indicate that the linear model does fit the data.

c. Upon reviewing the entire dataset, the 6 points are all days in November and December – months in which sales are different due to the holiday season.

13.79 a.

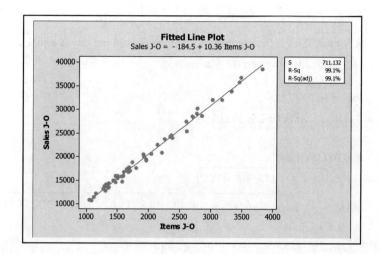

b. The data for January through October follow very closely to a straight line, linear, relationship. There do not appear to be any ordered pairs different from the others.

c. The relationship between the number of items purchased per day and the total daily sales is a strong positive linear relationship. As the number of items purchased per day increased so did the total daily sales. The corresponding correlation coefficient of 0.995 is significant as noted below with its p-value:

    Pearson correlation of Items J-O and Sales J-O = 0.995
    P-Value = 0.000 and $\hat{y}$ = -184.5 + 10.3555x

d.

Step 1: a. The slope β_1 of the line of best fit for the population of number of items purchased per day and the total daily sales.

    b. H_o: $\beta_1 = 0$ (no value)

    H_a: $\beta_1 \neq 0$

Step 2: a. random sample, normality assumed for y at each x
    b. t, df = 50   c. $\alpha = 0.05$

Step 3: a. n = 52, b_1 = 10.3555, s_{b1} = 0.1405

    b. t = $(b_1 - \beta_1)/s_{b1}$

     t* = (10.3555 - 0)/0.1405 = 73.68

Step 4: -- using p-value approach --------------------

    a. **P** = 2P(t > 73.68|df = 50);

     Using Table 6, ES11-p719: **P** < 0.01
     Using Table 7, ES11-p720: **P** = 0+
     Using a computer: **P** = 0.000

    b. **P** < α

   -- using classical approach ------------------
    a. critical region: t ≤ -2.01, t ≥ 2.01
    b. t* falls in the critical region

Step 5: a. Reject H_o.

    b. The slope is significantly different than zero, at the 0.05 level of significance.

d. Using the equation:
When x = 3000, then $\hat{y}$ = -185 + 10.3555(3000) = 30881.5

Step 1: $y_{x=3000}$, the daily total sales if the number of items purchased per day is equal to 3000.

Step 2: a. random sample, normality assumed for y at each x
b. t   c. $1 - \alpha = 0.95$

Step 3: n = 52, x_0 = 3000, $\bar{x}$ = 1991.02,
s_e = 711.132, SS(x) = 25600216.98

Step 4: a. $\alpha/2 = 0.05/2 = 0.025$; df = 50;
t(50, 0.025) = 2.01
b. $E = t(n-2, \alpha/2) \cdot s_e \cdot \sqrt{1 + (1/n) + [(x_0 - \bar{x})^2/SS(x)]}$
$E = (2.01)(711.132) \cdot \sqrt{1 + (1/52) + (3000 - 1991.02)^2 / 25600216.98}$
c. $\hat{y} \pm E = 30881.5 \pm 1470.94$

Step 5: $29410.60 to $32352.40, the 0.95 interval for $y_{|x=3000}$

CHAPTER EXERCISES

13.81 a. Always.

b. Never. r = 0.99 only indicates a strong linear correlation. It never indicates cause-effect.

c. Sometimes. An r value greater than zero indicates that as x increases, y tends to increase. However, there may be a few high x-values with low y-values.

d. Sometimes. The two coefficients measure two completely different concepts. Their signs are unrelated.

e. Always.

13.83 Step 1: a. The linear correlation coefficient for the population, ρ.
b. $H_o: \rho = 0.0$
$H_a: \rho > 0.0$

Step 2: a. random sample, assume normality for y at each x
b. df = n - 2 = 45 - 2 = 43
c. $\alpha = 0.05$

Step 3: a. n = 45
b. r* = 0.69

Step 4: -- using p-value approach --------------------
a. **P** = P(r > 0.69)
Using Table 11, ES11-p729: **P** < 0.005
b. **P** < α
-- using classical approach ------------------
a. critical region: r ≥ 0.257   (r ≥ 0.29 with interpolation)
b. r* falls in the critical region

Step 5: a. Reject H_o

b. There is sufficient reason to conclude that the correlation coefficient is positive, at the 0.05 level of significance.

13.85 a. Step 1: a. The linear correlation coefficient for the population, ρ.

b. H_o: ρ = 0.0

H_a: ρ ≠ 0.0

Step 2: a. random sample, assume normality for y at each x

b. df = n - 2 = 17 - 2 = 15

c. α = 0.05

Step 3: a. n = 17

b. r* = 0.61

Step 4: -- using p-value approach --------------------

a. **P** = P(r > 0.61)

Using Table 11, ES11-p729: **P** < 0.01

b. **P** < α

-- using classical approach -----------------

a. critical region: r ≤ -0.482, r ≥ 0.482

b. r* falls in the critical region

Step 5: a. Reject H_o.

b. Yes, the correlation coefficient is significantly different from zero, at the 0.05 level of significance.

b. ŷ = 1.8(50) + 28.7 = <u>118.7</u>

13.87 a.

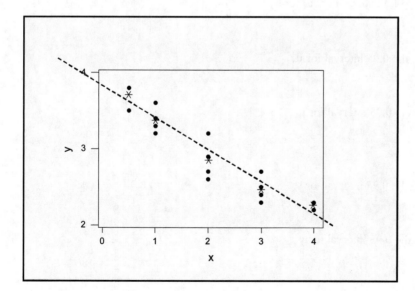

b. See dashed line on graph in (a).

c. See *'s on graph. The *'s seem to follow a curved path, not a straight line. The *'s are above the line at the ends and below the line in the middle.

d. Summary of data: $n = 18$, $\Sigma x = 37.5$, $\Sigma y = 52.7$, $\Sigma x^2 = 104.75$, $\Sigma xy = 98.75$, $\Sigma y^2 = 159.49$

$SS(x) = 104.75 - (37.5^2/18) = 26.625$
$SS(xy) = 98.75 - [(37.5)(52.7)/18] = -11.041666667 = -11.0417$

$b_1 = -11.041666667/26.625 = -0.4147104851 = -0.4147 = -0.415$
$b_0 = [52.7 - (-0.4147104851)(37.5)]/18 = 3.791757955 = 3.79$

$\hat{y} = \underline{3.79 - 0.415x}$

e. Number of decimal places makes a difference in the standard deviation of y about the regression line calculation.

Using Minitab output gives $s_e = 0.196697$

Using Formula 13.8 gives:

$s_e^2 = [159.49 - (3.791757955)(52.7) - (-0.4147104851)(98.75)]/16$
$= 0.0385635109$

$s_e = 0.1963759429$

f. $s_{b1} = \sqrt{s_e^2/SS(x)} = \sqrt{0.196697^2 / 26.625} = \underline{0.038}$

$-0.415 \pm 2.12(0.038) = -0.415 \pm 0.081$

$\underline{-0.496 \text{ to } -0.334}$, the 0.95 interval for β_1

g. At $x = 3.0$:
$\underline{2.43 \text{ to } 2.67}$, the 0.95 interval for $\mu_{y|x=3}$

At $x = 3.5$:
$\underline{2.19 \text{ to } 2.49}$ the 0.95 interval for $\mu_{y|x=3.5}$

h. At $x = 3.0$:
$\underline{2.12 \text{ to } 2.98}$, the 0.95 interval for $y_{x=3}$

At $x = 3.5$:
$\underline{1.90 \text{ to } 2.78}$ the 0.95 interval for $y_{x=3.5}$

13.89 Summary of data: $n = 21$, $\Sigma x = 1177$, $\Sigma y = 567$, $\Sigma x^2 = 70033$, $\Sigma xy = 32548$, $\Sigma y^2 = 15861$

$SS(x) = 70033 - (1177^2/21) = 4064.95$
$SS(xy) = 32548 - [(1177)(567)/21] = 769.0$
$SS(y) = 15861 - (567^2/21) = 552.0$

$r = 769.0/\sqrt{(4064.95)(552.0)} = 0.5133$

a. Step 1: a. The linear correlation coefficient for the number of stamens and the number of carpels in a particular species of flowers, ρ.
   b. $H_o: \rho = 0.0$
   $H_a: \rho \neq 0.0$

Step 2: a. random sample, assume normality for y at each x
   b. $df = n - 2 = 21 - 2 = 19$
   c. $\alpha = 0.05$

Step 3: a. $n = 21$
   b. $r^* = 0.513$

Step 4: -- using p-value approach --------------------
   a. $P = 2P(r > 0.513)$
      Using Table 11, ES11-p729: $0.01 < P < 0.02$
      Using computer: $P = 0.017$
   b. $P < \alpha$
   -- using classical approach ------------------
   a. critical region: $r \leq -0.433$, $r \geq 0.433$
   b. r^* is in the critical region

Step 5: a. Reject H_o.
   b. There is sufficient reason to conclude that there is linear correlation.

b. $b_1 = 769.0/4064.95 = 0.1892$
$b_0 = [567 - (0.1892)(1177)]/21 = 16.3958$

$\hat{y} = \underline{16.40 + 0.189x}$

c. Step 1: a. The slope β_1 of the line of best fit for the population of number of stamens per flower and the corresponding number of carpels.
   b. $H_o: \beta_1 = 0$ (no value)
   $H_a: \beta_1 > 0$

Step 2: a. random sample, assume normality for y at each x
   b. t, $df = 19$ c. $\alpha = 0.05$

Step 3: a. $n = 21$, $b_1 = 0.1892$,
   $s_e^2 = [15861 - (16.3958)(567) - (0.1892)(32548)]/19$
   $s_e^2 = 21.3947$
   $s_{b1} = \sqrt{s_e^2/SS(x)} = \sqrt{21.3947/4064.95} = 0.07255$
   b. $t = (b_1 - \beta_1)/s_{b1}$

$$t^* = 0.189/0.07255 = 2.61$$

Step 4: -- using p-value approach --------------------
  a. $P = P(t > 2.61 | df = 19)$;
  Using Table 6, ES11-p719: $0.005 < P < 0.01$
  Using Table 7, ES11-p720: $0.007 < P < 0.009$
  Using computer: $P = 0.0086$
  b. $P < \alpha$
  -- using classical approach -----------------
  a. critical region: $t \geq 1.73$
  b. t^* is in the critical region

Step 5: a. Reject H_0.
  b. The slope is significantly greater than zero, at the 0.05 level of significance.

d. Step 1: $y_{|x=64}$, the number of carpels found in a mature flower if the number of stamens is 64.
  Step 2: a. t b. $1 - \alpha = 0.95$
  Step 3: a. $n = 21$, $x_0 = 64$, $\overline{x} = 1177/21 = 56.05$
  $s_e = \sqrt{21.3947}$
  b. $\hat{y} = 16.40 + 0.189(64) = 28.50$
  Step 4: a. $\alpha/2 = 0.05/2 = 0.025$; $df = 19$;
  $t(19, 0.025) = 2.09$
  b. $E = t(n-2, \alpha/2) \cdot s_e \sqrt{1+(1/n)+[(x_0-\overline{x})^2/SS(x)]}$
  $E = (2.09)(\sqrt{21.3947})\sqrt{1+(1/21)+[(64-56.05)^2/4064.95]}$
  $E = 9.97$
  c. $\hat{y} \pm E = 28.50 \pm 9.97$
  Step 5: 18.53 to 38.47, the 0.95 interval for one $y_{x=64}$

13.91 Summary of data: $n = 24$ $\Sigma x = 1560$, $\Sigma y = 554$, $\Sigma x^2 = 114000$, $\Sigma xy = 40930$, $\Sigma y^2 = 15186$

$SS(x) = 114000 - (1560^2/24) = 12600$
$SS(xy) = 40930 - [(1560)(554)/24] = 4920$
$SS(y) = 15186 - (554^2/24) = 2397.83$

$r = 4920/\sqrt{(12600)(2397.83)} = 0.895$

a. Step 1: a. The linear correlation coefficient for the pounds of fertilizer and the pounds of wheat harvested, ρ.
  b. $H_0: \rho = 0.0$
  $H_a: \rho > 0.0$
  Step 2: a. random sample, assume normality for y at each x
  b. $df = n - 2 = 24 - 2 = 22$
  c. $\alpha = 0.05$
  Step 3: a. $n = 24$
  b. $r^* = 0.895$
  Step 4: -- using p-value approach --------------------

a. $P = P(r > 0.895)$
   Using Table 11, ES11-p729: $P < 0.005$
   Using computer: $P = 0.000$
b. $P < \alpha$
-- using classical approach ------------------
a. critical region: $r \geq 0.360$; $r \geq 0.34$ with interpolation
b. r^* is in the critical region

Step 5: a. Reject H_0.
b. There is sufficient reason to conclude that there is a positive linear correlation, at the 0.05 level of significance.

b. $b_1 = 4920/12600 = 0.39048 = 0.39$
   $b_0 = [554 - (0.39)(1560)]/24 = -2.298 = -2.30$

   $\hat{y} = -2.30 + 0.39x$

Step 1: $\mu_{y|x=50}$, the mean yield that could be expected if 50 pounds of fertilizer were used per plot.
Step 2: a. t   b. $1 - \alpha = 0.98$
Step 3: a. $n = 24$, $x_0 = 50$, $\bar{x} = 1560/24 = 65$
   $s_e^2 = [15186-(-2.298)(554)-(0.39048)(40930)]/22 = 21.667$
   $s_e = \sqrt{21.667} = 4.655$
   b. $\hat{y} = -2.298 + 0.3905(50) = 17.23$
Step 4: a. $\alpha/2 = 0.02/2 = 0.01$; df = 22;
   $t(22, 0.01) = 2.51$
   b. $E = t(n-2, \alpha/2) \cdot s_e \cdot \sqrt{(1/n)+[(x_0-\bar{x})^2/SS(x)]}$
   $E = (2.51)(4.655)\sqrt{(1/24) + [(50-65)^2/12600]}$
   $E = 2.85$
   c. $\hat{y} \pm E = 17.23 \pm 2.85$
Step 5: 14.38 to 20.08, the 0.98 interval for $\mu_{y|x=50}$

c. Step 1: $\mu_{y|x=75}$, the mean yield that could be expected if 75 pounds of fertilizer were used per plot.
Step 2: a. t   b. $1 - \alpha = 0.98$
Step 3: a. $n = 24$, $x_0 = 75$, $\bar{x} = 1560/24 = 65$
   $s_e = 4.655$
   b. $\hat{y} = -2.298 + 0.3905(75) = 26.99$
Step 4: a. $\alpha/2 = 0.02/2 = 0.01$; df = 22;
   $t(22, 0.01) = 2.51$
   b. $E = t(n-2, \alpha/2) \cdot s_e \cdot \sqrt{(1/n)+[(x_0-\bar{x})^2/SS(x)]}$
   $E = (2.51)(4.655)\sqrt{(1/24) + [(75-65)^2/12600]}$
   $E = 2.60$
   c. $\hat{y} \pm E = 26.99 \pm 2.60$
Step 5: 24.39 to 29.59, the 0.98 interval for $\mu_{y|x=75}$

13.93 Exercises 13.78 (customers vs items) and 13.92 (customers vs sales) both show a linear relationship but have several ordered pairs that appear to be different from the others. Therefore the fit is not as strong as in Exercise 13.79 (items vs sales). There is a stronger relationship with number of items purchased and the total daily sales. If more items are purchased, the sales have to also increase. When customers are involved the relationship is not as strong due to the fact that you can have a customer that does not make a purchase. This would effect the number of items purchased as well as the total daily sales.

13.95 Summary of data: $n = 5$, $\Sigma x = 16$, $\Sigma y = 38$, $\Sigma x^2 = 66$, $\Sigma xy = 145$, $\Sigma y^2 = 326$

$SS(x) = 66 - (16^2/5) = 14.8$
$SS(xy) = 145 - [(16)(38)/5] = 23.4$
$SS(y) = 326 - (38^2/5) = 37.2$

$b_1 = 23.4/14.8 = 1.5811$
$r = 23.4/\sqrt{(14.8)(37.2)} = \underline{0.9973}$ [Formula 13.3]
$r = 1.5811\sqrt{14.8/37.2} = \underline{0.9973}$ [Formula in exercise]

13.97 Yes. $s\sqrt{1/n} = s/\sqrt{n}$ and that is the estimate for the standard error of the mean.

CHAPTER 14 ∇ ELEMENTS OF NONPARAMETRIC STATISTICS

Chapter Preview

Chapter 14 introduces the concept of nonparametric statistics. Up to this point, especially in chapters 8, 9 and 10, the methods used were parametric methods. Parametric methods rely on the normality assumption through knowledge of the parent population or the central limit theorem. In nonparametric (distribution-free) methods, few assumptions about the parent population are required yet the methods are only slightly less efficient than their parametric counterparts. Chapter 14 will demonstrate nonparametric methods for hypothesis tests concerning one mean, two independent means, two dependent means, correlation and randomness.

A survey conducted by NFO Research, Inc. on the attitudes of teenagers toward social and moral values is used in this chapter's opening section "How Teenagers See Things".

SECTION 14.2 EXERCISES

14.1 a. The Sign Test only involves the counts of plus and minus signs.
  b. The population median can be tested using the sign test. By nature of a median, half of the data is above the median and half of the data is below the median.

Sign test (1 - α) Confidence Interval for M - the population median

1. Arrange data in ascending order (smallest to largest)
2. Assign the notation: x_1(smallest), x_2, x_3 ... x_n(largest) to the data.
3. Critical value = k (value from Table 12 using n and α)
4. (1 - α) confidence interval extends from the data values: x_{k+1} to x_{n-k}

14.3   Ranked data:
33 34 35 36 38 |39| 40 40 42 45
46 46 46 46 |47| 47 48 54 59 65

For n = 20 and 1 - α = 0.95, the critical value from Table 12 is k = 5.
$x_{k+1} = x_6 = 39$ and $x_{n-k} = x_{15} = 47$

<u>39 to 47</u>, the 0.95 interval for median M

14.5
-30 -16 -14 -14 -13 -12 -12 -11 -10 -10
 -9   -9   -8   -8   -8 |-7| -6   -6   -5   -4
 -4   -4   -3   -3   -2   -1   -1   0 |1| 2
  2   5   6   6   6   6   6   9   12   12
 13   16   18   19

For n = 44 and 1 - α = 0.95, the critical value from Table 12 is k = 15

$x_{k+1} = x_{16} = -7$ and $x_{n-k} = x_{29} = 1$

<u>−7 to 1</u> points of change in scores is the 95% confidence interval for the median.

The Sign Test

The Sign Test is used to test one mean (median) or the median difference between paired data (dependent samples). Refer to the Five-Step Hypothesis Test Procedure in: ES11-pp371&388, IRM-pp343&357, if necessary. The only changes are in:
  1. **the hypotheses**
      a. null hypothesis
         H_o: M = # <u>or</u> H_o: p(preference) = 0.5 <u>or</u> H_o: p(sign) = 0.5
      b. possible alternative hypotheses:
         H_a: M ≠ # <u>or</u> H_o: p(preference) ≠ 0.5 <u>or</u> H_o: p(sign) ≠ 0.5

> H_a: M < # or H_0: p(preference) < 0.5 or H_0: p(sign) < 0.5
> H_a: M > # or H_0: p(preference) > 0.5 or H_0: p(sign) > 0.5
> 2. **the critical value of the test statistic**,
>   a. for sample sizes ≤ 100, x; Table 12 (Appendix B, ES11-p730)
>   b. for sample sizes > 100, z; Table 4 (Appendix B, ES11-p718)
> 3. **the calculated test statistic**,
>   a. for n ≤ 100;   x = the number of the less frequent sign
>   b. for n > 100;   x = n(sign of preference)
>
> $$z^* = (x' - (n/2))/(½\sqrt{n}) \text{ where}$$
>
> x' = x - ½, if x > (n/2) or x' = x + ½, if x < (n/2)

14.7 a. H_0: Median = 18 vs. H_a: Median < 18

   b. H_0: Median = 32 vs. H_a: Median < 32

   c. H_0: Median = 4.5 vs. H_a: Median ≠ 4.5

14.9 a. **P** = P(x ≤ 3 for n = 18) = 0.01

   b. **P** = P(x ≤ 30 for n = 78) = ½(0.05 < P < 0.10): 0.025 < **P** < 0.05

   c. **P** = P(x ≤ 10 for n = 38) = ½(0.01) = 0.005

   d. **P** = 2P(z < -2.56) = 2[0.0052] = 0.0104

> Table 12, Critical Values of the Sign Test, gives the maximum allowable number of the less frequent sign, k, that will cause rejection of H_0.
> k is based on n (the total number of signs, excluding zeros) and α.
> Therefore, if x ≤ k, reject H_0 and if x > k, fail to reject H_0.

> **Notation for p-value approach and Table 12**
> If the hypothesis test for a sign test is two-tailed, use notation with 2 times the probability (Ex. **P** = 2P(x ≤ 40|n = 100)). A two-tailed test involves both sides and the Sign Test is only taking the less frequent sign (one side) as its test statistic. Table 12 is already set up to provide critical values for a two-tailed test, so no extra measures need to be taken when using the table.

14.11 Step 1: a. Median age of the population of all leukemia patients who receive stem cell transplants.
   b. H_0: Median = 42 years

   H_a: Median ≠ 42 years

Step 2: a. Assume sample is random. Age is continuous.
   b. x = n(least frequent sign)
   c. α = 0.05

Step 3: a. + = over 42 years of age; n = 100

$n(+) = 40$, $n(0) = 0$, $n(-) = 60$
  b. $x = n(+) = 40$
Step 4: -- using p-value approach ---------------
  a. $P = 2P(x \leq 40 | n = 100)$;
    Using Table 12, Appendix B, ES11-p730:
    $0.05 < P < 0.10$
  b. $P > \alpha$
  -- using classical approach -------------
  a. critical region: n(least freq sign) ≤ 39
  b. The test statistic is not in the critical region.

Step 5: a. Fail to reject H_0.
  b. The evidence is not sufficient to show the median age is not equal to 42 years, at the 0.05 level of significance.

14.13 Step 1: a. Proportion of boys who wear protective clothing
  b. H_0: $P(+) = 0.5$
    H_a: $P(+) \neq 0.5$
Step 2: a. Assume random sample. x is binomial and approximately normal.
  b. $x = n$(least freq sign)
  c. α is unspecified
Step 3: a. + = correct solution; n = 75
===
a. b. $x = n(+) = 20$
Step 4: -- using p-value approach ---------------
  a. $P = 2P(x \leq 20 | n = 75)$;
    Using Table 12, Appendix B, ES11-p730:
    $P < 0.01$
----
b. b. $x = n(+) = 27$
Step 4: -- using p-value approach ---------------
  a. $P = 2P(x \leq 27 | n = 75)$;
    Using Table 12, Appendix B, ES11-p730:
    $0.01 < P < 0.05$
----
c. b. $x = n(+) = 30$
Step 4: -- using p-value approach ---------------
  a. $P = 2P(x \leq 30 | n = 75)$;
    Using Table 12, Appendix B, ES11-p730:
    $0.10 < P < 0.25$
----
d. b. $x = n(+) = 33$
Step 4: -- using p-value approach ---------------
  a. $P = 2P(x \leq 33 | n = 75)$;
    Using Table 12, Appendix B, ES11-p730:
    $P > 0.25$
===
  b. If $P < \alpha$

Step 5: a. Reject H_o.

b. The evidence does show that the proportion is significantly different than one-half.

----

b. If $P > \alpha$

Step 5: a. Fail to reject H_o.

b. The evidence does not show that the proportion is significantly different than one-half.

For testing the median difference between paired data, subtract corresponding pairs of data and use the signs of the differences. Hypotheses can be written in three forms.

null hypothesis:
H_o: No difference between the pairs or H_o: M = 0 or H_o: p(+) = 0.5

possible alternative hypotheses:
H_a: There is a difference between the pairs

   H_a: M ≠ #   or H_a: p(+) ≠ 0.5

H_a: One of the pairs is greater than the other

   (subtract greater - smaller)   or

   H_a: M > #   or H_a: p(+) > 0.5

H_a: One of the pairs is less than the other

   (subtract smaller - greater)   or

   H_a: M < #   or H_a: p(+) < 0.5

14.15 a.

Nation	1999	2003	2007	d1=03-99	Sign1	d2=07-03	Sign2
Bulgaria	518	479	470	-39	-	-9	-
Cyprus	460	441	452	-19	-	11	+
Hong Kong	530	556	530	26	+	-26	-
Hungary	552	543	539	-9	-	-4	-
Iran, Islamic Republic	448	453	459	5	+	6	+
Japan	550	552	554	2	+	2	+
Korea, Republic of	549	558	553	9	+	-5	-
Lithuania	488	519	519	31	+	0	0
Romania	472	470	462	-2	-	-8	-
Russian Federation	529	514	530	-15	-	16	+
Singapore	568	578	567	10	+	-11	-
United States	515	527	520	12	+	-7	-

b. **d1 = 2003 − 1999**   *7 countries improved and 5 had lower scores:*

   $x^* = n(-) = 5$

Step 1:   a. Median change in science scores.
   b. H_o: M = 0   There is no difference between the 1999 and 2003 science scores.
      H_a: M > 0   The science scores improved.
Step 2:   a. Assume sample is random. Test scores are numerical.
   b. x = n(least frequent sign)
   c. $\alpha = 0.05$
Step 3:   a. + = positive, − = negative; n = 12
      n(+) = 7, n(0) = 0, n(−) = 5
   b. x = n(−) = 5
Step 4:   -- using p-value approach ---------------
   a. **P** = P(x ≤ 5|n = 12);
      Using Table 12, Appendix B, ES11-p730: **P** > 0.125
      Using a computer: **P** = 0.387
   b. **P** > α
   -- using classical approach -------------
   a. critical region: n(least freq sign) ≤ 2
   b. The test statistic is not in the critical region.

Step 5:   a. Fail to reject H_0.
   b. Insufficient evidence to support the claim that the science scores show a significant overall increase throughout the world, at the 0.05 level of significance.

c. **d2 = 2007 − 2003**   *1 country had no change, 4 countries improved and 7 had lower scores:*

   $x^* = n(+) = 4$

Step 1:   a. Median change in science scores.
   b. H_o: M = 0   There is no difference between the 2003 and 2007 science scores.
      H_a: M > 0   The science scores improved.
Step 2:   a. Assume sample is random. Test scores are numerical.
   b. x = n(least frequent sign)
   c. $\alpha = 0.05$
Step 3:   a. + = positive, − = negative; n = 12
      n(+) = 4, n(0) = 1, n(−) = 7
   b. x = n(+) = 4
Step 4:   -- using p-value approach ---------------
   a. **P** = P(x ≤ 4|n = 11);
      Using Table 12, Appendix B, ES11-p730: **P** > 0.125
      Using a computer: **P** = 0.887
   b. **P** > α
   -- using classical approach -------------
   a. critical region: n(least freq sign) ≤ 2
   b. The test statistic is not in the critical region.

Step 5:   a. Fail to reject H_0.

b. Insufficient evidence to support the claim that the science scores show a significant overall increase throughout the world, at the 0.05 level of significance.

14.17 Step 1: a. Preference for the taste of a new cola.
b. H_o: There is no preference; $p = P(\text{prefer}) = 0.5$
H_a: There is a preference for the new; $p > 0.5$

Step 2: a. x is binomial and approximately normal
b. z   c. $\alpha = 0.01$

Step 3: a. + = prefer new; n = 1228;
$n(+) = 645, n(0) = 272, n(-) = 583$
b. $x = n(+) = 645$; $x' = 644.5$
$z = (x' - (n/2))/(\frac{1}{2}\sqrt{n})$
$z* = (644.5 - (1228/2))/(\frac{1}{2}\sqrt{1228})$
$= (644.5 - 614)/17.5214 = 1.74$

Step 4: -- using p-value approach ---------------
a. $P = P(z > 1.74) = 1.0000 - 0.9591 = 0.0409$
b. $P > \alpha$
-- using classical approach -------------
a. critical region: $z \geq 2.33$
b. The test statistic is not in the critical region.

Step 5: a. Fail to reject H_o.
b. The evidence does not allow us to conclude that there is a significant preference for the new cola, at the 0.01 level of significance.

14.19 Step 1: a. Proportion of college students that wear 'eyeglasses' as their type of correctives lenses.
b. H_o: $P(+) = 0.5$
H_a: $P(+) > 0.5$

Step 2: a. Assume random sample. x is binomial and approximately normal.
b. x = n(least frequent sign)   c. $\alpha = 0.05$

Step 3: a. n = 1500

Step 4: a. $-z(0.05) = -1.65$
b. $z = (x' - (n/2))/(\frac{1}{2}\sqrt{n})$
$-1.65 = (x' - (1500/2))/(\frac{1}{2}\sqrt{1500})$
$-1.65 = (x' - 750)/19.3649$
$x' = 718.048$; critical value is __718__

SECTION 14.3 EXERCISES

14.21 a. Difference between two independent means
b. The actual size of the data is not used, only its rank.

The Mann-Whitney U Test

The Mann-Whitney U Test is used to test the difference between two independent means. Refer to the Five-Step Hypothesis Test Procedure in: ES11-pp371&388, IRM-pp343&357, if necessary. The only changes are in:

1. **the hypotheses**
    a. null hypothesis
       H_o: The average value is the same for both groups.
    b. possible alternative hypotheses:
       H_a: The average value is not the same for both groups.
       H_a: The average value of one group is greater than that of the other group.
       H_a: The average value of one group is less than that of the other group.

2. **the critical value of the test statistic**,
    a. for sample sizes ≤ 20, U; Table 13 (Appendix B, ES11-p731)
    b. for sample sizes > 20, z; Table 4 (Appendix B, ES11-p718)

3. **the calculated test statistic**,
    a. for $n \leq 20$; $U^* =$ smaller of U_a and U_b, where

    $$U_a = n_a \cdot n_b + \frac{(n_a)(n_b + 1)}{2} - R_b \quad \text{and}$$

    $$U_b = n_a \cdot n_b + \frac{(n_a)(n_{a+1})}{2} - R_a$$

    $R_a =$ sum of ranks for sample A, $R_b =$ sum of ranks for sample B

    b. for $n > 20$; $U^* =$ smaller of U_a and U_b

    $$z^* = (U - \mu_u)/(\sigma_u) \quad \text{where}$$

    $$\mu_u = \frac{n_a \cdot n_b}{2} \qquad \sigma_u = \sqrt{\frac{n_a n_b (n_a + n_b + 1)}{12}}$$

14.23 a. H_o: The distributions are the same for both groups
         H_a: The distributions are different for the groups

   b. H_o: The average value is the same for both groups
       H_a: The average value is not the same for the groups

   c. H_o: The distribution of blood pressure is the same for both groups
       H_a: The distribution of blood pressure for group A is higher than for group B

14.25 a. $P > 0.05$
   b. $P < 0.05$
   c. $P = 0.0089$

Table 13, Critical Values of U in the Mann-Whitney Test, gives only critical values for the left-hand tail. $U(n_1, n_2, \alpha)$ is based on the two sample sizes and the amount of α for a one or two-tailed test.

If $U^* \leq U(n_1, n_2, \alpha)$, reject H_o and if $U^* > U(n_1, n_2, \alpha)$, fail to reject H_o.

14.27 a. Critical region: $U \leq 88$

   b. Critical region: $z \leq -1.65$

Combine both samples in ascending order (lowest to highest). Be sure to retain a label identifying which sample each data value is from.
Rank each piece of data from 1(smallest) to $(n_a + n_b)$(largest).
R_a = sum of the ranks for sample A
R_b = sum of the ranks for sample B

Notation for p-value approach and Table 13

If the hypothesis test for a Mann-Whitney test is two-tailed, use notation with 2 times the probability (Ex. $P = 2P(U \leq 23 | n_b=6, n_g=8)$;). A two-tailed test involves both sides and the Mann-Whitney Test is only taking the smaller of the U statistics (one side) as its test statistic. Table 13 is already set up to provide critical values for a two-tailed test, so no extra measures need to be taken when using the table.

14.29 MINITAB verify -- answers given in exercise.

14.31 Step 1: a. Number of preoperative glaucoma medications for patients receiving combined cataract surgery and those receiving just the surgery.
   b. H_o: Number of medications is the same for both groups.
   H_a: Number of medications is not the same for both groups.
   Step 2: a. Independent samples and number values are numerical.
   b. U   c. $\alpha = 0.05$
   Step 3: a. $n_{cs} = 6$; $n_s = 5$
   b. $U^* = 12.5$
   Step 4: -- using p-value approach ---------------
   a. $P = 2P(U \leq 12.5 | n_{cs}=6, n_s=5)$;
      Using Table 13, Appendix B, ES11-p731:
      $P > 0.10$
      Using a computer: $P = 0.7066$
   b. $P > \alpha$
   -- using classical approach -------------
   a. $U(6,5,0.05)$; Crit. reg. ≤ 3

b. The test statistic is not in the critical region.

Step 5: a. Fail to reject H_o.

b. The evidence does allow us to reject the null hypothesis that the two groups are the same with respect to number of medications, at the 0.05 level of significance.

14.33 Excel verify – answers given in exercise.

14.35 Step 1: a. Rainfall amounts based on cloud unseeding and cloud seeding.

b. H_o: Rainfall amount is the same for the two methods.

H_a: Rainfall amount is higher with cloud seeding.

Step 2: a. Independent samples and rainfall amounts are numerical.

b. U   c. $\alpha = 0.05$

Step 3: a. $n_U = n_S = 25$; $R_U = 503$, $R_S = 772$, $U_U = 178$, $U_S = 447$

$\mu_u = (25 \cdot 25)/2 = 312.5$,

$\sigma_u = \sqrt{[(25)(25)(25+25+1)]/12} = 51.54$

b. $U^* = 178$

c. $z^* = (U - \mu_u)/\sigma_u$

$z^* = (178 - 312.5)/51.54 = -2.61$

Step 4: -- using p-value approach ---------------

a. $P = P(z < -2.61) = 0.0045$

b. $P < \alpha$

-- using classical approach ------------

a. critical values: $-z(0.05) = -1.65$

b. The test statistic is in the critical region.

Step 5: a. Reject H_o.

b. The evidence does allow us to conclude that there is a significant increase in the average amount of rainfall with cloud seeding, at the 0.05 level of significance.

SECTION 14.4 EXERCISES

The Runs Test

The Runs Test is used to test the randomness of a set of data. Refer to the Five-Step Hypothesis Test Procedure in: ES11-pp371&388, IRM-pp343&357, if necessary. The only changes are in:

1. **the hypotheses**

    H_o: The data occurred in a random order

    H_a: The data is not in random order

2. **the critical value of the test statistic**,

    a. for sample sizes ≤ 20, V; Table 14 (Appendix B, ES11-p732)

    b. for sample sizes > 20, z; Table 4 (Appendix B, ES11-p718)

> 3. **the calculated test statistic**,
>   a. for $n \leq 20$;   V = the number of runs
>   b. for $n > 20$;   V = the number of runs
>
> $$z^* = (V^* - \mu_V)/(\sigma_V) \quad \text{where}$$
>
> $$\mu_V = \frac{2n_1 \cdot n_2}{n_1 + n_2} + 1 \qquad \sigma_V = \sqrt{\frac{(2n_1 \cdot n_2)(2n_1 \cdot n_2 - n_1 - n_2)}{(n_1 + n_2)^2(n_1 + n_2 - 1)}}$$

14.37   a. H_o: The data did occur in a random order
  H_a: The data did not occur in a random order

  b. H_o: Sequence of odd/even is in random order
  H_a: Not in random order.

  c. H_o: The order of entry by gender was random
  H_a: The order of entry was not random

> Table 14, Critical Values for Total Number of Runs, gives two critical values, V, for each sample size category.
> If $V^* \leq$ the smaller V given or if $V^* >$ the larger V given, reject H_o.

14.39   a.   Critical regions: $V \leq 9$ or $V \geq 22$

  b.   Critical regions: $z \leq -1.96$ or $z \geq +1.96$

14.41   Step 1:   a. Randomness; P(women) and P(men).
  b. H_o: The hiring sequence is random.
  H_a: The hiring sequence is not of random order
  Step 2:   a. Each data fits one of two categories.
  b. V   c. $\alpha = 0.05$
  Step 3:   a. n(M) = 15, n(F) = 5
  b. $V^* = 9$
  Step 4:   -- using p-value approach ---------------
  a. Using Table 14, Appendix B, ES11-p732
  P > 0.05
  b. **P > α**
  -- using classical approach -------------
  a. Critical regions: $V \leq 4$ or $V \geq 12$
  b. The test statistic is not in the critical region.

Step 5: a. Fail to reject H_o.

b. The evidence is not significant, we can not conclude that this sequence is not random, at the 0.05 level of significance.

14.43 Step 1: a. Randomness; P(heads) and P(tails).

b. H_o: The results are randomly ordered.

H_a: The results are not randomly ordered.

Step 2: a. Each data fits one of two categories.

b. V   c. $\alpha = 0.05$

Step 3: a. n(H) = 13, n(T) = 12

b. V* = 21

Step 4: -- using p-value approach ---------------

a. Using Table 14, Appendix B, ES11-p732

$P < 0.05$

b. $P < \alpha$

-- using classical approach -------------

a. Critical regions: $V \leq 8$ or $V \geq 19$

b. The test statistic is in the critical region.

Step 5: a. Reject H_o.

b. The evidence is significant, we can conclude that the results are not random, at the 0.05 level of significance.

14.45 a. Median = 3.55; V = 7

b. Step 1: a. Randomness; P(occurrence).

b. H_o: Random reported average number of students per computer above and below median.

H_a: The data did not occur randomly.

Step 2: a. Each data fits one of two categories.

b. V   c. $\alpha = 0.05$

Step 3: a. n(a) = 7, n(b) = 7

b. V* = 7

Step 4: -- using p-value approach ---------------

a. Using Table 14, Appendix B, ES11-p732

$P > 0.05$

b. $P > \alpha$

-- using classical approach -------------

a. Critical regions: $V \leq 3$ or $V \geq 13$

b. The test statistic is not in the critical region.

Step 5: a. Fail to reject H_o.

b. The evidence is not significant, we are unable to conclude that this sequence lacks randomness, at the 0.05 level of significance.

Note: When the median is one of the data, the two categories are 'above the median' and 'below or equal to the median.'

14.47 a. MINITAB verify -- answers given in exercise.

b. $z^* = \underline{-3.76}$
$P = 2P(z < -3.76) = 2(0.00008) = \underline{0.00016}$

c. Yes, reject the hypothesis of random runs above and below the median.

d.

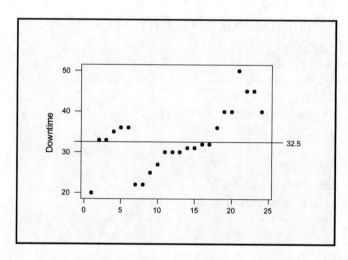

14.49 a. Step 1: a. Randomness of data above and below the median.
b. H_o: Randomness in number of absences.
H_a: The data did not occur randomly.

Step 2: a. Each data fits one of two categories.
b. V   c. $\alpha = 0.05$

Step 3: a. $\tilde{x} = 10.5$; n(above) = 13, n(below) = 13
b. $V^* = 9$

Step 4: -- using p-value approach ---------------
a. Using Table 14, Appendix B, ES11-p732
$P > 0.05$
b. $P > \alpha$
-- using classical approach -------------
a. Critical regions: $V \leq 8$ or $V \geq 20$
b. The test statistic is not in the critical region.

Step 5: a. Fail to reject H_o.

b. The evidence is not significant, we can not conclude that this sequence is not random, at the 0.05 level of significance.

b. Step 1: a. Randomness of data above and below the median.
b. H_o: Randomness in number of absences.

425

H_a: The data did not occur randomly.

Step 2: a. Each data fits one of two categories.

b. V c. $\alpha = 0.05$

Step 3: a. $\tilde{x} = 10.5$; n(above) = 13, n(below) = 13

b. V* = 9

c. n(1) = 13, n(2) = 13

$$\mu = [(2n_1n_2)/(n_1 + n_2)] + 1$$
$$\mu_V = [(2)(13)(13)/(13+13)] + 1 = 14$$

$$\sigma_V = \sqrt{\frac{(2n_1n_2)(2n_1n_2 - n_1 - n_2)}{(n_1 + n_2)^2(n_1 + n_2 - 1)}}$$
$$= \sqrt{[(2)(13)(13)][2(13)(13)-13-13]/(13+13)^2(13+13-1)} = 2.498$$

$z = (V - \mu_V)/\sigma_V$

$z^* = (9 - 14)/2.498 = -2.00$

Step 4: -- using p-value approach ---------------

a. $P = 2P(z < -2.00)$

Using Table 3, Appendix B, ES11-pp716-717

$P = 2(0.0228) = 0.0456$

b. $P < \alpha$

-- using classical approach -------------

a. Critical values: $z(0.025) = \pm 1.96$

b. The test statistic is in the critical region.

Step 5: a. Reject H_0.

b. The evidence is significant, we can conclude that this sequence is not random, at the 0.05 level of significance.

14.51 a. By comparing the number of actual occurrences to the expected number of occurrences using a multinomial or contingency table test the relative frequency of occurrences can be tested.

b. The runs test will test the order, or sequence, of occurrence for the numbers generated.

c. The correlation will test the independence of side-by-side outcomes to be sure there is no influence of one part of a game with another part of the same game.

d. When testing for randomness, it is the null hypothesis that states random, thereby making the "fail to reject" decision the desired outcome. The probability associated with that result is $1 - \alpha$, not the level of significance, and $1 - \alpha$ is known as the level of confidence.

SECTION 14.5 EXERCISES

The Spearman Rank Correlation Test

The Rank Correlation Test is used to test for the correlation or relationship between two variables. Refer to the Five-Step Hypothesis Test Procedure in: ES11-pp371&388, IRM-pp343&357, if necessary. The only changes are in:

1. **the hypotheses**
   a. null hypothesis
      H_o: There is no correlation or relationship between the two variables <u>or</u> $\rho_s = 0$
   b. possible alternative hypotheses:
      H_a: There is a correlation or relationship between the two variables <u>or</u> $\rho_s \neq 0$.
      H_a: There is a positive correlation <u>or</u> $\rho_s > 0$.
      H_a: There is a negative correlation <u>or</u> $\rho_s < 0$.

2. **the critical value of the test statistic**,
   a. two-tailed test, $\pm r_s$; Table 15 (Appendix B, ES11-p733)
   b. one-tailed test, $+r_s$ or $-r_s$; Table 15 (Appendix B, ES11-p733)

3. **the calculated test statistic**,

   $$r_s^* = 1 - \frac{6[\sum(d_i)^2]}{n(n^2 - 1)}$$

14.53 a. H_o: There is a no relationship between the two rankings
            H_a: There is a relationship between the two rankings

   b. H_o: The two variables are unrelated
       H_a: The two variables are related

   c. H_o: The is no correlation between the two variables
       H_a: There is positive correlation

   d. H_o: Refrigerator age has no effect on monetary value
       H_a: Refrigerator age has a decreasing effect on monetary value

---

Table 15, Critical Values of Spearman's Rank Correlation Coefficient, gives positive critical values based on sample size and the level of significance.
For a two-tailed test, add a plus and minus sign to the table value. For a one-tailed test, double the level of significance, then apply a plus or minus sign, whichever is appropriate.

14.55   a.   b.

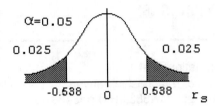
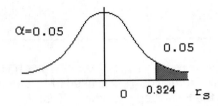

c.

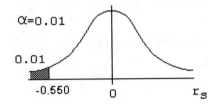

14.57 a. The formula for the rank correlation coefficient: 0.133

$$r_s = 1 - \frac{6\sum (d_i)^2}{n(n^2-1)}: \qquad r_s = 1 - \frac{6(143)}{10(99)} = 1 - .867 = 0.133$$

b.
Step 1: a. Correlation between the overall rating and the street price.

  b. $H_o: \rho_s = 0$

  $H_a: \rho_s > 0$

Step 2: a. Assume random sample of ordered pairs, one ordinal variable and one numerical variable

  b. r_s c. $\alpha = 0.05$

Step 3: a. $n = 10$, $\Sigma d^2 = 143$

  b. $r_s\text{*} = 0.133$

Step 4: -- using p-value approach ----------------

  a. Using Table 15, Appendix B, ES11-p733

  P > 0.05

  Using computer: **P** = 0.3652

  b. **P** > α

  -- using classical approach -------------

  a. critical region: $r_s \geq 0.564$

  b. $r_s\text{*}$ is not in the critical region

Step 5: a. Fail to reject H_o.

  b. There is not sufficient evidence presented by these data to enable us to conclude that there is any relationship between overall performance ratings of 17-inch computer monitors and their street price, at the 0.05 level of significance.

14.59 Summary of data: $n = 12$, $\Sigma d^2 = 70.5$

$r_s = 1 - [(6)(70.5)/(12)(12^2-1)] = \underline{0.753}$

14.61 Step 1: a. Correlation between undergraduate GPA and GPA at graduation from a graduate nursing program.

b. $H_o: \rho_s = 0$

$H_a: \rho_s > 0$

Step 2: a. Assume random sample of ordered pairs, both variables are numerical

b. r_s c. $\alpha = 0.05$

Step 3: a. $n = 10$, $\Sigma d^2 = 43.5$

b. $r_s = 1 - [(6)(\Sigma d^2)/(n)(n^2-1)]$

$r_s^* = 1 - [(6)(43.5)/(10)(99)] = 0.736$

$r_s^* = 0.732$ (using MINITAB)

Step 4: -- using p-value approach ---------------

a. Using Table 15, Appendix B, ES11-p733

$0.01 < P < 0.025$

Using computer: $P = 0.016$

b. $P < \alpha$

-- using classical approach -------------

a. critical region: $r_s \geq 0.564$

b. r_s^* is in the critical region

Step 5: a. Reject H_o.

b. There is sufficient reason to conclude there is a positive relationship, at the 0.05 level of significance.

14.63 a.

Airline	Comfort	Service	Web	Comfort, Rank	Service, Rank	Web, Rank
Midwest	23	22	18	13.0	12.5	4.5
Virgin America	23	24	23	13.0	14.0	13.5
JetBlue	23	22	22	13.0	12.5	11.5
Alaska	17	20	21	11.0	10.0	10.0
Hawaiian	16	19	19	9.0	9.0	7.0
Continental	15	17	22	7.0	7.0	11.5
Southwest	16	21	23	9.0	11.0	13.5
Frontier	16	18	16	9.0	8.0	3.0
AirTran	14	15	18	6.0	6.0	4.5
Delta	13	13	19	5.0	4.5	7.0
American	12	13	20	3.5	4.5	9.0
United	12	12	19	3.5	3.0	7.0
US Airways	11	10	15	1.5	1.5	2.0
Spirit	11	10	14	1.5	1.5	1.0

b. Step 1: a. Correlation between airlines ratings for comfort and service.

b. $H_0: \rho_s = 0$

$H_a: \rho_s \neq 0$

Step 2: a. Assume random sample of ordered pairs, both variables are numerical

b. r_s c. $\alpha = 0.05$

Step 3: a. $n = 14$, $\Sigma d^2 = 9$

b. $r_s = 1 - [(6)(\Sigma d^2)/(n)(n^2-1)]$

$r_s^* = 1 - [(6)(9)/(14)(195)] = 0.980$

$r_s^* = 0.980$ (using MINITAB)

Step 4: -- using p-value approach ---------------

a. Using Table 15, Appendix B, ES11-p733

P < 0.01

Using computer: **P = 0.0000**

b. **P < α**

-- using classical approach -------------

a. critical region: $r_s \leq -0.538$ and $r_s \geq 0.538$

b. r_s^* is in the critical region

Step 5: a. Reject H_0.

b. There is sufficient reason to conclude there is a correlation, at the 0.05 level of significance.

c. Step 1: a. Correlation between airlines ratings for comfort and Web.

b. $H_0: \rho_s = 0$

$H_a: \rho_s \neq 0$

Step 2: a. Assume random sample of ordered pairs, both variables are numerical

b. r_s c. $\alpha = 0.05$

Step 3: a. $n = 14$, $\Sigma d^2 = 205.5$

b. $r_s = 1 - [(6)(\Sigma d^2)/(n)(n^2-1)]$

$r_s^* = 1 - [(6)(205.5)/(14)(195)] = 0.548$

$r_s^* = 0.540$ (using MINITAB)

Step 4: -- using p-value approach ---------------

a. Using Table 15, Appendix B, ES11-p733

$0.02 < P < 0.05$

Using computer: **P = 0.046**

b. **P < α**

-- using classical approach -------------

a. critical region: $r_s \leq -0.538$ and $r_s \geq 0.538$

b. r_s^* is in the critical region

Step 5: a. Reject H_0.

b. There is sufficient reason to conclude there is a correlation, at the 0.05 level of significance.

d. Step 1: a. Correlation between airlines ratings for service and Web.

　　　　　　b. $H_0: \rho_s = 0$

　　　　　　　$H_a: \rho_s \neq 0$

　Step 2: a. Assume random sample of ordered pairs, both variables are numerical

　　　　　b. r_s　　　c. $\alpha = 0.05$

　Step 3: a. $n = 14$, $\Sigma d^2 = 166$

　　　　　b. $r_s = 1 - [(6)(\Sigma d^2)/(n)(n^2-1)]$

　　　　　　　$r_s^* = 1 - [(6)(166)/(14)(195)] = 0.635$

　　　　　　　$r_s^* = 0.631$ (using MINITAB)

　Step 4: -- using p-value approach ---------------

　　　　　a. Using Table 15, Appendix B, ES11-p733

　　　　　　　$0.01 < P < 0.02$

　　　　　　　Using computer: $P = 0.016$

　　　　　b. $P < \alpha$

　　　　　-- using classical approach -------------

　　　　　a. critical region: $r_s \leq -0.538$ and $r_s \geq 0.538$

　　　　　b. r_s^* is in the critical region

　　　　　--

　Step 5: a. Reject H_0.

　　　　　b. There is sufficient reason to conclude there is a correlation, at the 0.05 level of significance.

e. Comfort, service and web are all correlated with each other. The strongest correlation is between comfort and service, whereas the lowest correlation is between comfort and web (yet still significant).

14.65 a.

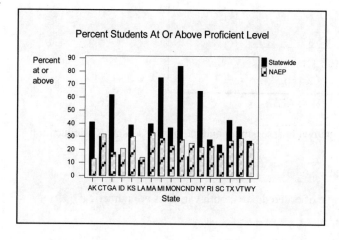

There appears to be very little relationship between the two sets of percentages. Some of the largest state percentages are paired with some of the lower NAEP percentages, while some of the lowest state percentages are also paired with some of the lowest NAEP percentages.

b.

	State	NAEP	Rstate	R NAEP	d	dsq	sumdsq
Arkansas	41	13	12	1	-11	121	594
Connecticut	30	32	7	16	9	81	
Georgia	62	18	14	3.5	-11	110.25	
Idaho	16	21	3	5	2	4	
Kansas	39	30	10	15	5	25	
Louisiana	12	14	1	2	1	1	
Massachusetts	40	33	11	17	6	36	
Michigan	75	29	16	13.5	-2.5	6.25	
Missouri	37	23	8	7.5	-0.5	0.25	
New York	65	22	15	6	-9	81	
North Carolina	84	28	17	12	-5	25	
North Dakota	15	25	2	9.5	7.5	56.25	
Rhode Island	28	23	6	7.5	1.5	2.25	
South Carolina	24	18	4	3.5	-0.5	0.25	
Texas	43	27	13	11	-2	4	
Vermont	38	29	9	13.5	4.5	20.25	
Wyoming	27	25	5	9.5	4.5	20.25	

c.

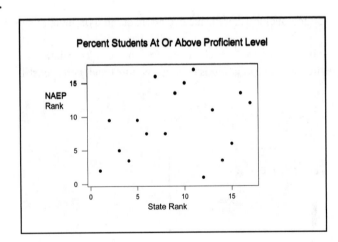

d.

Step 1: a. Correlation between statewide assessment and national assessment of education.

b. $H_o: \rho_s = 0$

$H_a: \rho_s \neq 0$

Step 2: a. Assume random sample of ordered pairs, both variables are numerical

b. r_s c. $\alpha = 0.05$

Step 3: a. $n = 17$, $\Sigma d^2 = 594$

b. $r_s = 1 - [(6)(\Sigma d^2)/(n)(n^2-1)]$

$r_s^* = 1 - [(6)(594)/(17)(288)] = 0.272$

$r_s^* = 0.270$ (using MINITAB)

Step 4: -- using p-value approach --------------

a. Using Table 15, Appendix B, ES11-p733

P > 0.10

Using computer: P = 0.294

b. **P** > α

-- using classical approach -------------

a. critical region: $r_s \leq -0.488$ and $r_s \geq 0.488$

b. r_s^* is not in the critical region

Step 5: a. Fail to reject H_o.

b. There is not sufficient reason to conclude there is a correlation between the two sets of percentages, at the 0.05 level of significance.

14.67 Summary of ranks: $n = 12$, $\Sigma x = 78$, $\Sigma y = 78$, $\Sigma x^2 = 650$, $\Sigma xy = 578.5$, $\Sigma y^2 = 649$

$SS(x) = 650 - (78^2/12) = 143$ $SS(y) = 649 - (78^2/12) = 142$
$SS(xy) = 578.5 - [(78)(78)/12] = 71.5$

$r_s = 71.5/\sqrt{(143)(142)} = \underline{0.502}$

CHAPTER EXERCISES

14.69 a.
```
Stem-and-leaf of Hydrogen   N = 52
Leaf Unit = 0.10
     3      2 078
    10      3 1356688
    18      4 13355689
   (12)     5 001112444589
    22      6 11137
    17      7 2388
    13      8 3378
     9      9 4479
     5     10 56
     3     11 1
     2     12 4
     1     13
     1     14
     1     15 2
```

b. skewed right

c. For $n = 52$ and $1 - \alpha = 0.95$, the critical value from Table 12 is $k = 18$.
$x_{k+1} = x_{19} = 5.0$ and $x_{n-k} = x_{34} = 6.3$

$\underline{5.0 \text{ to } 6.3}$, the 0.95 interval for median M

14.71 a. Using the Sign Test (one median):
   Step 1: a. Median score on exam.
            b. H_o: Median = 50
               H_a: Median ≠ 50
   Step 2: a. Assume sample is random. Exam score is continuous.
            b. x = n(least frequent sign)
            c. $\alpha = 0.05$
   Step 3: a. + = above 50, - = below 50, 0 = 50; n = 30
               n(+) = 10, n(0) = 2, n(-) = 20
            b. x = n(+) = 10
   Step 4: -- using p-value approach --------------
            a. **P** = 2P(x ≤ 10|n = 30);
                  Using Table 12, Appendix B, ES11-p730:
                  P ≈ 0.10
                  Using computer: P = 0.0987
            b. **P** > α
            -- using classical approach -------------
            a. critical region: n(least freq sign) = x ≤ 9
            b. The test statistic is not in the critical region.

   Step 5: a. Fail to reject H_o.
            b. The sample evidence is not sufficient to justify the claim that median is different than 50, at the 0.05 level of significance.

b. Step 1: a. Median score on exam.
            b. H_o: Median = 50
               H_a: Median < 50
   Step 2: a. Assume sample is random. Exam score is continuous.
            b. x = n(least frequent sign)
            c. $\alpha = 0.05$
   Step 3: a. + = above 50, - = below 50;
               n = 30, n(+) = 10, n(-) = 20
            b. x = n(+) = 10
   Step 4: -- using p-value approach --------------
            a. **P** = P(x ≤ 10|n = 30);
                  Using Table 12, Appendix B, ES11-p730:
                  P ≈ 0.05
                  Using computer: P = 0.0494
            b. **P** ≤ α
            -- using classical approach -------------
            a. critical region: n(least freq sign) = x ≤ 10
            b. The test statistic is in the critical region.

   Step 5: a. Reject H_o.
            b. The sample evidence is sufficient to justify the claim that median is less than 50, at the 0.05 level of significance.

14.73 Using the Sign Test (dependent samples):
  Step 1: a. Time required to run 220 yd sprint on two tracks.
    b. H_o: No difference between average times (no faster)
    H_a: Average time on B is less than on A (B is faster)
  Step 2: a. Assume sample is random. Time is continuous.
    b. x = n(least frequent sign)
    c. $\alpha = 0.05$
  Step 3: a. + = A time is greater; n = 10
    n(+) = 8, n(0) = 0, n(-) = 2
    b. x = n(-) = 2
  Step 4: -- using p-value approach ---------------
    a. $P = P(x \leq 2 | n = 10)$;
      Using Table 12, Appendix B, ES11-p730: $P \approx 0.125$
      Using computer: $P = 0.0547$
    b. $P > \alpha$

    -- using classical approach -------------
    a. critical region: n(least freq sign) = $x \leq 1$
    b. The test statistic is not in the critical region.

  Step 5: a. Fail to reject H_o.
    b. The evidence is not sufficient to justify the claim that track B is faster, at the 0.05 level of significance.

14.75 Reject for $U \leq 127$

14.77 Using Mann-Whitney U Test (independent samples):
  Step 1: a. Line width.
    b. H_o: No difference in line width.
    H_a: There is a difference in line width.
  Step 2: a. Independent samples and line widths are numerical.
    b. U   c. $\alpha = 0.05$
  Step 3: a. ranked data and ranks: (underlined = normal group)

    27.5 28.0 28.5 29.5 <u>30.5</u> <u>30.6</u> 30.7 <u>30.9</u> <u>32.9</u> <u>35.1</u>
     1   2   3   4   5   6   7   8   9   10
    $R_n = 38$, $R_m = 17$

    b. $U_n = n_n \cdot n_m + [(n_m)(n_m+1)/2] - R_m$
      $U_n = (5)(5) + [(5)(5+1)/2] - 17 = 23$
    $U_m = n_m \cdot n_n + [(n_n)(n_n+1)/2] - R_n$
      $U_m = (5)(5) + [(5)(5+1)/2] - 38 = 2$; $U^* = 2$
  Step 4: -- using p-value approach ---------------
    a. $P = 2P(U \leq 2 | n_n = 5, n_m = 5)$;
      Using Table 13, Appendix B, ES11-p731:
      $P \approx 0.05$

Using computer: P = 0.0367

b. $P \leq \alpha$

-- using classical approach -------------

a. $U(5,5,0.05) = 2$; $U \leq 2$

b. The test statistic is in the critical region.

Step 5: a. Reject H_0.

b. There is a significant difference in line width, at the 0.05 level of significance.

14.79 a.

Team, AL	w/ ERA, low is best		w/ BA, high is best	
	AVG, AL	AVG, AL Rank	ERA, AL	ERA, AL Rank
Baltimore Orioles	0.268	10.0	5.15	14.0
Boston Red Sox	0.270	11.0	4.35	7.0
Chicago White Sox	0.258	1.5	4.14	2.0
Cleveland Indians	0.264	8.0	5.06	13.0
Detroit Tigers	0.260	4.5	4.29	5.0
Kansas City Royals	0.259	3.0	4.83	12.0
Los Angeles Angels	0.285	14.0	4.45	9.0
Minnesota Twins	0.274	12.0	4.50	11.0
New York Yankees	0.283	13.0	4.26	3.5
Oakland Athletics	0.262	6.0	4.26	3.5
Seattle Mariners	0.258	1.5	3.87	1.0
Tampa Bay Rays	0.263	7.0	4.33	6.0
Texas Rangers	0.260	4.5	4.38	8.0
Toronto Blue Jays	0.266	9.0	4.47	10.0

Team, NL	AVG, NL	AVG, NL Rank	ERA, NL	ERA, NL Rank
Arizona Diamondbacks	0.253	4.0	4.42	11
Atlanta Braves	0.263	12.0	3.57	3
Chicago Cubs	0.255	5.0	3.84	5
Cincinnati Reds	0.247	2.0	4.18	7
Colorado Rockies	0.261	10.0	4.22	8
Florida Marlins	0.268	14.0	4.29	9
Houston Astros	0.260	9.0	4.54	13
Los Angeles Dodgers	0.270	15.5	3.41	1
Milwaukee Brewers	0.263	12.0	4.83	15
New York Mets	0.270	15.5	4.45	12
Philadelphia Phillies	0.258	7.5	4.16	6
Pittsburgh Pirates	0.252	3.0	4.59	14
San Diego Padres	0.242	1.0	4.37	10
San Francisco Giants	0.257	6.0	3.55	2
St. Louis Cardinals	0.263	12.0	3.66	4
Washington Nationals	0.258	7.5	5.00	16

b. <u>Batting Averages:</u>

Step 1: a. Batting averages.
b. H_o: Batting averages in AL are not higher. ($\leq$)
H_a: Batting averages in AL are higher. (>)

Step 2: a. Independent samples and batting averages are numerical.
b. U   c. $\alpha = 0.05$

Step 3: a. $n_{AL} = 14$, $n_{NL} = 16$; W = 268 using Minitab, therefore $U_1 = 163$, $U_2 = 61$
b. $U^* = 61$ (Check: $163 + 61 = 224 = 14 \times 16$)

Step 4: -- using p-value approach ---------------
a. $P = P(U < 61)$
Using Table 13, Appendix B, ES11-p731: $P < 0.025$
Using Minitab: $P = 0.0179$
b. $P < \alpha$
-- using classical approach -------------
a. critical region: $U \leq 71$
b. The test statistic is in the critical region.

Step 5: a. Reject H_o
b. The American League batting average in 2009 was higher than the National League at the 0.05 level of significance.

<u>Earned Run Averages:</u>

Step 1: a. Earned run averages
b. H_o: Earned run average for NL is not lower. ($\geq$)
H_a: Earned run average for NL is lower. (<)

Step 2: a. Independent samples and earned run averages are numerical.
b. U   c. $\alpha = 0.05$

Step 3: a. $n_{AL} = 14$, $n_{NL} = 16$; W = 215.5 using Minitab, therefore $U_1 = 79.5$, $U_2 = 144.5$
b. $U^* = 79.5$ (Check: $79.5 + 144.5 = 224 = 14 \times 16$)

Step 4: -- using p-value approach ---------------
a. $P = P(U < 79.5)$
Using Table 13, Appendix B, ES11-p731: $P < 0.025$
Using Minitab: $P = 0.0917$
b. $P > 0.05$
-- using classical approach -------------
a. critical region: $U \leq 71$
b. The test statistic is not in the critical region.

Step 5: a. Fail to reject H_o
b. The National League earned run average in 2009 is not lower than the American League at the 0.05 level of significance.

14.81 Using the Runs Test:
Step 1: a. Randomness in sequence of occurrence of defective and nondefective parts.

b. H_0: Random order.

H_a: Lack of randomness.

Step 2: a. Each data fits into one of two categories

b. V c. $\alpha = 0.05$

Step 3: a. n(n) = 20, n(d) = 4

b. V* = 9

Step 4: -- using p-value approach --------------

a. $P = P(V \leq 9)$

$P > 0.05$

b. $P > \alpha$

-- using classical approach -------------

a. critical region: $V \leq 4$ and $V \geq 10$

b. the test statistic is not in the critical region

Step 5: a. Fail to reject H_0.

b. The sample results do not show a significant lack of randomness, at the 0.05 level of significance.

14.83 a. Median = 22.5

Company		Job Growth	Company		Job Growth
1	a	26	11	a	23
2	a	54	12	b	13
3	a	34	13	b	17
4	b	10	14	a	23
5	a	31	15	b	9
6	a	48	16	b	3
7	a	26	17	b	15
8	b	22	18	b	11
9	a	24	19	b	1
10	b	10	20	a	122

Runs above: 6 Runs below: 5

b.

Step 1: a. Randomness of job growth rate percentages.

b. H_0: The job growth rate percentages are listed in a random sequence

H_a: Lack of randomness.

Step 2: a. Each data fits into one of two categories

b. V c. $\alpha = 0.05$

Step 3: a. $n_a = 9$ $n_b = 11$

b. V* = 11

Step 4: -- using p-value approach --------------

a. $P = P(V \leq 11)$;

$P > 0.05$

b. $P > \alpha$

-- using classical approach -------------

a. critical region: $V \leq 6$ and $V \geq 16$

b. The test statistic is not in the critical region

Step 5: a. Fail to reject H_o.

c. Conclusion: There is not sufficient evidence to reject the null hypothesis that the job growth rate percentages are listed in a random sequence at the 0.05 level of significance. Based on this sample evidence, a higher job growth rate does not imply a higher rank in attractiveness.

14.85 Using Spearman's Rank Correlation:
Step 1: a. Correlation between two daily high temperatures.
b. $H_o: \rho_s = 0$ (Independence)
$H_a: \rho_s > 0$ (Positive correlation)
Step 2: a. Assume random sample of ordered pairs, numerical variables
b. r_s c. $\alpha = 0.05$
Step 3: a. $n = 18$, $\Sigma d^2 = 116.5$
b. $r_s = 1 - [(6)(\Sigma d^2)/(n)(n^2-1)]$
$r_s^* = 1 - [(6)(116.5)/(18)(18^2-1)] = 0.880$
Step 4: -- using p-value approach ---------------
a. Using Table 15, Appendix B, ES11-p733
P < 0.005
Using computer: **P = 0.000**
b. **P < α**
-- using classical approach -------------
a. critical regions: $r_s \geq 0.401$
b. r_s^* is in the critical region

Step 5: a. Reject H_o
b. There is a significant amount of correlation shown between the two sets of temperatures, at the 0.05 level of significance.

Stopped to say hey but you were asleep. Hope you feel better in the morning.

— Jack

P.S. Lock your door